DISCARDED

A LAND BETWEEN WATERS

Latin American Landscapes

Series Editors
Christopher R. Boyer and Lise Sedrez

Editorial Board
Guillermo Castro Herrera
José Augusto Drummond
Stefania Gallini
Stuart McCook
John R. McNeill
Shawn Miller
Cynthia Radding
John Soluri

A Land Between Waters
Environmental Histories of Modern Mexico

EDITED BY CHRISTOPHER R. BOYER

The University of Arizona Press
TUCSON

 THE UNIVERSITY OF
ARIZONA PRESS
© 2012 The Arizona Board of Regents
All rights reserved

www.uapress.arizona.edu

Library of Congress Cataloging-in-Publication Data

Land between waters : environmental histories of modern Mexico / edited by Christopher R. Boyer.
  p. cm. — (Latin American landscapes)
Includes bibliographical references and index.
ISBN 978-0-8165-0249-3 (cloth : acid-free paper)
1. Mexico—Environmental conditions. 2. Landscape changes—Mexico—History. 3. Nature—Effect of human beings on—Mexico—History. 4. Environmental degradation—Mexico—History. 5. Political ecology—Mexico—History. 6. Environmental policy—Mexico—History. 7. Mexico—History—1810– 8. Mexico—Politics and government—1810– I. Boyer, Christopher R. (Christopher Robert)

GE160.M58L36 2012
304.20972—dc23

2012000982

Publication of this book is made possible in part by the proceeds of a permanent endowment created with the assistance of a Challenge Grant from the National Endowment for the Humanities, a federal agency.

Manufactured in the United States of America on acid-free, archival-quality paper containing a minimum of 30% post-consumer waste and processed chlorine free.

17 16 15 14 13 12   6 5 4 3 2 1

Contents

Acknowledgments	vii
1. *The Cycles of Mexican Environmental History* **Christopher R. Boyer**	1
2. *Downslope and North: How Soil Degradation and Synthetic Pesticides Drove the Trajectory of Mexican Agriculture through the Twentieth Century* **Angus Wright**	22
3. *Mexico's Breadbasket: Agriculture and the Environment in the Bajío* **Martín Sánchez Rodríguez**	50
4. *Nature as Subject and Citizen in the Mexican Botanical Garden, 1787–1829* **Rick A. López**	73
5. *Besieged Forests at Century's End: Industry, Speculation, and Dispossession in Tlaxcala's La Malintzin Woodlands, 1860–1910* **José Juan Juárez Flores**	100

6. *Water and Revolution in Morelos, 1850–1915*   124
   Alejandro Tortolero Villaseñor

7. *King Henequen: Order, Progress, and Ecological Change in Yucatán, 1850–1950*   150
   Sterling Evans

8. *Class and Nature in the Oil Industry of Northern Veracruz, 1900–1938*   173
   Myrna I. Santiago

9. *Parables of Chapultepec: Urban Parks, National Landscapes, and Contradictory Conservation in Modern Mexico*   192
   Emily Wakild

10. *The Illusion of National Power: Water Infrastructure in Mexican Cities, 1930–1990*   218
    Luis Aboites Aguilar

11. *Episodes of Environmental History in the Gulf of California: Fisheries, Commerce, and Aquaculture of Nacre and Pearls*   245
    Mario Monteforte and Micheline Cariño

12. *Conclusion: Of the "Lands in Between" and the Environments of Modernity*   277
    Cynthia Radding

About the Contributors   297

Index   301

Acknowledgments

This book is the first time a transnational collection of environmental histories about modern Mexico has been published on either side of the US-Mexican border. Although historians have written for decades about the relationship between people and their environment in Mexico, only recently has the field achieved the dimensions that make a publication such as *A Land Between Waters* viable. This book therefore signals the advent of this important new field and, I hope, will stimulate an ongoing scholarly conversation about Mexico's past and the challenges it faces in the present. Bringing Mexican and North American voices together not only required an extraordinary degree of patience from the twelve other authors in this volume, whose diligence and good humor I gratefully recognize, but also demanded the work of many others as well.

Matthew Janzen and Elizabeth Janzen translated chapters 3 and 5, the contributions by Martín Sánchez and José Juan Juárez, respectively. Their thoroughly professional work will make these authors' important scholarship available to an English-speaking audience for the first time. Chapter 6, by Alejandro Tortolero, was translated by Alfonso Camargo Caballero and Judith Ostria Velázquez. I am responsible for the translation of chapter 10, by Luis Aboites. His contribution is a revised version of a chapter published in Spanish in *Ciudades mexicanas del siglo XX: Siete estudios históricos*, edited by Carlos Lira Vásquez and Ariel Rodríguez Kuri (Mexico City: El Colegio de México/Universidad Autónoma Metropolitana–Azcapotzalco, 2009). He thanks Diana Birrichaga for her severe criticism of an earlier version, as well as Edith Kauffer for her comments after he presented a later version at the Seminario Agua y Tierra, held in Aguascalientes in September 2007. He also thanks the editor of the present collection for suggestions in preparing the English-language version.

Rick López would like to thank the participants of the "Nuevas Fronteras: New Trends and Transformations in Modern Mexican History" conference held at Yale University, June 5, 2009, for their comments on a previous version of his chapter. I would like to thank the participants of the 2011 Tenth Annual Graduate Conference of Stony Brook University's Latin American and Caribbean Studies Center for their thoughtful comments on a draft version of the introduction to this book. Alejandro Tortolero's research was supported by CONACYT grant number H-43960. Martín Sánchez's work received economic support from the Consejo Nacional de Ciencia y Tecnología for the broader project entitled "Patrones históricos de uso y manejo del agua en la cuenca Lerma Chapala Santiago" (Historical Patterns in the Use and Management of Water in the Lerma-Chapala-Santiago Basin). He would like to thank Marco Antonio Hernández for designing the maps in his chapter. The map in the chapter by Sterling Evans was prepared by Wenonah Fraser-Van Heyst, of Brandon University's Department of Geography.

Producing and editing a book—particularly a binational one—requires substantial time and resources. I am fortunate to have received a fellowship from the University of Illinois at Chicago Institute for the Humanities during the 2009–2010 academic year, which gave me the time and focus to organize this volume. The UIC Department of History also contributed a subvention to the University of Arizona Press, which I gratefully acknowledge. I also thank editors Kristen Buckles and her predecessor, Patti Hartman, for their enthusiastic support of this project and the edited series Latin American Landscapes, of which it forms a part. Sally Bennett edited the text with extraordinary care and an enviable attention to style. Finally, I must once again thank my wife, Amy Shannon, and son, Isaac Boyer, for their graceful tolerance of my distracted wanderings (both corporeal and mental) over the years it took to bring this book to press.

A LAND BETWEEN WATERS

CHAPTER ONE

The Cycles of Mexican Environmental History

Christopher R. Boyer

The native Mexica who ruled the Valley of Mexico at the eve of the Spanish conquest called their homeland "Anáhuac," a term usually translated as "the land by the water" or "the land between the waters." Their cities lay alongside a labyrinth of shallow lakes and manmade channels that covered much of the valley floor. As the Aztec empire grew, the term *Anáhuac* was sometimes used to denote all of central Mexico, though it never lost its original meaning.[1] Today, the Valley of Mexico is no longer dominated by lakes. Centuries of civil engineering projects have drained most of the water from the basin, and in its place rises one of the world's greatest cities. Ancient concerns about floods and an overabundance of water have given way (in most seasons) to fears about potentially crippling water shortages. Although pipes and canals now divert water into the valley from distant rivers, most of the supply comes from drilling ever deeper into the aquifer. Pumping the subterranean vestiges of the lakes has taken a toll, however. Some of Mexico City's most beloved landmarks have subsided as the water table inches lower, as if to remind modern Mexicans that their fate still remains tied to the proper balance of land and water.

The nation as a whole likewise dwells between the waters. Seacoasts and rivers demarcate nearly all the borders of contemporary Mexico, from the Pacific Ocean on the west to the Gulf of Mexico on the east and from the Río Bravo (known as the Rio Grande in the United States) in the north to the Suchiate and Usumacinta to the south.[2] The seas in particular have provided a means of trade and cultural exchange over the centuries, initially by connecting native cities and empires, and later by placing Mexico within the global exchange of commodities flowing between Asia, Europe, Africa, and the Americas. The seas have also harbored natural and human threats, from hurricanes to flotillas of invaders to oil spills. As if shunning these dangers, most of the population has historically lived in the central highlands far from the coasts. These lands also exist between the waters, albeit in a temporal rather than a geographic sense. In most parts of the country, rain falls during one or two sharply defined seasons, the most common of which lasts from May to October. Farmers must endure long dry spells punctuated in some regions by unpredictable torrential downpours, until the reliable spring rains begin to bathe the fields more consistently. The succession of aridity and plenty inspired ingenious systems for managing and storing water, from the dikes designed by native rulers, to seasonal reservoirs developed during the colonial era, to massive irrigation projects in the nineteenth and twentieth centuries.[3]

The climate of the central plateau ranges from the subhumid tropics of the plains to the more temperate, forested, and sometimes harsh conditions of the mountains that tower over most major cities. Some of these lands bear crops readily. Maize, first domesticated in Mexico around ten thousand years ago and now the world's most ubiquitous crop, grows promiscuously in a range of soils, altitudes, and climates.[4] Other grains, chilies, fruits, and legumes complemented maize before the Spanish conquest of 1521, when large domestic animals became available to provide traction and a reliable source of meat. This is an oftentimes unforgiving land where volcanoes brooded and smoked and sometimes appeared from nowhere, as in the case of El Jorullo (1759) and Paricutín (1943), the only New World volcanoes born in historical times. Earthquakes also threatened to topple buildings, destroy lives, and visit death and disease, though we only have records of relatively recent ones in Huamuxtitlán, Guerrero (1845), Colima (1932), and Mexico City (1985). Droughts regularly threatened to choke farms with dust, as they still do in the arid north. They appear to have aggravated the social unrest that undermined Teotihuacan in the sixth century and contributed to the independence movement of 1810–21 and the revolution of 1910–15.[5]

Native people did not necessarily use the land sustainably before the conquest—indeed, some scholars believe deforestation, erosion, and mismanagement of water resources had pushed the Valley of Mexico to the verge of calamity on the eve of the conquest—but nothing could have prepared them for the devastations of colonialism. The conquistadors arrived not only with Toledo steel and warhorses but also with Old World sicknesses such as smallpox, diphtheria, typhus, and influenza. Microbial diseases assailed Mexico's indigenous population, which had numbered around 25 million in the central plateau, and reduced their numbers by approximately 90 percent between the 1520s and the 1650s.[6] The biotic transfer that Alfred Crosby has dubbed the "Columbian exchange" continued as cattle, pigs, sheep, goats, donkeys, and horses arrived to the Americas, along with plants such as wheat and oats and (more ominously) sugarcane and new species of weeds such as the dandelion. Meanwhile, New World staples such as corn, potatoes, and beans diffused throughout the rest of the globe.[7] Silver strikes in Zacatecas, Guanajuato, San Luis Potosí, and Pachuca ushered in socially and ecologically devastating extractivist regimes in the mid-sixteenth century that resulted in locally significant deforestation, erosion, and accumulations of toxic mining tailings.[8] Agricultural estates grew up to meet the demand for food in cities, and mining centers also wrought environmental change as landowners converted forests into croplands and pasture, introduced livestock into the arid north and the torrid south, and diverted water for irrigation.[9]

Generations of historians have studied environmental revolution associated with colonialism, but they have largely ignored the more recent and ambiguous political ecology of modernity.[10] Yet the relationship between people and the landscape has undergone fundamental transformations in the past two centuries as new technologies have revolutionized resource extraction, organic fuels have ceded to fossil fuels, and Mexico has become an urban nation. This does not add up to a one-way history of relentlessly increasing exploitation of natural resources and a secular trend of environmental decline, however. On the contrary, a close examination of Mexico's modern environmental history leads to a substantially more complicated periodization.

Intensive and Extensive Cycles in Modern Mexico

Over the past 250 years, Mexico has oscillated between phases of relative stability associated with economic growth and periods of political volatility accompanied by economic stagnation. Political stability has historically spurred economic expansion in ways that reinforced state authority. It has typically led landowners, mining interests, and others to invest in technologies that make relatively intensive uses of land and other resources both in renewable sectors such as agriculture and in extractive sectors such as mining and petroleum, as well as in those that combine elements of each, such as forestry and fisheries. It is hardly surprising that most mechanization, technological development, and investment in hydrological projects—and eventually the deployment of Green Revolution technologies such as pesticides and fertilizers—have occurred during times of relative social peace and economic advancement. These conditions have usually buoyed government revenues and allowed leaders at the state and national levels to extend their bureaucratic reach. The result has normally been to increase state capacities to promote and regulate the use of nature and to elaborate initiatives intended to promote modernization and continued growth. Periods of stability have also favored population growth, which can place new pressures on the land and, eventually, urban infrastructures. Finally, they have tended to overlap with heightened scientific study of the natural world and political interest in conservation.

These expansionist phases have been followed by periods of political instability and disinvestment during which state authority has crumbled. The resulting economic chaos and social dislocation made investment more risky, leading most landowners, entrepreneurs, and rural people to prefer extensive rather than intensive productive techniques, often in ways that the state could do little or nothing to regulate. Intervals of revolution, warfare, and crisis crippled the state's regulatory apparatus and capacity to enact developmentalist initiatives. Population growth usually declined or reversed, often in the context of migration to the cities or abroad. Scientific inquiry became difficult or impossible. In contrast to periods of stability, local authorities and private interests were left on their own to monitor the effects of natural resource use.

We can trace these transitions from what can be called the political ecology of centralization to that of decentralization over the course of modern Mexican history. For obvious reasons, they closely track the trajectory of Mexican political history more generally (see table 1.1). Centralization occurred between 1765 and 1808 during the culminating stage of the Bourbon reforms, then reappeared during the Porfirio Díaz administration (often called the Porfiriato) between 1876 and 1910 and again during the long phase of economic expansion known as the Mexican Miracle, from the mid-1940s to the 1980s. These phases differ from each other in several crucial regards, most notably that Bourbon centralization occurred in the context of colonialism, powered by organic forms of energy such as animal traction and wood, whereas fossil fuels incompletely displaced these sources during the nineteenth- and twentieth-century phases of centralization.[11] Following each of these periods came phases of decentralization and relatively extensive resource use, the first of which stretched from 1810 to the 1860s as a result of Mexico's protracted war for independence and the

Table 1.1 Phases of Mexico's political ecology, 1765–2011

Phase	Historical period	Mode of political ecology
1765–1810	Bourbon reforms	Intensive/centralized
1810–1876	Independence; age of caudillos	Extensive/decentralized
1876–1910	Porfiriato	Intensive/centralized
1910–1940s	Revolution and reconstruction	Extensive/decentralized
1940s–1982	Mexican Miracle	Intensive/centralized
1982–2011	Neoliberal era	Savage/decentralized

ensuing episodes of political instability, fiscal meltdown, civil war, and foreign invasion. Decentralization reappeared between 1910 and the 1940s on the heels of the Mexican Revolution, the 1918 influenza pandemic, and the 1929 global economic crisis. State power all but disappeared during some of these episodes, leaving rural communities and landowners to their own devices while choking off investment in agriculture and most extractive industries. The revolution also broke the back of landowners' political power and opened the way for a process of agrarian reform that culminated with massive land redistribution during the administration of Lázaro Cárdenas (1934–40), hence the radical extension of land cultivated through low-tech (though not necessarily low-impact) agricultural practices. The most recent phase of decentralization began with the neoliberal turn of the 1980s, when economic restructuring spurred migration (particularly from rural areas), while violence associated with narcotrafficking vitiated government authority in many cities and substantial parts of the coastline and borderlands.

No simple correlation exists between centralizing phases and the degradation of ecosystems on the one hand, and decentralization and environmental recovery on the other. Without a doubt, the relatively intensive practices associated with the political ecology of centralization *can* be devastating to the environment: mining, the exhaustive and unsustainable use of forests, new construction of irrigation systems and (eventually) dams and water pumping infrastructure, and the quest to make land more productive, all took a heavy toll on the natural world. Yet these phenomena typically occurred in the context of heightened, sometimes authoritarian, resource management policies that provided at least some buffers against environmental devastation. The central government created management policies, regulatory bureaucracies, and (eventually) environmental recovery programs that blunted some of the worst effects of "development." Conversely, the disappearance of state authority during periods of crisis, instability, and economic decline usually made it attractive for landowners and rural people to spread their crops and animals farther into the countryside. In some instances, it opened the door to the outright plunder of natural resources by corporations eager to take advantage of lax oversight by an enfeebled or distracted state apparatus. In other words, the usefulness of periodizing Mexico's environmental history in terms of centralizing versus decentralizing political ecologies lies not with these concepts' ability to predict environmental harm in a general sense but rather with their capacity to focus our attention on the complicated link between the nation's shifting political ecology and people's use of nature. It emphasizes the variations and change over time in resource use, while complicating "declension" narratives of secular environmental decline that describe the relationship between people in nature as the relentless expansion of modernity or of humans' unquenchable thirst for more and more natural resources.[12]

Toward an Environmental History of Mexico

Mexico's modern environmental progression began with the centralization prompted by the Bourbon reforms, a loosely related set of policies initiated in 1700 and bolstered by Charles III in 1765 after Spain's ignominious setbacks during the Seven Years' War. The Crown ordered a reorganization of the army and militias, moved to secularize political power (including the expulsion of the Jesuit order in 1764), and streamlined administration and the collection of taxes. The Crown declared "free trade" for most of the empire in 1778 and expanded it to Mexico (then New Spain) in 1789, thus breaking Cádiz's monopoly on trade and sparking a short-lived economic boom within the empire. The revitalization of colonial trade meant renewed demand for Mexican dyewoods, silver, cochineal, and even sugar, all of which had predictably negative consequences for the environment.[13] The Bourbon policies also unleashed new pressures on—and competition over—natural resources. Cattle ranches appeared in the northern frontier states of Nuevo Santander (Tamaulipas), Baja California, and parts of Texas, leading to wholesale changes in the landscape. Ranching also expanded in the Yucatán, putting new pressure on the Maya homeland and encouraging some native people to emulate the colonists and establish herds on land owned by *cofradías* (religious confraternities).[14] A similar integration of European and native productive strategies took place in at least some parts of central Mexico, where deepening commodification of corn and timber resources sparked conflicts between hacienda owners and indigenous communities. Yet the rebounding native population of the central valleys contended with a series of droughts probably caused by El Niño weather patterns, which cut short the rains during the 1780s and 1790s and led some Nahua villagers to migrate to nearby townships or invade hacienda lands.[15] It is not surprising in these circumstances that water became an increasingly central point of contention, not only in the far north—where the 1789 Plan de Pitic (modern-day Hermosillo, Sonora) set a precedent for the administration of water rights—but also in central parts of the nation such as Puebla, where Sonya Lipsett-Rivera detected the rise of a Creole "water monopolist class" in the late colonial period. These elites slowly chipped away at the water rights of indigenous communities and rival hacienda owners before the establishment of a formal water bureaucracy in 1808.[16]

The reinvigoration of colonial rule sparked innovations in agriculture and inspired scientific inquiry on a level not seen since the first decades of European exploration. Religious and secular scholars began to work out increasingly elaborate procedures to collect and categorize Mexican flora and fauna. They used their knowledge to combat disease through sanitary measures such as relocating grave sites outside of urban churchyards.[17] By the eve of independence in 1821, recognizably modern descriptions of natural history flowed from the pens of foreign observers such as Alexander von Humboldt, the Creole naturalist José Antonio de Alzate, and the director of the Mexico City Royal Botanical Garden, Martín Sessé y Lacasta.[18]

Independence undermined this dynamic. It removed indigenous communities from the Crown's tutelage and paternalist protections, decentralized political and economic structures, and opened the new nation to the global economy. Perhaps the greatest casualty of independence was political stability itself. Four decades of intermittent warfare and palace "revolutions" presaged the 1858–61 War of the Reform. The nation also struggled against occupations by the United States in the 1840s and France in the 1860s. These upheavals vitiated the authority of the central state and pummeled the economy but left the autonomy of regional power holders and many rural communities largely intact. Political insecurity wracked the countryside and ushered in a period of ecological decentralization. Haciendas confronted a credit crunch and repeated raids on their products to such an extent that they never fully recovered in many parts of the West, and some landowners simply split up their property and sold it to family farmers (*rancheros*) who could rarely make investments in irrigation or other modern technologies.[19] Historians know little about the environmental consequences of this process, though we may suppose that rancheros in particular cleared the land to make way for crops and cattle. We do know that local governments (*municipios*) decided questions of water rights and forest use without the paternalist royal bureaucracy to raise objections to the dispossession of native lands and resources.[20]

The disintegration of colonial authority also opened the way for a new kind of largely unregulated extractivism, especially by foreign interests. Silver mines had been devastated during the wars of independence as trade routes disappeared and workers abandoned the pumps that kept water from seeping into the mine shafts; by the 1840s, silver output had decreased by over 80 percent. Investors arrived in the north during the 1830s with new technology such as Cornish pumps and inefficient first-generation steam engines hungry for wood fuel, but even the miracle of steam could not revive the industry.[21] At the opposite end of the nation, British lumber interests greeted the waning official presence and the 1848 Caste War of the Yucatán by snapping up untold amounts of mahogany illegally logged in Mexico.[22] Foreign investment in industries such as mining did attempt to intensify the use of natural resources, but the social and political conditions after independence forced them to abandon most of these plans. Capital flight and the degradation of state authority left people in most parts of the country on their own to negotiate access to land, water, and forests. These conditions probably eased the overexploitation of ecosystems around the now-failing mines and the core fields of former haciendas but at the same time allowed entrepreneurs to open up new resource frontiers for export commodities.

Porfirio Díaz's modernizing regime reversed most of these trends between 1876 and 1910. Díaz used a combination of political cunning, foreign investment, and self-perpetuation in office to forge the most stable regime Mexico had seen since the end of Spanish rule. His administration worked to stem banditry while paring down the army in a bid to keep it out of politics. He and his technocrat-advisors, known as *científicos*, demanded the privatization of communally owned lands while shoring up the rights of large landowners and other employers. Domestic and foreign capital began to flow into agriculture, mining, heavy industry, railroads, and the timber sector, all of which shifted the nation back to a political ecology of centralization. Both federal and state governments granted concessions to railroad companies, giving rise to a transportation revolution that created unprecedented demand for wood used as ties and fuel for steam engines.[23] Mining also made a comeback thanks to British, North American, and Mexican capital, as well as new technologies such as dynamite and pneumatic drills. The largest precious metals mines adopted cyanide refining, which demanded less fuel than the older *patio* amalgamation process but came at a predictably high environmental cost in terms of polluted waterways and public health.[24] Nor was this the only extractive industry that dealt in toxins. The petroleum industry that developed in the 1890s in the Huasteca region of Veracruz remade the environment by clearing jungle to make way for oil derricks and holding tanks, many of which posed serious hazards to the health of Mexican workers and the environment in which they lived.[25]

Porfirian centralization coincided with unprecedented levels of state support for economic development, in particular through the tendering of concessions that allowed private companies privileged access to natural resources in exchange for capital investment. In theory, concessions also functioned to outsource management functions, insofar as concession holders typically pledged to conserve the resources over which they gained dominion. In fact, they had more ambiguous effects. Concessions for pearl oyster collection in Baja California, for example, led one foreign-owned corporation to pillage oyster beds in a quest for short-term profits, while another responded by developing a pathbreaking process of aquaculture that produced impressive harvests of pearls and performed a host of environmental services in the Gulf of California. Moreover, Porfirian political and fiscal stability underwrote a renaissance in sciences such as cartography, agronomy, meteorology, and forestry.[26]

The Porfiriato is probably best known as the triumphant culmination of nineteenth-century liberalism, a period when discriminatory but market-oriented legislation and pliant functionaries facilitated the massive transfer of land from rural communities to hacienda owners, rancheros, and wealthy villagers. What these new owners did with their property varied from place to place, but the development of reliable markets and transportation networks typically created incentives for landowners to intensify agricultural practices on at least a portion of their holdings. Planters in thickly settled regions such as the Bajío and the Valley of Mexico attempted to redouble production using tractors, irrigation pumps, and improved varieties of crops and livestock to open rich new croplands to commercial agriculture. Hacienda owners in the Chaco and the Michoacán hotlands used imported technology to drain the shallow lakes on which villagers depended.[27] Agro-industry connected to global markets also appeared in formerly marginal regions on the national frontiers. Planters dispossessed Maya communities in Yucatán to plant the henequen for cordage that wheat farmers in the US and Canadian heartlands fed to their McCormick reapers. Cotton bloomed in the desertlike Laguna district of the Coahuila-Durango border thanks to irrigation and water pumps. Cattle ranches stretched far into the plains of Chihuahua once the Apache threat disappeared in the 1880s. All the while, railroads threaded deeper into the countryside and made it possible to transport agricultural goods to distant markets; their arrival boosted the value of property close to railways and led in some instances to social conflict.[28] The conjuncture of these factors spurred sugar plantations in Morelos to irrigate more and more of what had become some of the most valuable sugar land in the world, touching off a scramble to control water on such a scale that it ultimately exacerbated social tensions in the homeland of Zapatismo.

The Revolution of 1910 had multiple and interconnected origins, but popular unrest caused by the massive commodification of the "lands, woods, and waters" (to quote the Plan of Ayala) clearly stoked the fires. Most mines, oil wells, logging companies, and haciendas tried to remain in operation until 1912 or so, but the violence and instability of the revolution eventually forced most of them to close down, at least temporarily. Some landowners took advantage of the chaos to extend their operations at the expense of neighboring estates and villages; in other instances, rural people invaded land that they had lost to private landowners over the previous decades or perhaps that they simply coveted. In either case, rural unrest decentered the nation's political ecology, particularly once the vicious warfare of 1914–15 forced extractive industries out of business in most of the country and imperiled agriculture on both the commercial and the village scales. This productive disruption took a devastating toll during the hungry years after 1916, which brought migration, dislocation, and death from the 1918 global influenza pandemic.[29] The population declined and streamed out of the countryside, while official surveillance all but disappeared and allowed those who remained behind to make their way on their own. This usually meant a more extensive regime of land use, as in the case of the Zapatistas' short-lived redistribution of sugar plantations in Morelos.

Postrevolutionary presidents tried desperately to jump-start the nation's economy from 1917 onward by encouraging new investment and reopening the door to extractive industries. Their efforts met with only partial success, because investors were spooked, not least because the agrarian reform authorized by the new constitution called the inviolability of private property into question. Land reform also placed the fields and forests into the hands of peasant communities that lacked the capital or technology to use them intensively, although many woodland communities immediately leased their woods to timber companies. The Cárdenas administration of 1934–40 sped the pace of redistribution and labored to provide peasant communities the tools to use their land more efficiently by making rural credit available, organizing producers' cooperatives, and using institutions such as the Banco

Ejidal as makeshift agricultural extension agencies. Emily Wakild and I have described these development-minded Cardenista policies as "social landscaping," that is, a holistic political project intended to manage rural society and nature together to rationalize the countryside.[30] Cardenismo combined elements of centralization and decentralization insofar as it sought to increase rural productivity and manage resource use in the context of a strengthening state, yet it decentralized many elements of rural political life by granting land from haciendas and agribusinesses to rural communities. One result of these contradictory trends was the creation of hybrid spaces such as parks and nature preserves intended both to protect nature and to open it to use by people from a variety of social classes.[31] Indeed, the Cardenista experiment in social landscaping might have offered an alternative to the cycles of centralization and decentralization if not for the realignment of political priorities that took place in the wake of World War II.

The period known as the Mexican Miracle dated from the mid-1940s to the debt crisis of 1982 and ushered in another round of centralization. Rapid industrial development and an average rate of economic growth that hovered around 6 percent per annum, migration to the cities, and a demographic explosion that tripled the nation's population from approximately 20 million to over 60 million. Declining infant mortality and improved healthcare made for a nation that was young, urban, and upwardly mobile. The official party, known as the Party of the Institutionalized Revolution (PRI), had monopolistic control of a well-funded and efficacious state apparatus. Political leaders used power over tariff policies and control of labor and campesino unions to embark on a program of import-substitution industrialization and agricultural modernization based on the use of pesticides and fertilizers.

Although investment in the land reform sector declined and the *ejidal* system began to wither, federal policies did target some rural areas for development.[32] "River basin" commissions were charged with building roads, dams, and irrigation systems, as well as providing basic human services, in the Tepalcatepec, Papaloapan, and other watersheds.[33] Extension services and Green Revolution technologies aimed primarily at highly capitalized agribusinesses succeeded in intensifying the use of commercial land, although at great ecological cost.[34] The state also began to create neoconcessions for logging companies and paragovernmental corporations in industry as well as in natural resource extraction.[35] These government-sponsored corporations (sometimes in the form of workers' cooperatives) also appeared in high-tech sectors such as pharmaceuticals, including a major push into steroid hormone production derived from an endemic species of yam, which soon became rare in the countryside.[36] Industrialization and the booming urban population also created new challenges, including pollution and an immense demand for running water and sewerage that the federal government struggled to provide during a half-century interval before capitulating. Since the 1980s, drinking water has effectively been privatized, while municipalities have inherited the nation's increasingly antiquated water infrastructure.

The period of neoliberalism from 1982 to the present bears many of the hallmarks of a decentralized political ecology, although with a few notable departures from previous phases. Neoliberal ideology and a globalizing economy rather than warfare or revolution curbed the reach of state power in this phase. Initially under the PRI, and later in the context of political pluralism, national leaders privatized the paragovernmental corporations, abandoned the model of state-led industrialization and commercial protectionism, ended land reform, and allowed markets to determine which industries would survive and which would perish in the global economy. Notwithstanding the 1994 Zapatista uprising in Chiapas, the state initially appeared capable of maintaining social order and even making new investments in highways, railroads, and urban infrastructure. It seemed as if the political ecology of centralization would endure even though the federal government had surrendered much of its control over economic and environmental regulation. By the end of the twentieth century, however, it had become clear that the neoliberal turn had catalyzed what can best be called a political ecology of "savage decentralization." Official abandonment of the popular classes and the loss of (oftentimes unproductive) industrial jobs provoked mass migration to the cities and abroad. The decline of federal authority and the PRI-centric status quo also helped to feed the rise of narcotrafficking networks that flourished in Mexico as a result of the United States' interdiction policies in South America and its insatiable demand for illegal narcotics. Drug lords virtually displaced state authority in some parts of the countryside, while deregulation in mining, agriculture, forestry, industry, and other sectors allowed corporations (now viewed as the providers of increasingly rare high-quality jobs) to displace the state and begin using natural resources and the environment without the inconvenience of "burdensome" government regulations.

Neoliberalism and the overall climate of insecurity have brought new pressures on the environment. The virtual abandonment of the countryside has left the rural poor to extend production wherever possible, above all by pushing the cattle frontier into marginal land, jungles, and forests. Others have turned to nontraditional crops such as marijuana or opium poppies. Many land reform beneficiaries have divided and privatized their land reform parcels through the federal PROCEDE program, as pressure on other landscapes has increased from agro-industry and mining, as well as the toxic industrial production of the northern *maquila* sector. Organized crime syndicates have appropriated remote forests and seacoasts to produce drugs and conceal their trade routes. Millions of people have simply abandoned the countryside in favor of a new life in the cities or the United States. Rural out-migration has led in some instances to the decommodification of nature, as overworked land is left fallow, unused forests regrow, and played-out silver mines become nothing more than toxic relics of a bygone era.[37] Yet migration can also spell trouble for cities in the form of more pollution, overtaxed mass transit systems, and increased demand on inadequate water infrastructure. In the case of the nation's capital, these challenges have taken place in the context of shortages so severe that officials have routinely been forced to ration water.[38]

Conceptualizing the environmental history of modern Mexico as a cycle of centralizing and decentralizing phases of resource use does not merely reveal the fundamental interconnectedness of political economy and the environment; it suggests that a systematic study of environmental history can contribute to contemporary debates about environmental policy. For example, proposals to create markets in carbon emissions credits and the United Nations' initiative on Reducing Emissions from Deforestation and Forest Degradation (REDD+) have led to lengthy policy debates about whether centralized governance structures protect the environment better than decentralized frameworks that might generate a higher degree of local participation in conservation initiatives.[39] The brief overview presented here suggests that the issue needs to be reframed, insofar as environmental degradation and ecosystem recovery have occurred during both decentralizing and centralizing phases of resource extraction. The political ecology of centralization has historically been associated with state policies that both promote and regulate resource use; by the same token, decentralized modes have historically arisen during periods of diminished state power, allowing communities and commercial interests alike greater latitude either to preserve or to despoil the resources under their control.

The Cycles of Mexican Environmental Historiography

If the periodization presented here seems anecdotal in places, one reason is that key historical questions about Mexico's past remain unexplored even though foundational works of environmental history began to appear over half a century ago. The first phase of research began in the late 1950s with Sherburne Cook's examination on soil erosion and demography in pre-Hispanic Mexico and culminated with Alfred Crosby's deeply influential work on the Columbian Exchange in 1972.[40] These books placed Mexican studies at the leading edge of a burgeoning scholarly interest in the history of human interaction with the natural world. Yet relatively few historians followed up on these promising beginnings. Only a handful of Mexican environmental histories appeared in the 1970s and 1980s, although the ones that did laid the foundations for most of the authors writing today.[41] The 1990s brought renewed interest as historians turned their attention to the politics of water and its role in shaping the agricultural landscape, especially during the colonial period.[42] Historians opened up other avenues of analysis as well. Elinor Melville's influential work on the introduction of sheep in the Valle del Mezquital revised Cook's arguments about erosion and described how untended animals grazed native plants down to the roots, leaving behind a virtual desert. Alejandro Tortolero published a volume of essays that provided the first general overview of environmental change in central Mexico. Cynthia Radding investigated the relationship between native people and the landscape in colonial Mexico (insights she later expanded into a comparative analysis with Bolivia), and Brigitte Boehm, along with her colleagues at El Colegio de Michoacán, launched a long-term project that interprets the history of the Lerma River basin as a closely linked social and ecological network.[43] Only recently, however, has a sufficiently large cohort of environmental historians emerged to maintain a scholarly conversation about the interaction between people and nature in the modern era.[44]

This volume integrates the scholarly voices that appeared in the 1980s and 1990s with those from the subsequent generation of historians. Rather than take an encyclopedic approach to the Mexican environment, it functions as an "invitation" to Mexican environmental history (to paraphrase Luis González) by presenting a snapshot of modern Mexican environmental historiography at this stage of its development. That does not mean it contains a mere hodgepodge of topics, however. The themes its authors take up reflect the ecological challenges that Mexico has faced in the past and continues to confront today. Unsurprisingly, water receives the most attention.[45] Martín Sánchez, Alejandro Tortolero, and Luis Aboites investigate its role in agriculture and urban development to demonstrate how it became an object of political and social struggle in modern Mexico. They show that the technologies for the retention and distribution of water captured the imagination of landowners, urbanites, and political leaders, many of whom regarded irrigation, dams, and (in the twentieth century) running water and sewage as emblems of modernity. Chapters by Angus Wright and Sterling Evans address the pivotal issue of land use, commercialization, and export agriculture in central Mexico and Yucatán. José Juan Juárez and Emily Wakild discuss the misuse and attempted conservation of forest landscapes in Porfirian Tlaxcala, in the carefully tended groves of Chapultepec Park, and beyond. Mario Monteforte and Micheline Cariño contribute a chapter that punctuates a discussion of the long-term decline of fisheries—specifically, the pearl oysters of the Gulf of California—with a surprising case study of a nineteenth-century experiment in aquaculture that seemed to offer a viable alternative. All of the authors emphasize the social embeddedness of the environment, but two authors pay particular attention to the scientific context and political economy of human relationships with nature: Rick López uses the Mexican Royal Botanical Garden as a window onto elite conceptions of the natural landscape, while Myrna Santiago describes the interplay of class and ecology in the oil fields of the Huasteca Veracruzana.

The breadth and depth of contributions demonstrate the fundamental place of water, land, forest, petroleum, and science in structuring the relationship between people and the natural world, but the book's silences attest to the scale of work that remains to be done. Despite the work of Wakild and Aboites, for instance, urban environmental history remains a largely undeveloped topic.[46] While Cariño and Monteforte have opened a path for the study of fisheries, we still know very little about their history either inland or in most coastal areas.[47] And notwithstanding the work of Wright, Evans, and others—not to mention the early work of Cook—we have just begun to explore the history of soils and land conversion in Mexico. Another obvious direction for new studies lies at the crossroads of environmental history and other fields: historians of the urban experience, gender, and science will find that environmental topics remain all but unknown. Equally significant is that historians clearly have tended to concentrate on periods of intensification that favored the expansion of irrigation and extractive industries, brought new development in agriculture and forestry, and facilitated advances in cartography, museology, and the natural sciences. We know significantly less about periods of decentralization and political flux, when economic turmoil and the retreat of the state resulted in a more extensive use of natural resources. It seems quite possible that concerted research on these crucial phases of state withdrawal and economic decline may complicate our understanding of environmental change and yield insights into the mechanisms of local environmental stewardship.

Studies such as the ones in this book come not a moment too soon. Mexico's current regime of deregulation and neoliberalism has coincided with the breakdown of state authority on a scale unseen since the 1910–17 revolution and is trending toward an increasingly decentralized and savagely extensive use of nature. At the same time, transnational environmental policies (whether in the form of a global market for carbon emissions, REDD+, or some other comprehensive climate initiative) appear likely to gain traction in Mexico and throughout the hemisphere, with significant implications for rural people—indeed, all people—in Mexico. While the insights that environmental history can offer in this context are not sufficient in themselves to guide national policies, it is nonetheless true that robust and engaged environmental histories of Mexico can provide invaluable insights into the relationship between the state, the economy, and the environment. In so doing, they might provide an outline that can prevent Mexicans from rushing headlong into an increasingly clouded future.

Notes

1. Literal translators such as the great cartographer Antonio García Cubas agree that *Anáhuac* means "the land near the water" or "next to [*junto a*] the waters," but today it is more commonly rendered as "between the waters." Historians may have picked up on the latter translation, first used by Mariano Fernández de Echeverría y Veytia in his *Historia Antigua de Méjico*, 2 vols., comp. C. F. Ortega (Mexico City: Imprenta Juan Ojeda, 1836), 1:1, precisely because it was used by the first great popularizer of Mexican history, William Prescott, who understood it to suggest that Anáhuac stretched from one ocean to the other.

2. Many observers believe that the term *Anáhuac* eventually came to refer to the Aztec empire, rather than the Valley of Mexico proper during the final years of the Aztec empire.

3. For an overview of the Mexican landscape, see Christopher R. Boyer, "Historical Geography," in *A Companion to Mexican History and Culture*, ed. William Beezley, 119–30 (Oxford, U.K.: Wiley-Blackwell, 2011). On irrigation and water management in the modern period, see chapters by Sánchez, Tortolero, and Aboites in this volume; see also Antonio Escobar Ohmstede, Martín Sánchez Rodríguez, and Ana María Gutiérrez Rivas, eds., *Agua y tierra en México, siglos XIX y XX*, 2 vols. (Zamora: El Colegio de Michoacán; San Luis Potosí: El Colegio de San Luis, 2008); and Brigitte Boehm Schoendube, "Historias del agua en zonas de alta inversión para el desarrollo en el centro occidente de México," in *El agua en la historia de México*, ed. Manuel Durán, Martín Sánchez, and Antonio Escobar, 33–59 (Guadalajara: Centro Universitario de Ciencias Sociales y Humanidades de la Universidad de Guadalajara; Zamora: El Colegio de Michoacán, 2005).

4. Arturo Warman, *La historia de un bastardo: Maiz y capitalismo* (Mexico City: Fondo de Cultura Económica, 1988).

5. See synoptic treatments by Enrique Florescano, *Breve historia de la sequía en México* (Mexico City: CONACULTA, 2000); and Georgina H. Enfield, *Climate and Society in Colonial Mexico: A Study in Vulnerability* (London: Blackwell, 2008).

6. The fundamental source on demography remains Sherburne F. Cook and Woodrow W. Borah, *The Indian Population of Central Mexico* (Berkeley: University of California Press, 1963).

7. Alfred W. Crosby, *The Columbian Exchange: Biological and Cultural Consequences of 1492* (Westport, Conn.: Greenwood, 1972).

8. Elinor G. K. Melville, "Disease, Ecology, and the Environment," in *The Oxford History of Mexico*, ed. Michael C. Meyer and William H. Beezley, 213–43 (New York: Oxford University Press, 2000). Information on the continued presence of toxins is from José Antonio Ávalos Lozano and Miguel Aguilar Robledo, "El impacto de la industria minero-metalúrgica sobre los paisajes de la Nueva España, siglos XVIII y XIX, estudio de caso en el Sitio Sagrado Natural de Wirikuta," presentation at the V Simposio de la Sociedad Latinoamericana y Caribeña de Historia Ambiental, La Paz, Mexico, 15–18 June 2010.

9. Major works that attend to the environmental dimension of colonialism include Felipe Castro Gutiérrez, *Los tarascos y el imperio español, 1600–1740* (Mexico City: UNAM; Morelia: Universidad Michoacana, 2004), esp. 305–44; Elinor G. K. Melville, *A Plague of Sheep: Environmental Consequences of the Conquest of Mexico* (Cambridge: Cambridge University Press, 1994); Cynthia Radding, *Wandering Peoples: Colonialism, Ethnic Spaces, and Ecological Frontiers in Northwestern Mexico, 1700–1850* (Durham, N.C.: Duke University Press, 1997); and Kevin Terraciano, *The Mixtecs of Colonial Oaxaca: Ñudzahui History, Sixteenth through Eighteenth Centuries* (Stanford: Stanford University Press, 2001).

10. Political ecology can roughly be defined as the study of how economic, political, social, and cultural structures transform and are transformed by the natural world. For a brief description, see James B. Greenberg and Thomas K. Park, "Political Ecology," *Journal of Political Ecology* 1, no. 1 (1994): 1–12.

11. Edmund Burke III, "The Big Story: Human History, Energy Regimes, and the Environment," in *The Environment and World History*, ed. Edmund Burke III and Kenneth Pomeranz, 33–53 (Berkeley: University of California Press, 2009).

12. See José Agusto Pádua, "As bases teóricas da história ambiental," *Estudos Avançados* 24, no. 68 (2010): 81–101.

13. John Fisher, "The Imperial Response to 'Free Trade': Spanish Imports from Spanish America, 1778–1796," *Journal of Latin American Studies* 17, no. 1 (May 1985): 35–78. See also Daviken Studnicki-Gizbert and David Schecter, "The Environmental Dynamics of a Colonial Fuel-Rush: Silver Mining and Deforestation in New Spain, 1522 to 1810," *Environmental History* 15 (January 2010): 94–119.

14. Nancy M. Farriss, *Maya Society under Colonial Rule: The Collective Enterprise of Survival* (Princeton: Princeton University Press, 1984), 33–35 and 267–80; Terry G. Jordan, *North American Cattle-Ranching Frontiers: Origins, Diffusion, and Differentiation* (Albuquerque: University of New Mexico Press, 1993), 123–56; Lucia Hernández, ed., *Historia ambiental de la ganadería en México* (Xalapa: Instituto de Ecología, 2001).

15. Alejandro Tortolero Villaseñor, *Notarios y agricultures: Crecimiento y atraso en el campo mexicano, 1780–1920* (Mexico City: Siglo XXI, 2008), 184–86, 201–11; Arij

Ouweneel, *Shadows over Anáhuac: An Ecological Interpretation of Crisis and Development in Central Mexico, 1730–1800* (Albuquerque: University of New Mexico Press, 1996), 59–100.

16. Sonya Lipsett-Rivera, *To Defend Our Water with the Blood of Our Veins: The Struggle for Resources in Colonial Puebla* (Albuquerque: University of New Mexico Press, 1999), 77; Michael C. Meyer, *Water in the Hispanic Southwest: A Social and Legal History, 1550–1850*, with a new afterword (Tucson: University of Arizona Press, 1996), esp. 30–45.

17. Jorge Cañizares-Esguerra, "From Baroque to Modern Colonial Science," in *Nature, Empire, and Nation: Explorations of the History of Science in the Iberian World* (Stanford: Stanford University Press, 2006); and Pamela Voekel, *Alone Before God: The Religious Origins of Modernity in Mexico* (Durham, N.C.: Duke University Press, 2002).

18. Mauricio Beuchot, *Filosofía y ciencia en el México dieciochesco* (Mexico City: UNAM, 1996). See also Rick López's contribution to this book.

19. Margaret Chowning, *Wealth and Power in Provincial Mexico: Michoacán from the Late Colony to the Revolution* (Stanford: Stanford University Press, 1999), 261–331. See also the classic article by Jan Bazant, "The Division of Some Mexican Haciendas during the Liberal Revolution," *Journal of Latin American Studies* 3, no. 1 (May 1971): 25–37.

20. Luis Aboites, *El agua de la nación: Una historia política de México (1888–1946)* (Mexico City: CIESAS, 1998), 26–45; Martín Sánchez Rodríguez, *"El mejor de los títulos": Riego, organización social y administración de recursos hidráulicos en el Bajío mexicano* (Zamora: El Colegio de Michoacán, 2005).

21. Cuauhtémoc Velasco Avila, Eduardo Flores Clair, Alma Parra, and Edgar Gutiérrez, *Estado y minería en México (1767–1910)* (Mexico City: Fondo de Cultura Económica, 1988); and José Alfredo Uribe Salas, "El distrito minero El Oro–Tlalpujahua entre dos siglos y el mercado internacional de tecnología," in *Five Centuries of Mexican History/Cinco siglos de historia de México: Papers of the VIII Conference of Mexican and North American Historians, San Diego, California, October 18–20, 1990*, ed. Virginia Gueda and Jaime E. Rodríguez O., 119–35 (Mexico City: Instituto Mora, 1992).

22. Jan de Vos, *Oro verde: La conquista de la Selva Lacandona por los madereros tabasqueños* (Mexico City: Fondo de Cultura Económica), 202–27.

23. In addition to Juárez's contribution to this book, see my forthcoming *Political Landscapes: Native Communities and Resource Management in Mexican Forests, 1880–2000* (Durham, N.C.: Duke University Press); and Andrew Matthews's *Instituting Nature: Authority, Expertise, and Power in Mexican Forests* (Cambridge, Mass.: MIT Press, 2011). For case studies of the ecological impact of railroads in San Luis Potosí, see Luz Carregha Lamadrid, "Tierra y agua para ferrocarriles en los partidos del Oriente potosino, 1878–1901," and Miguel Ángel Solís Esquivel, "Ferrocarriles y recursos naturales: La construcción del ramal San Bartolo/Rioverde, 1899–1902," both in *Entretejiendo el mundo rural en el "oriente" de San Luis Potosí, siglos XIX y XX*, ed. Antonio Escobar Ohmstede and Ana María Gutiérrez Rivas, 177–204 and 205–36, respectively (San Luis Potosí: El Colegio de San Luis; Mexico City: CIESAS, 2009).

24. I would like to thank Ted Beatty for sharing his ongoing research on the uses of cyanide during the Porfiriato. See also John Mason Hart, *The Silver of the Sierra Madre: John Robinson, Boss Shepherd, and the People of the Canyons* (Tucson: University of Arizona Press, 2008), 163–65.

25. Myrna I. Santiago, *The Ecology of Oil: Environment, Labor, and the Mexican Revolution, 1900–1938* (Cambridge: Cambridge University Press, 2006).

26. Lane Simonian, *Defending the Land of the Jaguar: A History of Conservation in Mexico* (Austin: University of Texas Press, 1995), 59–84. On sciences in particular, see Raymond Craib, *Cartographic Mexico: A History of State Fixations and Fugitive Landscapes* (Durham, N.C.: Duke University Press, 2004); and Joseph Cotter, *Troubled Harvest: Agronomy and Revolution in Mexico, 1880–2002* (Westport, Conn.: Praeger Publishers, 2003), 19–48.

27. Alejandro Tortolero, *De la coa a la máquina de vapor: Actividad agrícola e innovación tecnológica en las haciendas de la región central de México, 1880–1914* (Mexico City: Siglo XXI, 1995); Gustavo Lorenzana Durán, "Las aguas del canal Porfirio Díaz: Una disputa entre la Compañía Constructora Richardson y los colonos de Cócorit, Bácum y San José, 1911–1912," in *Negociaciones, acuerdos y conflictos en México, siglos XIX y XX: Agua y tierra*, ed. Aquiles Omar Ávila Quijas, Jesús Gómez Serrano, Antonio Escobar Ohmstede, and Martín Sánchez Rodríguez, 225–44 (Zamora: El Colegio de Michoacán, 2009); Paul Friedrich, *Agrarian Revolt in a Mexican Village*, rev. ed. (Chicago: University of Chicago Press, 1977), 43–49; Alfredo Pureco Ornelas, *Empresarios lombardos en Michoacán: La familia Cusi entre el porfiriato y la postrevolución (1884–1938)* (Zamora: El Colegio de Michoacán, 2010), 183–221.

28. John Coatsworth, *Growth against Development: The Economic Impact of Railroads in Porfirian Mexico* (DeKalb: University of Northern Illinois Press, 1981), 149–74.

29. Leticia Gamboa Ojeda, "La epidemia de influenza de 1918: Sanidad y política en la ciudad de Puebla," *Quipu* 8 (1991): 91–109.

30. Christopher R. Boyer and Emily Wakild, "Social Landscaping in the Forests of Mexico: An Environmental Interpretation of Cardenismo, 1934–1940," *Hispanic American Historical Review* 92, no. 1 (February 2012): 73–106.

31. See Wakild, *Revolutionary Parks: Conservation, Social Justice, and Mexico's National Parks, 1910–1940* (Tucson: University of Arizona Press, 2011); Simonian, *Defending the Land*, 85–110; Boyer, "Revolución y paternalismo ecológico: Miguel Ángel de Quevedo y la política forestal, 1926–1940," *Historia Mexicana* 57, no. 1 (July–September 2007): 91–138. For discussions of the use of water in this period, see Luis Aboites Aguilar, *La irrigación revolucionaria: Historia del sistema nacional de riego del Río Conchos, Chihuahua, 1927–1938* (Mexico City: SEP/CIESAS, 1987); Aquiles Omar Ávila Quijas, "La organización para la distribución del agua de la presa de Malpaso y el Estado mexicano, 1909–1922," in *Negociaciones, acuerdos y conflictos en México, siglos XIX y XX: Agua y tierra*, ed. Aquiles Omar Ávila Quijas, Jesús Gómez Serrano, Antonio Escobar Ohmstede, and Martín Sánchez Rodríguez (Zamora: El Colegio de Michoacán, 2009); Matthew Vitz, "'The Lands with Which We Shall Struggle': Land Reclamation, Revolution, and Development in Mexico's Lake Texcoco Basin, 1910–1950," *Hispanic American Historical Review* 92, no. 1 (February 2012): 41–71; and Mikael Wolfe, "Conflicto por un cambio de régimen de aguas en La Laguna: La 'construcción social' de la primera gran presa en el río Nazas, 1900–1936," *Buenaval* 2 (Summer 2006): 1–35.

32. Fundamental discussions of this process include Armando Bartra, *Los herederos de Zapata: Movimientos campesinos postrevolucionarios en México, 1920–1980* (Mexico City: Era, 1985); John Gledhill, *Casi Nada: A Study of Agrarian Reform in the Homeland of Cardenismo* (Albany: SUNY Institute of Mesoamerican Studies; Austin: University

of Texas Press, 1991); and Arturo Warman, *El campo mexicano en el siglo XX* (Mexico City: Fondo de Cultura Económica, 2001).

33. David Barkin and Timothy King, *Regional Economic Development: The River Basin Approach in Mexico* (Cambridge: Cambridge University Press, 1970).

34. See also Angus Wright, *The Death of Ramón González: The Modern Agricultural Dilemma*, 2nd ed. (Austin: University of Texas Press, 2005); and Cotter, *Troubled Harvest*, 179–279.

35. Dan Klooster, "Conflict in the Commons: Commercial Forestry and Conservation in Mexican Indigenous Communities," PhD diss., University of California, Los Angeles, 1997, pp. 189–93; Francisco Chapela, "Indigenous Community Forest Management in the Sierra Juárez, Oaxaca," in *The Community-Managed Forests of Mexico: The Struggle for Equity and Sustainability*, ed. David Barton Bray, Leticia Merino-Pérez, and Deborah Barry, 91–110 (Austin: University of Texas Press, 2005); Andrew Salvador Mathews, "Suppressing Fire and Memory: Environmental Degradation and Political Restoration in the Sierra Juárez of Oaxaca, 1887–2001," *Environmental History* 8, no. 1 (January 2003): 77–108.

36. Christopher R. Boyer, "Contested Terrain: Forestry Regimes and Community Responses in Northeastern Michoacán, 1940–2000," in *The Community Forests of Mexico*, ed. David Barton Bray, Leticia Merino-Pérez, and Deborah Barry, 27–48 (Austin: University of Texas Press, 2005); Gabriela Soto Laveaga, *Jungle Laboratories: Mexican Peasants, National Projects, and the Making of the Pill* (Durham, N.C.: Duke University Press, 2009).

37. The literature on land conversion is too vast to gloss here. An important starting point for such studies, however, is Jacques M. Chevalier and Daniel Buckles, *A Land Without Gods: Process Theory, Maldevelopment, and The Mexican Nahuas* (London: Zed Books, 1995). For an overview of the link between migration and afforestation, see Dan Klooster, "Beyond Deforestation: The Social Context of Forest Change in Two Indigenous Communities in Highland Mexico," *Journal of Latin American Geography* 26 (2000): 47–59.

38. Diane E. Davis, "The Social Construction of Mexico City," *Journal of Urban History* 24, no. 3 (March 1998): 364–413; Joel Simon, *Endangered Mexico: An Environment on the Edge* (San Francisco, Calif.: Sierra Club Books, 1997), 60–90.

39. For a recent debate on these themes in Latin America, see "Summary of the Oaxaca Workshop on Forest Governance, Decentralization and REDD+ in Latin America and the Caribbean, 31 August–3 September 2010," *Linkages: ISD Reporting Services* 130, no. 1 (6 September 2010).

40. Crosby's work (*The Columbian Exchange*) built upon a previous generation of historical demography that examined the holocaust wrought by European diseases and wholesale environmental change. This literature includes, most notably, Sherburne Cook, *Soil Erosion and Population in Central Mexico* (Berkeley: University of California Press, 1949); Sherburne F. Cook and Woodrow W. Borah, *The Indian Population of Central Mexico* (Berkeley: University of California Press, 1963); and Carl O. Sauer, *The Early Spanish Main* (Berkeley: University of California Press, 1966).

41. Fernando Ortiz Monasterio, ed., *Tierra profanada: Historia ambiental de México* (Mexico City: INAH/Secretaría de Desarrollo Urbano y Ecología, 1987); and Michael C. Meyer, *Water in the Hispanic Southwest: A Social and Legal History, 1550–1850* (Tucson: University of Arizona Press, 1984). A year earlier, Clifton B. Kroeber published

Man, Land, and Water: Mexico's Farmlands Irrigation Policies, 1885–1911 (Berkeley: University of California Press, 1983). Although not explicitly an environmental history, John Tutino's *From Insurrection to Revolution in Mexico: Social Bases of Agrarian Violence, 1750–1940* (Princeton: Princeton University Press, 1986) is a valuable resource.

42. Luis Aboites, *El agua de la nación*; Lipsett-Rivera, *To Defend Our Water*; Exequiel Ezcurra, *De las Chinampas a la megalopolis: El medio ambiente en la cuenca de México* (Mexico City: Fondo de Cultura Económica, 1991).

43. Alejandro Tortolero Villaseñor, ed., *Tierra, agua y bosques: Historia y medio ambiente en el México central* (Mexico City: CEMCA, Instituto de Investigaciones Dr. José María Luis Mora; Guadalajara: Universidad de Guadalajara, 1996); Melville, *Plague of Sheep*; Cynthia Radding, *Wandering Peoples and Landscapes of Power and Identity: Comparative Histories in the Sonoran Desert and the Forests of Amazonia from Colony to Republic* (Durham, N.C.: Duke University Press, 2005). Boehm's work grew out of her earlier interest in the ecological history of the Lake of Chapala. Her tragic death in 2005 was a major setback for the project, which nevertheless continues to develop and has thus far led to the publications of two major collected works: Brigitte Boehm Schoendube, Juan Manuel Durán Juárez, Martín Sánchez Rodríguez, and Alicia Torres Rodríguez, eds., *Los estudios del agua en la cuenca Lerma-Chapala-Santiago*, vols. 1 and 2 (Zamora: El Colegio de Michoacán; Guadalajara: Universidad de Guadalajara, 2002 and 2005).

44. This resurgence of Mexican environmental history was part of a flourishing interest in environmental history throughout Latin America in the 1990s and 2000s. Recent overviews of this process include Stefania Gallini, "Historia, ambiente, política: El camino de la historia ambiental en América Latina," *Nómadas* 30 (April 2009): 92–102; Lise Sedrez, "Latin American Environmental History: A Shifting Old/New Field," in *The Environment and World History*, ed. Edmund Burke III and Kenneth Pomeranz, 255–75 (Berkeley: University of California Press, 2009); and on Mexico specifically, Emily Wakild, "Environmental History," in *A Companion to Mexican History and Culture*, ed. William H. Beezley, 518–37 (Oxford, U.K.: Wiley-Blackwell, 2011).

45. See also Juan Manuel Durán, Martín Sánchez, and Antonio Escobar, eds., *El agua en la historia de México* (Guadalajara: Universidad de Guadalajara; Zamora: El Colegio de México, 2005).

46. See Matthew Vitz, "Revolutionary Environments: The Politics of Nature and Space in the Valley of Mexico, 1890–1950," PhD diss., New York University, 2010.

47. Again, there are some exceptions. See José Adán Cháirez A., *Historia de la pesca del atún en México* (Ensenada: Editorial Chairez, 1996); Graciela Alcalá, *Políticas pesqueras en México (1946–2000): Contradicciones y aciertos en la planificación de la pesca nacional* (Mexico City: El Colegio de México, 2003).

CHAPTER TWO

Downslope and North
How Soil Degradation and Synthetic Pesticides Drove the Trajectory of Mexican Agriculture through the Twentieth Century

Angus Wright

The physical landscape of Mexico was transformed by the way the Mexican government responded to the nation's age-old problem of soil degradation. The response involved a commitment to large-scale irrigation works and the use of synthetic pesticides and fertilizers. This transformation of the landscape has multiple economic, political, and cultural implications not only for Mexican society but also for the entire planet. Mexican leaders and technical advisers from the United States justified their chosen path of agricultural development based on a view of Mexico's soils as naturally poor and severely degraded by use and abuse. The strategy was supported by the Rockefeller Foundation and the United States government both because it was seen as desirable for Mexico and because the Mexican case was seen as an experimental testing ground for American policy that could prove favorable to American commercial and foreign policy goals. As such, it was perhaps the most significant example of a much larger effort to promote the purposes of the United States government and private interests around the world through the use of technological and scientific expertise exercised, at least in theory, in the mutual interests of all involved. Although it had deeper roots, the design of the specific strategy emerged from the political situation in Mexico at the beginning of World War II and became the template for what would come to be called the Green Revolution, which transformed economies and landscapes around the world as the dominant model for international agricultural development and a central element of Cold War American foreign policy. It was a major factor in the great worldwide exodus of rural people to cities, but it also had other results for the movement of people and transformation of landscapes. As in Mexico, throughout the tropical and subtropical world one result was the tendency to drive economic development downslope into humid lowlands and desert valleys, with a variety of political, cultural, and economic consequences.[1]

Seeing the traditional areas of Mexican agriculture as naturally inadequate and exhausted by use, Mexican leaders and their foreign advisors in the twentieth century looked north toward desert valleys where state-financed large-scale irrigation projects could take advantage of land that had been relatively lightly used and little damaged, as well as to humid tropical lowlands farther south. The farming technologies developed to take advantage of these northern desert valley and humid tropical soils were integrally dependent on the use of newly invented synthetic fertilizers and pesticides. This is the story of how the approach taken to deal with soil degradation in Mexico became intertwined with pesticide use in a way that shaped the fate of Mexico's countryside and rural people.

The new synthetic pesticides, in addition to being a key part of what would come to be called "the package" of Green Revolution technologies, also seemed to offer an apparently simple solution to insect-borne Old World human diseases, particularly malaria and yellow fever. These diseases had been introduced into Mexico at the time of the Conquest and had tended to push dense human populations in humid lowlands upslope, away from mosquito-infested humid tropical areas. If both tropical crop pests and epidemic illness among the agricultural labor force could be controlled, a new era would open for the humid tropics.

Based on this view of Mexico's problems with soils and disease, by the mid-twentieth century, synthetic agrichemicals had become an important tool in overcoming a variety of barriers that stood in the way of a fuller use of Mexico's land resource and its expanding human population. Exhausted soils could be left behind and land put into production in developing economic frontiers.

The strategy of opening new frontiers was to be significantly undermined by a combination of ecological, economic, and health problems inherent in the newly adopted crop production technologies. It supported rapid economic growth by exploiting and in some ways widening the already deep inequalities in Mexican society. The beginning of the twenty-first century found many Mexicans looking for new political arrangements while at the same time searching for more sustainable solutions to the ancient problem of maintaining the health of the soil and, with it, the health of human society.

The Pre-Columbian and Colonial Background

The people of ancient Mesoamerica were intensely interested in the problem of soil erosion. As the inventors of many of the world's crop plants, most notably maize, Mesoamerican farmers had a long and complicated relationship with cultivated land.[2] For example, pre-Columbian people built terraces designed to maximize production while protecting against erosion on slopes. Some of these are still in use. There is also a good deal of evidence of the difficulties of controlling soil erosion and of dramatic failures to do so.[3] Archaeologists have identified major episodes of soil loss coincident with the early florescence of maize-based agriculture.[4] Success and failure alike left a heritage of knowledge and techniques among traditional Mexican farmers directed at protecting soils from erosion and maintaining or restoring lost fertility. Recent research, for example, suggests that the ancient Maya experienced episodes of severe erosion at the beginning of the Classic period but apparently learned to reduce erosion rates as population increased, rather than, as many previously thought, moving toward greater rates of erosion throughout the Classic period until collapse.[5]

Whatever may have occurred before their arrival, the Spanish were deeply impressed by the productive capabilities of the cultures they encountered: "An . . . obvious determination by the Spanish intruders was that the mainland landscapes of cultivation and settlement in Mexico were aesthetically pleasing, intricately managed, and manifestly prosperous. The pre-Conquest Indian inhabitants of Mexico had painstakingly constructed intensive, often irrigated, systems of productive horticulture tailored for each of the major ecological regions: the coastal lowlands, the piedmont, and the semi-arid, broad, volcano-surrounded basins of the high plateau. So successful were these systems of cultivation that the inhabitants clearly produced abundant and even surplus food for themselves."[6] The methods used to accomplish this impressive result reflected an evolved response to the opportunities and challenges offered by Mexico's landscape, whose varied physical geography has created an extraordinary diversity of biomes, species, and soils. Volcanic activity has created some soils that are extraordinarily fertile, being particularly rich in phosphate and potassium though often deficient in nitrogen. Volcanoes have also created many soils that are not suitable for crops, and even in the twentieth century, volcanoes in some places simply demolished fertile land under burdens of unworkable rock or toxic debris. High primary productivity deriving from the combination of high rainfall and intense tropical sunlight contributes large amounts of organic matter to soils in many areas, but torrential storms are capable of causing rapid soil erosion. Heavy rainfall can also leach nutrients from the soil, especially when they are put to agricultural use. The particular granular structure of soils in many regions, including, for example, the Mixtec highlands of Oaxaca, makes them peculiarly erosion prone. Potentially fertile desert soils are often the product of millions of years of alluvial deposition and bloom extravagantly with irrigation. These soils, however, tend to be subject to problems of salinity and perched water tables (where impermeable surfaces under the topsoil lead to poor drainage and the accumulation of salty water in the root zone) when converted to intensive agriculture. In much of Mexico, rugged topography alone virtually ensures significant erosion problems. At the same time, sediments delivered by precipitation running down mountainous slopes can sometimes provide new nutrients to valley soils.[7]

A seeming paradox of Mesoamerican history prior to the Spanish invasion derives from the relationship of sophisticated cultures with the characteristics of Mexican soils. The largely highland areas under Aztec sway alone are thought to have had a population in the range of 14 million people.[8] The paradox is that remarkably high population densities could be supported in agricultural societies on a land base that from the perspective of many later observers seemed to be incapable of doing so. The resolution of the paradox lies in the high potential fertility of a land that was also very vulnerable to degradation—a vital key to understanding the nation's history.

The work of geologist Sherburne Cook, historians Woodrow Wilson Borah and Lesley Bird Simpson, and geographer Carl Sauer in the mid-twentieth century originally shaped this key to use the history of soil erosion combined with the documentary record to unlock the story of the demographic collapse in Mexico (and, by extension, to the New World as a whole), opening our eyes to entirely new interpretations of the history of Mexico and all of the Americas.[9] This in turn helped clear the way for the social history of François Chevalier and "ethnohistory" of Charles Gibson, William Taylor, James Lockhart, and others.[10] The pathbreaking work of Gibson's student Elinor Melville in one of the most important works of the new wave of environmental history returned us to the themes developed by Cook and Borah while also giving us a much fuller and more detailed picture of the effect of the Conquest on the Mexican landscape and its peoples.[11]

Despite the strength of this current of historical study, historians have not always recognized the complex role of soil erosion. This is partly because of their inability to fully understand the effectiveness of the intensive management techniques used by pre-Columbian farmers and the damage done by the Conquest. More fatefully for Mexico and much of the world, agronomists and agricultural scientists mostly failed to achieve this understanding. Ideological predispositions and the political purposes of observers in the twentieth century regarding the perceived inability of Mexico to reliably feed itself have colored the view taken of Mexico's past.

As we have seen, the Spanish colonizers' initial assessment of indigenous agriculture was one of frank admiration for the "manifest prosperity" that Mexican peoples had achieved. The example that was most evident and impressive to the Spanish was the *chinampa* agriculture on which Tenochtitlan (the modern Mexico City), then one of the world's largest cities, relied for most of its food, medicinal plants, and flowers. The artificially constructed fields of the chinampas set in the shallow bed of Lake Texcoco created one of the most productive forms of agriculture ever invented, rivaled only by intensively cultivated rice paddy agriculture of Asia and by the most intensive chemically based modern systems. Even immediately prior to World War II, chinampa agriculture provided most of the food and flowers for the then one million people of Mexico City despite severe degradation from drainage, diversion, and pollution.[12] Much of lowland tropical Mexico and Guatemala also relied on raised bed agriculture in lake or swampy environments.[13] In Mexico's dry land, indigenous people had built irrigation canals to adapt food production to aridity and drought.[14] Although losses to soils and fertility occurred both prior to and after the Spanish Conquest (as discussed below), Mexican farmers before European contact were engaged in highly sophisticated and usually successful management of soil resources.[15]

There were dramatic exceptions, but the sometimes spectacular subsistence crises of pre-Hispanic Mexico rarely occurred because of indigenous peoples' ignorance of methods to control soil erosion; rather, we surely need to look for combinations of natural and social factors that set limits to what farmers could successfully accomplish. The decades-long argument over the possible role of soil exhaustion in the collapse of Classical Mayan culture, for example, demonstrates repeatedly that while degradation of the soil may have been a major factor in the collapse, the case does not become convincing unless we consider social factors that would explain why these obviously very capable farmers failed to respond to signs of trouble. Major social disruptions likely occurred as a result of soil degradation in pre-Conquest Mesoamerica, but Mesoamerican cultures had learned a great deal about how to manage soil fertility and erosion.[16]

The issues are complex and cannot be based on an Edenic view of pre-Conquest Mexico and its inhabitants. Admiration for indigenous technique should not obscure its failures or the problems built into some of its successes. One of the most important examples of soil management in pre-Conquest Mexico occurred in the Mixteca in what is now the state of Oaxaca. Pre-Conquest Mixtec society is thought to have been "the most highly stratified society in Meso-America."[17] Their agricultural methods reflected sharp differences between rulers and ruled. Bottomland was controlled by a hereditary nobility. Around 30 percent of those not in the nobility performed work obligations for the noble families. The remaining commoners farmed the steep slopes. As rich soils formed under pine and oak forests were cleared and farmed, soil inevitably moved downslope toward the bottomlands of the nobility. The nobles responded by directing laborers to build check dams in the valleys and canyons to capture the soil. These check dams took on the character of valley-land irrigated terraces as they grew into carefully designed systems of soil and water retention. The rich valleys of the nobility became broader and the soils deeper. Commoners became progressively impoverished and desperate as their laboriously cleared fields were washed out from under them, forcing them to higher and steeper ground. The nobility grew richer in soils and labor from the deepening problems of the poor.[18]

The question arises whether the Mixtec system, founded on exploitation and inequality, could be considered stable. The Spanish Conquest avoided a test of that question because it immediately threw the entire system into radical disarray. The introduction of domesticated grazing animals onto the fragile slopes, the use of the plow, and a shift in the mix of cultivars stressed the system, accelerating erosion. European diseases reduced the population of the Mixteca to perhaps 15 percent of its pre-Conquest level within a century. As erosion rates increased from grazing and plowing, the labor gangs required to maintain the terraces and check dams were difficult to assemble. In addition, the Spanish had different purposes in mind for the region, above all, the provision of mules for mines to the north. Although the pre-Conquest human population was not reached again in the rural Mixtec until approximately 1950, it doubled by the mid-1970s. Arable land had been reduced by erosion by at least 75 percent compared with that available at the Conquest.[19]

Other regions suffered similarly disastrous consequences. The one we know most about is the Valle del Mezquital north of Mexico City. In her classic study, Elinor Melville demonstrated that what had once been a rich agricultural region was fundamentally altered for the worse, primarily by a series of changes resulting from a "plague of sheep." Spanish domesticated grazing animals caused generalized soil erosion and hydrologic modifications, drying up springs and wells. Melville points out that contrary to Crosby's view, the Spanish intrusion did not create a "neo-European" landscape. The Valle del Mezquital had more closely resembled the European ideal of a productive agricultural region at Conquest; the European invasion itself was what "transformed it into something often perceived as archetypical of the 'naturally' poor Mexican regions." Melville explains that "[i]n the process the Otomí were displaced, alienated, and marginalized, their history and that of their region mystified. The Otomí are identified with the alien conquest landscape, not with the fertile, productive landscapes of contact. Their skills as cultivators were forgotten, their reputation as eaters of beetles, bugs, and the fruit of the nopal cactus confirmed."[20]

Perhaps one of the most profound transformations in Mexico involved the chinampas of Lake Texcoco that had provisioned Tenochtitlan and Mexico City. Though they admired chinampa productivity, the Spanish, not without reason, saw the waters of Lake Texcoco as the source of the "miasmas" that bred Old World diseases. They misidentified the waters of the lake as a major cause of catastrophic flooding of the city, rather than understanding the effects of their own mismanagement of regional land and water. The relentless and largely deliberate destruction of the chinampas from the colonial era to the present forced agriculture onto less productive land, itself undergoing the destruction described by Melville.[21]

Soil erosion and degradation has been an important theme in Mexican history for two or three thousand years. Where erosion was in some sense under control, as in the Mixteca, the system of control sometimes came at a heavy human price and was likely not sustainable in the long run. Recent assertions that incidents of erosion in the highlands of Michoacán dwarf the problems of those created by the Conquest are not convincing, however, because we do not know enough about the chain of causation.[22] Further, what we know about the damage to Mexican soils after the Conquest is that it was generalized, pervasive, and persistent through centuries. Had that been the case in the pre-Conquest era, it is difficult to imagine that the Spanish would have ever encountered the densely settled and "manifestly prosperous" civilizations that so impressed them. The fact that Mexico by the twentieth century was frequently characterized as a poor country in terms of its agricultural potential resulted to a considerable extent from centuries of European domination, both as it destroyed the land and as it formed attitudes that interpreted the devastated land as a natural heritage rather than a human creation. It might be said that the Spanish did not so much discover but rather created a New World and, from the point of view of agriculture, a poorer one.

A Deepening Crisis

The exploitive boom during the first century after the Conquest gave way to a prolonged economic depression that lasted through the seventeenth century. Agriculture and mining continued to create massive erosion.[23] A great thirst for stability helped shape the institution of the inwardly focused hacienda.[24]

Haciendas, missions, and straightforwardly commercial operations such as sugar plantations absorbed some indigenous communities and eliminated others.[25] By the early nineteenth century, the relentless dispossession of indigenous communities that had gone on during the three centuries of the colonial period was not complete—such communities continued to control about half the arable land in Oaxaca, for example, at independence.[26] The liberal land laws of the nineteenth century led to a more rapid process of dispossession. Estimates vary, but outside of inaccessible mountain regions as little as 4 to 10 percent of agricultural land was in the hands of smallholders and indigenous communities by the beginning of the 1910 revolution.[27]

We do not know in detail how the land itself fared between 1600 and 1910. We know of haciendas with excellent conservation practices, thus ensuring the security that was the ideal of such estates. Where such conservation practices succeeded, they probably relied on indigenous knowledge combined with Spanish traditional land practices. Spanish traditions helped to craft the colonial pastoral arrangements that often made intelligent use of seasonal changes across vast landscapes to deliver livestock to mines and population centers with little damage to pastures.[28] The exquisite adaptation to local conditions and the intricate ways in which some indigenous techniques used in the twentieth century are integrated with culture and local social organization strongly suggest continuous development over a period of centuries if not millennia.[29]

The landscape of Mexico and its soil endowment, however, continued to undergo serious degradation. Mexico's loss of a large share of its territory in the war with the United States robbed it of land that Economic Minister Lucas Alamán had hoped would counterbalance the pressure on what was seen as increasingly stressed land in the central highlands. Porfirio Diaz's advisors saw development of land in what remained of the north as essential for the same reasons. In the twentieth century, Mexico became a much-cited example of a country worn out by damage to its soils and in need of an agricultural strategy of recuperation.

Revolutionary Choices or Counterrevolutionary Choices Made by Revolutionaries?

The leaders who came to national power as a result of the Mexican Revolution were not much concerned with redistribution of land to a disenfranchised peasantry, much as that had been among the major themes of revolutionary struggle. In many states, governors pursued aggressive agrarian reform. At the national level, however, the portions of the 1917 Constitution that governed Mexican agrarian law until 1992 by no means ensured that serious agrarian reform would occur. The peasant leader Emiliano Zapata was assassinated in 1919. Francisco Villa, seen by some as a leader of the disenfranchised rural peasantry, was assassinated in 1924. General and Governor Álvaro Obregón of the northern border state of Sonora became President Obregón in 1920, and his successor was another Sonoran, Plutarco Elías Calles. "The Sonoran Dynasty" controlled the Mexican government until 1934. This group of political associates had a new vision for Mexican agriculture, but it contrasted starkly with Zapata's vision of a communal peasantry.[30]

Obregón and Calles were demonized in the United States for their willingness to confront the United States government in disputes arising out of the revolutionary Constitution. This has tended to obscure the much more significant fact that both of these Sonoran Dynasty leaders were partners with North American investors in the development of irrigated agriculture and associated industrial enterprises. Rail lines and port facilities built in the Sea of Cortez during the Porfiriato linked Sonora to the American Southwest and therefore to the entire North American market. The Sonorans saw that they were well positioned to take advantage of the distinct commercial advantages offered by access to US markets. The United States was opening its West and Southwest to commercial agriculture through the 1902 Reclamation Act, which provided a rich new source of federal financing for irrigation systems. The Mexican government could do the same for the desert valleys of Mexico. The road to wealth led to the north, not south to Mexico's populous central highlands.[31]

Though he had grown up as a poor orphan, Obregón had become a wealthy farmer, inventor, and businessman before 1910. He bought out a US trading firm and became known as "the garbanzo king" by monopolizing the trade.[32] He contracted with Herbert Hoover to supply the American Relief Administration with garbanzos for European famine relief. Calles, before his presidency a proven failure at managing his father's business, which supplied commercial farmers, was able to use his position in the Sonoran Dynasty and as president to enrich himself through agribusiness enterprises.[33]

The Sonoran Dynasty had little sympathy for Zapata's agrarian radicalism. Obregón and Calles sometimes manipulated peasant demands as a tool to cut wealthy opponents down to size through expropriation of their lands for redistribution to landless peasants. Their redistributions affected about 3 percent of Mexico's agricultural land. The Sonorans were opposed to excessively large landholdings to the degree that they led to inefficiency and lack of entrepreneurial zeal but not on principles of justice or fairness. In the guise of a radical, anticlerical, and anti-imperialist government, much of the Sonoran Dynasty program consisted of the promotion of North American agribusiness with participation by "revolutionary generals" and assorted cronies: "The low level of the country's economic and financial resources led the government to give priority to raising levels of production, including agricultural production, and made them reluctant to break up agricultural estates, especially the relatively efficient holdings oriented to commercial production." In 1930 Calles "announced that the agrarian reform program was a mistake, that the peons did not know how to use their land, and that production of foodstuffs was steadily declining." Calles and Dwight Morrow, the American ambassador, agreed that agrarian reform had already gone too far at a time when it had hardly gone anywhere at all. The Sonorans were interested in colonization of newly opened land, largely in the North, that represented a return to nineteenth-century policies that had encouraged Mexican and foreign investment in the desert states.[34] In 1926, Calles set up new legal and institutional mechanisms to support colonization of land that was largely located in irrigated or soon-to-be-irrigated parts of the North and Northwest.[35]

The underlying assumptions of the Sonoran Dynasty were widely held by Mexican elites. The young Daniel Cosío Villegas, who was to become one of the nation's most prominent intellectuals, wrote in 1924, "The industry of agriculture in our country is deficient not only because our methods of cultivation are backward . . . but because the soil itself is poor. In order that our agriculture shall be able to satisfy our necessities . . . costly engineering works, especially of irrigation, will be necessary throughout the nation. We cannot expect anything as a gift of Nature; everything in Mexico depends upon the activity and ingenuity of man. It is for this reason I say 'economically we are poor'; but more than this, the origin of our economic poverty is our natural poverty."[36] As Melville noted, the poverty of the Mexican soil has been taken as its original condition rather than as a consequence of centuries of misuse and the specific impact of European animals, technologies, and purposes. The internalization of this attitude among Mexican intellectuals was a significant factor in what Mexico would choose to do about its land, in an ahistorical "man versus nature" opposition that serves a superb example of twentieth-century high modernism at work.

Calles, who had continued to indirectly control the presidents who served after him from 1928 through 1934, believed that he would be able to continue the policies of the Sonoran Dynasty after the presidential election of 1934. The new president, General Lázaro Cárdenas, governor of Michoacán, was able to set a new course and drive Calles into exile. One of the most important elements of the Cárdenas vision for Mexico was a major redistribution of land as envisioned by the 1917 Constitution. Cárdenas expropriated about 49 million acres of land and created thousands of *ejidos* (land grants to communities) and restored a modest number of indigenous common lands. Between a quarter and a half of the Mexican population became members of ejido families. The Cárdenas government encouraged the formation of credit, marketing, and machine cooperatives and offered government credit to *ejidatarios* (land reform beneficiaries).

The Cárdenas administration saw agricultural education, including soil conservation, as central to the success of the agrarian reform program. The administration sought a variety of other creative ways to promote both the culture and the practice of modern conservation.[37] At the same time, the national government and popular writers and artists tried to deepen Mexico's awareness of its indigenous past and its remaining indigenous people. There was a reawakening of a modest degree of interest in indigenous crops and agricultural methods.[38]

Cárdenas had a serious commitment to conservation similar to Franklin Roosevelt's. The emphasis was on conservation for the sake of improving economic performance and reducing poverty. Soil conservation was still seen largely as an aspect of forest protection as it had been in the past, and new laws and institutions gave the government more power to effect forest conservation. Despite the focus on forests, the linkage of Cárdenas's conservation programs with land redistribution began to associate soil erosion more closely with issues of equity.

Cárdenas was determined to build the power of the ruling party on a base of popular enthusiasm. Land reform was part of this project, as were programs that extended government authority—and thereby, patronage—into many new areas. His use of extended trips into provincial areas and of radio and newspaper publicity brought national politics and his own personality into a prominence that was becoming the mark of modern politicians everywhere. Cárdenas's view was distinctly nationalist; the most enduring concrete legacy of this perspective was his nationalization of Mexico's large petroleum industry. This in turn led to confrontations with the United States and England, home to the main owners of the expropriated oil facilities. Some in the US government argued for armed invasion of Mexico, but consistent with Roosevelt's Good Neighbor policy seeking Latin American allies in the coming war with Germany and Italy, cooler heads prevailed.

While war was ruled out, political factions in Mexico and the United States began to plan for a presidential succession in the 1940 Mexican election that would blunt Cárdenas's populist nationalism. Businessmen and landowners saw the 1940 election as a crucial test of their ability to survive. Cárdenas had made the ruling party much more popular and powerful, and opposing its dominance was not seen as an option. The battle would be over internal control of the party. The conservative factions were able to prevail by selecting Manuel Ávila Camacho as the presidential candidate, and he was duly elected to office.[39]

In his inaugural speech, Ávila Camacho embraced "the vital energy of private initiative" and promised to "increase the protection given to private agricultural properties, not only defending those that exist, but also forming new properties in the vast regions not now cultivated." Agriculture would be seen not as a means for achieving greater equality but rather as the foundation of "industrial greatness."[40] Beginning with the Ávila Camacho administration and continuing through the rest of the twentieth century, land reform, in those occasions when large acreages were distributed, was often aimed rather obviously at reducing rural unrest in certain regions and slowing the migration rate to the cities. Much of the land was only marginally productive, but significant parts of it were important in the restoration of forestlands to indigenous communities. Despite occasional land redistributions, agrarian reform would not be seen, as it was by Cárdenas, as a primary means to achieve the agricultural and economic development of the nation.

Agriculturally significant land distributions occurred mostly in newly opened irrigation districts in the northern deserts, with financing for the dams and canals from the Mexican government and, later, the World Bank. The distributions on good irrigated land went mostly to, or ended up in the control of, what were envisioned in the 1917 Constitution as *pequeños propietarios* (private smallholders rather than ejidos and ejidatarios). These smallholders, many of whom would come to own large extents of land, formed the bedrock for the expansion of entrepreneurial, commercial agriculture precisely on the model envisioned by the Sonoran Dynasty. Foreign capital participated heavily in financing many of these operations.

In the Culiacán and Fuerte Valleys in Sinaloa and Sonora, ejidos gained land rights during the Cárdenas administration and later. These ejidos, however, would be subjected to the systematic corruption of the intentions of land reform, sometimes through illegal leasing or buyouts of ejidal land involving the use of state and private violence against the recalcitrant holdouts (which, for example, continued in the Fuerte Valley into the 1970s). Genuine ejidos were sometimes displaced by commercial farmers who operated under the legal guise of an ejido grant.[41]

The Mexican Green Revolution Package

Sitting beside Ávila Camacho at his inauguration was the recently elected vice president of the United States, Henry Wallace, previously Roosevelt's secretary of agriculture. Wallace had inherited the legacy of his father as corn breeder and owner of one of the world's largest seed companies. Although known in the United States as a left-wing politician, Wallace's role in the restoration of conservative forces in Mexico showed another side. Wallace grandiloquently commented that he had done nothing less than help avoid a revolution in Mexico. But the most important thing that Wallace did during his weeks in Mexico was to help negotiate a research program that would support the country's new emphasis on agricultural productivity as opposed to land reform.[42]

The election of Ávila Camacho put Mexico's government solidly behind an agricultural policy that reasserted the vision of the Sonoran Dynasty. The Rockefeller Foundation, which had begun to widen its concerns beyond its Mexican child welfare programs of the 1920s, provided critical funding and leadership for a research program aimed at improving agricultural productivity in Mexico, with an eye to an extension of successes in Mexico to the rest of the tropical and subtropical world.[43] With the approval of the program by the US Department of Agriculture, the Rockefeller Foundation, and the Mexican government, work began in 1941.

The first step was a research trip to be undertaken by leading US agricultural scientists to study Mexico's agricultural problems and recommend appropriate research strategies. The foundation recruited a team led by Richard Branfield of Cornell, Paul C. Mangelsdorf of Harvard, and Elvin Stakman of the University of Minnesota. This survey commission began its five-month journey in July 1941. The story of the commission's journey is told in the book *Campaigns Against Hunger*, published when the Rockefeller Foundation and the US government were eagerly promoting what they saw as their Mexican success in India, Pakistan, and much of the rest of the world. The scientists of the commission revealed concern for Mexicans and some appreciation of their culture and intelligence, but a smug, self-congratulatory paternalism prevails, based on the assumption that it was scientists like the authors themselves who held all the keys to ending hunger and bringing prosperity to Mexico and all poorer nations. The fact that all were trained in the sciences, with virtually no input from social scientists or historians, reflected the unexamined assumption that the solutions to hunger were all to be found in technology. There was no serious attempt to analyze the agrarian reform program.[44]

The recommendations from the survey commission began with a focus on improving "soil management and tillage practices." It might have been expected that what would follow would have been an effort to improve soil management and tillage practices on the agricultural land then under cultivation, but the logic went elsewhere. Better soil management instead would be pursued as a product of improved varieties to be developed by plant and animal breeders, by expanded large-scale irrigation works, and by the "more rational and effective control of plant diseases and insect pests."[45] While the team recognized the genius in chinampas, they concluded that Mexicans could never again hope to have such a productive agriculture. Little attention was given to other traditional Mexican practices for soil management, and relatively little study was given to what specifically should be done on existing acreages. Mexican soils were held to be rich in most nutrients but chronically poor in nitrogen. The Haber-Bosch process and the availability of commercially produced ammonia fertilizers were the solution. "Improved" crop varieties would be designed to maximize nitrogen uptake. This would require abundant and reliable water delivery that could only be had by the great expansion of irrigated acreages through dams that would supply water to what were regarded as virtually empty frontiers of lowland and desert land.[46] The scientists were proudly recommending a bold and transformative approach. It was posed as the alternative to another possible bold and transformative choice: to focus policy and scientific effort more intensely on the protection, improvement, and restoration of land currently farmed by land reform beneficiaries and others.

Their recommendations constituted a remarkable and surely not coincidental meeting of minds. In 1941, Ávila Camacho had lectured the nation that

> [t]he future of agriculture lies in the fertile lands of the coast. A march to the sea will relieve congestion in our central plateau, where worn-out lands must be devoted to crops which colonial policy denied them with the result that the traditional maize culture of the indigenous population has continued to be dominant. The fertility of the coastal plains will make it uneconomical to raise many products in the central plateau. But the march to the sea requires . . . sanitary and health measures, the opening of communications, the reclamation and drainage of swamps, and, to make such projects possible, the expenditure of large sums of money. It will be necessary to organize a new kind of tropical agriculture, which, because of the very nature of its production, cannot be the small-scale type.[47]

Given as prescription and prophecy, no more prescient vision of the future of the Green Revolution in Mexico and elsewhere could have been given. It is also noteworthy that, contrary to continual denials of proponents of the Green Revolution from the 1940s to the present, for Ávila Camacho it was a clear rejection of small-scale agriculture.[48]

In the latter half of the 1940s, all of this planning started to bear fruit in several different ways. The Mexican government created river basin commissions to provide irrigation for agriculture and electricity for industry, modeled to a considerable degree on the Tennessee Valley Authority. The success of these commissions varied widely. The most successful, measured in terms of various economic and welfare measures, were those in the North, especially the North Pacific (Sonora, Sinaloa, Baja California Norte and Sur, and Nayarit). By 1960, the North Pacific accounted for 43 percent of the Mexican government's large-scale irrigation efforts in terms of acreage. The northern states as a whole account for more than three-quarters. The only competition is in the central region, with 14.5 percent, mainly in one of Mexico's most enduring agricultural regions: the Bajío of Guanajuato, Querétaro, and adjoining states. The newly irrigated states experienced rapid growth in investment, agricultural production, and industry during this period, with outstanding improvement in per capita income and overall welfare indices.[49] Without counting the obvious and hidden costs, the Green Revolution was clearly at least a short-term production success, although it has been convincingly demonstrated that about two-thirds or more of the gains could have been achieved on the same irrigated acreage with the use of traditional crop varieties that depended less on synthetic agrochemicals, rather than the Green Revolution varieties. What might have been achieved with a focus on innovation and improvement on existing farmland has not been estimated.[50]

Although the focus on large-scale irrigation in the northern desert was often discussed as the settlement of empty frontiers, the frontier was far from empty; agriculture had been practiced for millennia in most of what was termed "new land."[51] What was new was technology and control. When Ávila Camacho created the first Soil Conservation Department in 1942, it was placed under the authority of the National Irrigation Commission rather than the Secretariat of Agriculture.[52] It was moved to agriculture in 1946, and the Secretariat of Agriculture was soon renamed the Secretariat of Agriculture and Water Resources (SARH—Secretaría de Agricultura y Recursos Hidráulicos). Employees frequently complained in the 1980s that it was "recursos hidráulicos" and not "agricultura" that dominated the agency's bureaucracy. Successive presidents gave lip service to soil conservation as an urgent cause, citing alarming statistics, but provided very little money. Government soil conservationists were shoveling against the tide of government priorities.

The Pesticide Revolution

As irrigated acreage grew, the most celebrated plant-breeding successes came from a new research center near the aptly named planned town of Ciudad Obregón, Sonora, successes for which Norman Borlaug would win the Nobel Prize. The new varieties could only succeed when the complete "package" was used. The new varieties had been designed to absorb and convert more nitrogen into grain, but more nitrogen would be toxic to the plants if not delivered with adequate water applied at the right times. In turn, the greater mass of plant material produced more densely in a more humid soil and field environment could be expected to attract more plant diseases and insect pests. This would require more use of the newly available synthetic pesticides.[53] Commercial agriculture became heavily dependent on regular fertilizer and pesticide use. Peasants sometimes used pesticides in very hazardous ways, but their dependence on pesticides was spotty, limited by costs when not by caution. Agricultural schools and extension agencies spent a significant part of their effort working with private farmers to maintain lists of recommended formulas for fertilizer and pesticide application in given regions and crops.

The seed, water, and chemical package was costly to develop, costly to support with needed infrastructure, and costly for the individual farmer to acquire, not to mention the expense of publicly funded irrigation works necessitated by the new seeds and agrochemicals. But the long-term price was higher still. The Green Revolution grain crops could seldom be produced competitively unless cultivated and harvested by machinery, which would only pay out with relatively large expanses of land. All of this meant that access to credit became more important. The Mexican government offered meaningful rural credit for about three decades (roughly from the 1950s through the early 1980s), but it was offered selectively and unreliably and was often used primarily as a means of exercising political influence. Successful commercial farmers relied more heavily on commercial credit, much of it foreign, into the 1980s. By then, government rural credit programs had been nearly eliminated.

Private and government banks typically mandated the scheduled use of agrochemicals to ensure successful repayment of large loans, even when farmers and agronomists were reluctant to apply chemicals so freely. Farmers also worked under contract farming, in which production costs were advanced by a marketing firm in exchange for delivery of the crop—these also typically required scheduled chemical use.

In Mexico, the nationalism of Cárdenas was quickly harnessed to the productivist model of Ávila Camacho, the Rockefeller Foundation, and the United States government. Most agrochemicals were made using petroleum and/or natural gas as feedstock. Pemex, the nationalized oil firm, became a major producer of agrochemicals through its subsidiary, Fertimex. By the 1980s, Fertimex had become the world's largest producer and major exporter of DDT even though most industrial countries had virtually banned its use. Fertimex produced an array of other pesticides as well and sometimes provided free fertilizers and pesticides to farmers in the 1980s, in the name of agricultural development. Cheap asphalt paving materials from Pemex provided the farm-to-market roads. Cheap fuels kept the trucks and tractors moving.[54] Much of the public investment in agriculture was justified using the argument that Mexico's political and economic independence relied on self-sufficiency in basic grains, until this argument was abandoned in the embrace of liberalized trade that set in during the 1990s, when Mexico's market was thrown open to international competition.[55]

The large investments in agriculture led to a more critical assessment of opportunities for profit. Selling bulk grains that were often produced more cheaply abroad and had very slim profit margins in the best of circumstances began to look less attractive. In the 1960s, farmers and creditors began to see the potential for greatly expanded markets for tropical and out-of-season fruits and vegetables in the United States and Europe. Based on the Mexican government's program to ensure an adequate stock of basic grains for the Mexican people, everything needed—cheap water, land, chemicals, road, communications, credit, and tightly controlled cheap labor—was available to shift a sizable portion of grain land into more valuable crops. These could be sold to foreign consumers with a virtually insatiable appetite for out-of-season luxury foods and more money than could be found in the domestic market.

The Culiacán Valley in the North Pacific state of Sinaloa was the most successful at seizing this opportunity. Good road and rail lines ran straight north six hundred miles to Nogales, Arizona. By the early 1980s, the Culiacán Valley was reliably providing one-third or more of all the "summer" vegetables (tomatoes, cucumbers, peppers, eggplant, peppers, and so on) sold between December and May in the United States. In accordance with the provisions of the 1917 Constitution (applicable until amended in 1992), these farms were at least nominally owned by Mexicans, but 90 percent of the financing came from the United States and with it much of the control over production and marketing. Other irrigated valleys of the northern states increased the export flow of high-value fruits, vegetables, and, eventually, even wines to US and Canadian and Mexican urban markets. Farther south, humid-zone agricultural areas such as Apatzingán, Michoacán, experienced the same kind of boom. In the Bajío, grain and livestock production continued to prevail as it had for centuries, but with a growing export horticultural sector, including, for example, former president Vicente Fox's strawberry operation. Not surprisingly, Mexico's self-sufficiency in grains, the supposed focus of Green Revolution research and policies, was increasingly undermined by the even more intensely pesticide-dependent agro-export boom based on the very investments made to ensure national food security.[56]

By the 1980s it had become clear that many Mexicans and much of the best Mexican land were beginning to pay a heavy price for the enormous expansion of commercial agriculture dependent on synthetic pesticides. In the Culiacán Valley, studies from the 1970s until at least the early years of the twenty-first century showed that the lavish use of pesticides resulted in the serious contamination of waterways and a systematic attack on the health of farmworkers and rural residents.

My own field study conducted in the winter of 1983–84 and with renewed visits until 1989 found that highly toxic combinations of pesticides were being used up to fifty times in a ten-month season. From January through May, growers very often applied an insecticide/fungicide combination twice a week. Growers and their supervising agronomists routinely ignored all industry-suggested safety precautions meant to protect farmworkers, rural residents, and the environment, as well as national laws and international standards. Local university professors and researchers reported that they were too frightened of job loss or violent reprisals to study the problem or speak out too loudly about it.

Farmworkers confronted a series of threats to their well-being and that of their families. They frequently experienced acute pesticide poisonings for which they were sometimes treated in local clinics or which they simply ignored for fear of being laid off. Local doctors attended farmworkers who had died from pesticide exposure and far more who had suffered pesticide-caused illness. Most doctors and nurses were not adequately trained to recognize or treat pesticide poisoning. Workers used backpack sprayers, which often leaked pesticides that were heavy skin-contact nerve toxins. Workers were directed to labor in the fields during and immediately after application of highly toxic mixtures. Airplanes applying pesticides sometimes sprayed directly over farmworkers, including women and children. Farmworker families lived on the field borders in open sheds designed for poultry operations, which inevitably allowed drift from ground and air pesticide applications to contaminate living space, bedding, and food. Pesticide containers were commonly used for drinking water and food storage. People bathed in contaminated irrigation canals.

Workers had widely varying ideas about the problem—some understood the dangers well but felt there was nothing that could be done about it because they needed the work to survive. Most had partial knowledge of the nature of the problem. Some referred to the pesticides as "medicines for the plants" and believed that these "medicines" would also be good for people.

The fact that most of the farmworkers suffered from many other illnesses—gastrointestinal diseases and tuberculosis were endemic—further complicated the problem. One doctor working in the area reported that nearly all the farmworker patients he had treated suffered from anemia. The consequences of pesticide poisoning were likely to be more severe when combined with other health problems. Moreover, the symptoms of pesticide poisoning are easily confused with the symptoms of many other diseases.

Work crews usually included women, some of them pregnant, and often children, who were known at the time to be especially subject to harm from the pesticides used. In 1983, three Mixtec women who worked in the

Culiacán Valley fields died of pesticide exposure while suffering miscarriages of fetuses they had not known they were carrying. We now know that the permanent harm to workers, but especially to children and pregnant women, is much more serious, more complicated, and sometimes more subtle than most experts thought in the 1980s. Present toxicological knowledge allows us to expect that a significant number of farmworkers working under the conditions found in Culiacán and elsewhere in Mexico and around the world suffer from a whole variety of pesticide-related mental and physical problems that probably will never be diagnosed as the consequence of pesticide exposure.[57]

The Culiacán field study revealed that agronomy professors, extension agents, and regulatory officials were systematically misinformed or dishonest about the kinds and quantities of pesticides applied, as observed in hundreds of field applications. We cannot know whether conditions have improved over the past decades, because growers have made entering fields and doing adequate field studies of pesticide use nearly impossible, ever since results of the studies of the 1980s and early 1990s have become public. The pervasive violence used in the drug trade and labor control in the region also has been used to intimidate scholars who wished to carry out research, even when they were working under international agreements between Mexico and the United States. We are forced to rely on less direct evidence. For example, biologists contracted by the shrimp industry that expanded into coastal lagoons receiving field drainage in the 1990s found consistently high levels of contamination from the same types of pesticides used commonly in the 1980s. Interviews with farmworkers in one study reveal what appear to be marginal improvements in conditions and training. In some regions, however, such as the tobacco fields of Nayarit, recent direct observation and systematic study show that the horrendous conditions found in the Culiacán Valley in the 1980s still prevail.[58] In the Fuerte Valley, studies of childhood exposure to pesticides have shown notable mental and physical developmental problems that appear to be connected to ubiquitous background pesticide exposure in environments where pesticides are heavily used but where the children are not known to have suffered from acute poisoning incidents.[59]

The problem of general background exposure to pesticides, as well as the acute symptoms that appear directly after exposure, is exacerbated by the fact that synthetic pesticides have been heavily used in much of Mexico for control of diseases carried by insect vectors. The discovery in 1938 that DDT was effective in killing insects and other arthropods but that it had low acute human toxicity led to its widespread use in World War II, particularly in the control of typhus. It soon emerged as a seemingly miraculous solution to malaria control, with DDT-based campaigns achieving spectacular rates of malaria reduction around the world. By the early 1960s, however, two problems had emerged. One was the rapid growth of resistance to DDT in mosquito populations, causing a strong resurgence of malaria in many regions. Public health officials often reacted by increasing DDT application rates or by shifting to other pesticides that were equally or more toxic to humans, a shift that happened for the same reasons in agriculture. The second problem was the accumulation of evidence that DDT was causing significant environmental problems. DDT and other pesticides of its class are fat soluble and are therefore stored in animal and human tissue.[60] They also persist in toxic form for long periods of time. These characteristics caused "biological magnification" of the pesticides, and therefore serious harm, among such organisms as raptors, pelicans, and seals living at high trophic levels. Evidence accumulated that while DDT caused acute poisoning symptoms in humans only at relatively high and rare dosage rates, it was probably carcinogenic and hazardous to human health in a variety of chronic and more subtle ways at low doses. Extraordinarily high levels of DDT and its toxic breakdown products were discovered in mother's milk in areas where DDT was being used for malaria control and for agricultural pest control, as in parts of Guatemala and lowland tropical Mexico.[61]

The widespread use of DDT in malaria control is one of the reasons that Mexico's nationalized chemical company, Fertimex, had become the world's largest manufacturer of DDT in the 1980s. By that time, public health officials in Mexico had begun to transition to other malaria control methods, including pesticides that were not fat soluble and persistent and were of relatively low human toxicity (such as malathion), in combination with "hygienic" measures to control mosquito reproduction. Mexico nearly eliminated DDT use for malaria control in the 1990s. However, DDT remains an important environmental contaminant as a chemical that is long-lived, has toxic breakdown products, and is stored in animal tissue. It is used illegally because it is cheap and easily obtained. We now also understand that DDT and its relatives are potent endocrine disruptors, meaning that they may have effects on fetal and childhood development and various adult diseases, such as infertility and endometriosis. In areas where pesticide exposure comes through both agriculture and public health uses, the problem of understanding the cumulative and synergistic effects of exposure to multiple pesticides is very difficult and little studied although the effects may be quite serious.

From One Crisis to Another

In the 1980s and 1990s, the majority of farmworkers subjected to the systematic and abusive use of pesticides were Mixtec- and Zapotec-speaking people from Oaxaca or neighboring states. Mixtecs were the most numerous. Mixtecs most commonly gave as the reason for migration the inability to make a living on their traditional lands. This was not surprising; the Food and Agriculture Organization of the United Nations had classified the Mixteca as one of the most seriously eroded regions of the world, with more than 75 percent of the agricultural land destroyed by soil erosion. The federal soil conservation service had worked with notably ineffectual results in the region for decades. The farmworkers who labored in northern Mexico in the latter part of the twentieth century were fleeing one environmental disaster to find work amidst another ongoing environmental calamity.[62]

The lavish and abusive use of pesticides has had predictable agronomic and ecological consequences that include the rapid development of pesticide resistance among target organisms, including insects, weeds, and plant pathogens. While plant breeders work on producing crop varieties that are in turn more resistant to pests, many farmers respond to pesticide-resistant pests primarily by increasing the frequency and dosage of pesticide applications. Many secondary pests have arisen as serious problems as a result of suppression of their predators by pesticide use against other organisms. Pesticides contaminate local water supplies, shrimp farms, and the Sea of Cortez.[63] Large-scale and intensive irrigation has also resulted in predictable consequences, namely, the serious increase in soil salinity and drainage problems that threaten the continued productivity of the "new lands."[64]

Many observers had anticipated that negotiations under the North American Free Trade Agreement (NAFTA) would result in better pesticide regulation in Mexico, as well as better environmental policy in general. Negotiations have indeed produced a body of pesticide regulations in Mexico that is as rigorous as the standards of the US Environmental Protection Agency. However, there is little evidence that this has significantly improved the situation in practice. Kevin Gallagher's study of the environmental consequences of NAFTA shows that in some areas there has been improvement in Mexico's environmental performance since the trade liberalization treaty went into effect. These improvements are largely independent of anything prompted by NAFTA, however. With regard to many other areas, there are better laws but continuing poor enforcement. Budgetary austerity in Mexico enacted as part of the overall concept of economic liberalization under NAFTA has cut enforcement budgets to the bone in most areas. Soil erosion rates in particular increased by 89 percent in the first decade after NAFTA.[65] In 2008, analysts at the Mexican Instituto Nacional de Ecología called attention to an "alarming situation" in which soil degradation affected 45 percent of the national territory and in which "it can be said that the institutional response to this situation continues to be weak and dispersed."[66]

The Persistence of a Vision: Culmination Rather Than Change?

A consistent vision has been held by the Sonoran Dynasty, Daniel Cosío Villegas, the Rockefeller Foundation, Ávila Camacho, and, with a few diversionary moments, the Mexican government to the present. This was what Cosío Villegas saw as solving the "natural" poverty of Mexico's soils through irrigation, what Ávila Camacho saw as "the march to the sea," and what Rockefeller-financed researchers saw as solving the degradation of Mexico's soils by opening new lands with new crops and new technologies. Zapata's agrarian radicalism was honored in highly contested experiments in some states in the 1920s and for a little more than four years of the Lázaro Cárdenas administration from 1935 through 1940 (and then in a spirit of social engineering that would not likely have pleased Zapata). It may be said to have been dishonored by decades of demagogic manipulation of the peasant cause by PRI politicians otherwise preoccupied with the "march to the sea." An alternative approach that would have linked soil conservation and productivity gains to agrarian reform was conceived but not effectively implemented for the long run by the Cárdenas administration.

The Green Revolution approach to soil erosion in Mexico meant the abandonment of Mexico's rural poor to their fate on what had been determined to be soils too "worn-out" to carry the force of Mexico's economic development. Irrigation of "new land" and pesticide use would enable the creation of a new class of agricultural entrepreneurs who oversaw the bulk of Mexico's agricultural growth as it tended downslope and north. What PRI politicians saw as a path to secure economic growth and opportunity was also a significant factor in the rise to power of a new political party, PAN (Partido de Acción Nacional), which has some of its strongest support among the entrepreneurial agribusiness leaders of the desert North. (One, a vegetable grower from Sinaloa, ran as the first serious PAN contender for president in 1988. The PRI sought unsuccessfully to undermine PAN support by running another Sinaloa entrepreneur in 2000.) Since the rise of the Sonoran Dynasty in 1920, the PRI may have been doggedly preparing the ground, literally and figuratively, for its own demise. What has been seen as a new beginning with the fall of the PRI and rise of the PAN, however, may be more culmination than change.

Is a New Beginning Possible?

Is an alternative story, a real new beginning, possible in Mexico's future? In the 1980s, as some Mixtec farmers were leaving their communities to work in the poisoned fields of Culiacán, others chose to stay home and began to experiment with traditional forms of Mixtec farming and erosion control. Using a combination of reforestation, terraces, ditches, and check dams, they have been able to restore significant quantities of land. For example, Leon Santos, a Mixtec farmer, won the Goldman Environmental Prize in 2008 for his work with the Center for Integral Small Farmer Development (CEDICAM) in the Mixtec region of northern Oaxaca. The organization works with more than 1,500 farmers in twelve communities. It claims to have planted more than a million trees and to have protected or restored about seven thousand hectares of land using stone terraces and walls and other techniques, leading to a 50 percent increase in agricultural production. Communities that were able to farm only 25 to 30 percent of their land now are reported to be using 80 percent. With improved water retention, springs are coming back to life. Farmers are transitioning away from synthetic pesticide and fertilizer use, adapting traditional techniques and seeds, and learning new methods of sustainable agriculture. Similar movements are taking place in other areas of Mexico.[67]

Are such people and organizations minor and transitory, or will they prove to have transformational importance as Mexico's agriculture adapts to its self-created problems and to global climate change? Will they ultimately need the massive state support that made the Green Revolution possible, or can they thrive on their own? Will they require or cause new shifts in political power? It would be very bold to suggest that the nascent movement for sustainable agriculture among Mexico's rural poor will become important in Mexican life. The full power of the Mexican state—supported by supportive alliances with US businesses, finance, and government—has run in the opposite direction since (and even from before) the Mexican Revolution. Any strategy for change will have to deal with that stubborn reality.

Notes

1. This story is told in much greater detail, but with different emphasis, in Angus Wright, *The Death of Ramón González: The Modern Agricultural Dilemma*, 2nd ed. (Austin: University of Texas Press, 2005).
2. Teresa Rojos Rabiela, *Agricultura indigena: Pasado y presente* (Mexico City: Ediciones de la Casa Chata, 1994).
3. Gene C. Wilken, *Good Farmers: Traditional Agricultural Resource Management in Mexico and Central America* (Berkeley: University of California Press, 1987).
4. Sarah Bunney, "Prehistoric Farming Caused Devastating Soil Erosion," *New Scientist* 125, no. 1705 (1990): 20; S. L. O'Hara, F. Alayne Street Parrott, and Timothy Burt, "Accelerated Soil Erosion around a Mexican Highland Lake Caused by Prehispanic Agriculture," *Nature* 362 (1993): 48–51; C. T. Fisher, Helen P. Pollard, Israde Alcántara, Victor H. Gardeño, Monroy Sabin, and K. D. Banerjee, "A Reexamination of Human-Induced Environmental Change within the Lake Pátzcuaro Basin, Michoacán, Mexico," *Proceedings of National Academy of Sciences* 100 (2003): 4957–62; Flavio S. Anselmetti, David S. Hodell, Daniel Ariztegui, Mark Brenner, and Michael F. Rosenmeier, "Quantification of Soil Erosion Rates Related to Ancient Mayan Deforestation," *Geology* 35, no. 10 (2007): 915–18.
5. Anselmetti et al., "Quantification of Soil Erosion."
6. John F. Richards, *The Unending Frontier: An Environmental History of the Early Modern World* (Berkeley: University of California Press, 2003), 335.
7. Angel Bassols Battalla, *Recursos naturales de Mexico: Teoria, conocimiento, y uso* (Mexico City: Editorial Nuestro Tiempo, 1982).
8. Richards, *Unending Frontier*, 340.
9. Carl O. Sauer, *Aboriginal Population of Northwestern Mexico*, Ibero-Americana no. 10 (Berkeley: University of California Press, 1930); Sherburne F. Cook and Lesley Bird Simpson, *The Population of Central Mexico in the Sixteenth Century*, Ibero-Americana no. 31 (Berkeley: University of California Press, 1948); Sherburne F. Cook, *Soil Erosion and Population in Central Mexico*, Ibero-Americana no. 34 (Berkeley: University of California Press, 1949); Sherburne F. Cook and Woodrow Borah, *The Indian Population of Central Mexico, 1531–1610*, Ibero-Americana no. 44 (Berkeley: University of California Press, 1960).
10. François Chevalier, *Land and Society in Colonial Mexico* (Berkeley: University of California Press, 1971); Charles Gibson, *The Aztecs under Spanish Rule* (Stanford: Stanford University Press, 1964); William B. Taylor, *Landlord and Peasant in Colonial Oaxaca* (Stanford: Stanford University Press, 1972).
11. Elinor G. K. Melville, *A Plague of Sheep: Environmental Consequences of the Conquest of Mexico* (Cambridge: Cambridge University Press, 1994).
12. Elisabeth Schilling, "Los 'jardines flotantes' de Xochimilco (1938)," in *La agricultural chinampera*, ed. Teresa Rojos Rabiela, 71–98 (Chapingo: Universidad Autónoma Chapingo, 1983).
13. Tim Beach, Sheryl Luzzader-Beach, N. Dunning, J. Jones, J. Lohse, T. Guderjan, S. Bozarth, S. Millspaugh, and T. Bhattacharya, "A Review of Human and Natural Changes in Maya Lowland Wetlands over the Holocene," *Quaternary Science Reviews* 28, no. 17–18 (March 2009): 1710–24.

14. William E. Doolittle, *Cultivated Landscapes of Native North America* (Oxford: Oxford University Press, 2000).

15. Teresa Rojas Rabiela, ed., *La agricultura chinampera* (Chapingo: Universidad Autónoma Chapingo, 1983).

16. Arthur A. Demarest, Prudence Rice, and Don S. Rice, eds., *The Terminal Classic in the Maya Lowlands: Collapse, Transition, and Transformation* (Boulder: University of Colorado Press, 2004).

17. Kent Flannery, "Precolumbian Farming in the Valleys of Oaxaca, Nochixtlán, Tehuacán, and Cuicatlán: A Comparative Study," in *The Cloud People: Divergent Evolution of the Zapotec and Mixtec Civilizations*, ed. Kent Flannery and Joyce Marcus, 323–39 (New York: Academic Press, 1983), 328.

18. Ronald Spores, *The Mixtecs in Ancient and Modern Times* (Norman: University of Oklahoma Press, 1984).

19. Wright, *Death of Ramón González*, chap. 5.

20. Melville, *Plague of Sheep*, 115.

21. Joel Simon, *Endangered Mexico: An Environment on the Edge* (San Francisco: Sierra Club Books, 1997), 64–72; Alaine Musset, *El agua en el Valle de México: Siglos XVI–XVII* (Mexico City: Centro de Estudios Mexicanos y Centroamericanos, 1992).

22. Bunney, "Prehistoric Farming"; O'Hara et al., "Accelerated Soil Erosion"; Anselmetti et al., "Quantification of Soil Erosion."

23. Peter J. Bakewell, *Silver Mining and Society in Colonial Mexico: Zacatecas, 1546–1700* (Cambridge: Cambridge University Press, 1971).

24. Chevalier, *Land and Society*; Richards, *Unending Frontier*, chap. 10.

25. Gibson, *Aztecs under Spanish Rule*.

26. Taylor, *Landlord and Peasant*.

27. John Hart, *Revolutionary Mexico: The Coming and Process of the Mexican Revolution* (Berkeley: University of California Press, 1987).

28. Karl W. Butzer and Elizabeth K. Butzer, "Transfer of the Mediterranean Livestock Economy to New Spain: Adaptation and Ecological Consequences," in *Global Land Use Change: A Perspective from the Columbian Encounter*, ed. B. L. Turner II, Antonio Gómez Sal, Fernando González Bernáldez, and Francesco di Castri, 151–93 (Madrid: Consejo Superior de Investigaciones Científicas, 1995).

29. Wilken, *Good Farmers*. Comparison of techniques described by Wilken with those discussed in Doolittle, *Cultivated Landscapes*, also strongly suggests the long provenance of techniques still in use.

30. Wright, *Death of Ramón González*, 166–67; Nora Hamilton, *The Limits of State Autonomy: Post-Revolutionary Mexico* (Princeton: Princeton University Press, 1982).

31. John Mason Hart, *Empire and Revolution: The Americans in Mexico since the Civil War* (Berkeley: University of California Press, 2002), esp. chaps. 3, 6, 7, 11, and 12; John J. Dwyer, *The Agrarian Dispute: The Expropriation of American Owned Rural Land in Post-Revolutionary Mexico* (Durham, N.C.: Duke University Press, 2008).

32. Linda Hall, *Alvaro Obregón: Power and Revolution in Mexico, 1911–1920* (College Station: Texas A&M Press, 1981).

33. Jürgen Buchenau, *Plutarco Elías Calles and the Mexican Revolution* (Lanham, MD: Rowman and Littlefield, 2006).

34. Hart, *Empire and Revolution*, part 2; Hamilton, *Limits of State Autonomy*, 96–97.

35. David Barkin and Timothy King, *Regional Economic Development: The River Basin Approach in Mexico* (Cambridge: Cambridge University Press, 1970), 53.

36. Villegas quoted in E. C. Stakman, Richard Bradfield, and Paul C. Mangelsdorf, *Campaigns Against Hunger* (Cambridge, Mass.: Harvard University Press, 1967), 2.

37. Simonian, *Defending the Land of the Jaguar: A History of Conservation in Mexico* (Austin: University of Texas Press, 1995), chap. 5.

38. Wright, *Death of Ramón González*, 168–71.

39. Hamilton, *Limits of State Autonomy*, 216–40.

40. Ávila Camacho quoted in Cynthia Hewitt de Alcantara, *Modernizing Mexican Agriculture: Socio-Economic Implications of Technological Change, 1940–1970* (Geneva: United Nations Research in Social Development, 1976), 21–22.

41. Detailed accounts of how this was done can be found in CEPAL, *Economía campesina y agricultura empresarial* (Mexico City: Siglo XXI, 1982); and David Mares, *Penetrating the International Market: Theoretical Considerations and a Mexican Case Study* (New York: Columbia University Press, 1987).

42. Wright, *Death of Ramón González*, 172–77.

43. John E. Farley, *To Cast Out Disease: A History of the International Health Division of the Rockefeller Foundation (1913–1951)* (Oxford: Oxford University Press, 2003); John H. Perkins, *Geopolitics and the Green Revolution: Wheat, Genes, and the Cold War* (Oxford: Oxford University Press, 1997); Wright, *Death of Ramón González*.

44. E. C. Stakman, Richard Bradfield, and Paul C. Mangelsdorf, *Campaigns Against Hunger* (Cambridge, Mass.: Belknap Press, 1967).

45. Ibid., 33.

46. Ibid., chaps. 1, 8, 9.

47. Ávila Camacho quoted in Barkin and King, *Regional Economic Development*, 54.

48. The applicability of the Green Revolution to smallholder agriculture was and remains an area of intense controversy and a large literature. In Mexico, the controversy focuses on the Rockefeller-financed Puebla Project, designed to show that the Green Revolution was appropriate to smallholders. See Bruce H. Jennings, *Foundations of International Agricultural Research: Science and Politics in Mexican Agriculture* (Boulder, Colo.: Westview, 1988). See also Wright, *Death of Ramón González*; and Perkins, *Geopolitics and the Green Revolution*.

49. Barkin and King, *Regional Economic Development*, 54–65.

50. Paul Lamartine Yates, *Mexico's Agricultural Dilemma* (Tucson: University of Arizona Press, 1981).

51. For an excellent picture of land practices in the northwestern desert to the mid-nineteenth century, see Cynthia Radding, *Wandering Peoples: Colonialism, Ethnic Spaces, and Ecological Frontiers in Northwestern Mexico, 1700–1850* (Durham, N.C.: Duke University Press, 1997); and Doolittle, *Cultivated Landscapes*.

52. Simonian, *Defending the Land*, 113.

53. Stakman, Bradfield, and Mangelsdorf, *Campaigns Against Hunger*, esp. chaps. 5 and 9.

54. Fertimex, *Plan de desarrollo de Fertimex en la producción, formulación y comercialización de pesticidas* (Mexico City: Gerencia General de Planeación, 1981).

55. Wright, *Death of Ramón González*, chaps. 6–9.

56. This and following pages are based on Wright, *Death of Ramón González*. For

policies, theory, and implications, see esp. chaps. 6–9 and afterword; for the field study, see the introduction, chaps. 1–4, and the afterword.

57. Wright, *Death of Ramón González*, afterword.

58. Patricia Díaz Romo and Samuel Salinas Álvarez, *Plaguicidas, tabaco, y salud: El caso de los jornaleros huicholes, jornaleros mestizos, y ejidatarios en Nayarit, México* (Oaxaca: Proyecto Huicholes y Plaguicidas, 2002); see also the video *Huichols and Pesticides*, available at www.huicholesyplaguicidas.org and at www.panna.org (also available in Spanish and twelve indigenous languages of Mexico).

59. Elizabeth Guillette, M. M. Meza, M. G. Águilar, A. D. Soto, and I. E. García, "An Anthropological Approach to the Evaluation of Preschool Children Exposed to Pesticides in Mexico," *Environmental Health Perspectives* 106, no. 6 (June 1998): 347–53.

60. Rachel Carson, *Silent Spring* (New York: Houghton Mifflin, 1962). Carson's work was little known and little heeded in Mexico, with concerns of some scientists largely dismissed on the grounds of the need for malaria control and economic development. By the time it became well known, Mexican farmers were mostly moving on to reliance on nonpersistent but far more toxic organophosphates and carbamates (whose dangers Carson had also discussed, but with less resonance among the public and government regulators) because of the buildup of resistance to DDT and other organochlorines among insect populations. This subject is discussed in Wright, *Death of Ramón González*, esp. 16–19.

61. Saul Franco Agudelo, *El Paludismo en America Latina* (Guadalajara: Universidad de Guadalajara, 1990).

62. Wright, *Death of Ramón González*, chap. 5.

63. Ibid., afterword.

64. Dina L. Umali, *Irrigation-Induced Salinity: A Growing Problem for Development and the Environment*, World Bank Technical Paper 215 (Washington, D.C.: World Bank, 1993).

65. Kevin Gallagher, *Free Trade and the Environment: Mexico, NAFTA, and Beyond* (Stanford: Stanford University Press, 2004); D. Campbell and L. Berry, "Land Degradation in Mexico: Its Extent and Impact," FAO, LADA Working Paper, 2003, available at http://lada.virtual.centre.org.

66. Helena Cotler, Esthela Sotelo, Judith Domínguez, Maria Zorilla, Sofía Cortina, and Leticia Quiñones, "La conservación de suelos: Un asunto de interés público," *Gaceta Ecológica* (Instituto Nacional de Ecologia) 30 (2007): 5–71 (also available at ine.gob.mx/publicaciones/gacetas/conservacion.html).

67. "Jesus Leon Santos, Goldman Prize 2008," http://www.goldmanprize.org/node/713.

CHAPTER THREE

Mexico's Breadbasket
Agriculture and the Environment in the Bajío

Martín Sánchez Rodríguez

For more than 250 years, the alluvial valleys of the Lerma River's middle basin were the scene of intense human efforts to convert them into cropland for grain production. The region, which is one of Mexico's principal river basins and stretches from the Valley of Toluca to its mouth in Santiago Ixcuintla, Nayarit, eventually became the breadbasket of New Spain and independent Mexico. This transformation was no simple task. The hydraulic conditions in the basin required not only the deployment of an ancient form of water management but also "surplus" land for production, capital investment, and the organization of a reliable labor force. The creation of an agricultural landscape in the Mexican Bajío came about largely because of a technique of irrigation and soil management based on water capture using *cajas de agua*, which were low-lying fields with artificial embankments that could be converted into temporary reservoirs.

In 1981, Eric Van Young published a study of the rural economy around Guadalajara that spanned a period of nearly 150 years. He concluded that the changes in agriculture between 1675 and 1810 had been essentially quantitative rather than qualitative, in the sense that the area under cultivation expanded through a process of internal colonization of lands that were already occupied, in contrast to the European pattern of increasing productivity through technological innovation. At the same time, he suggested that the increase in land values during Mexico's colonial period was not solely the result of inflation but also resulted from the investment of capital generated by commerce and mining.[1] Most historians now agree that the most important technological advances in New Spain's agriculture derived from the administration of water and the expansion of irrigation. What did this mean in environmental terms? What were the social and environmental consequences of the construction of a hydraulic infrastructure? What technical differences are there between the use of perennial water supplies and seasonal ones? During what stages or periods were these distinct types of water use developed in Mexican agriculture, and what role did they play in the creation of new agricultural landscapes?

The Precolonial Landscape

In 1521, Spanish conquerors first set foot on the lands ruled by the Cazozin, lord of the Purépechas, and took their first look at the territory known as Michoacán. At that time the Tarascan empire occupied approximately the same area as the present-day state. Toward the east, it was bordered by territory ruled by the Aztecs; to the south, the Tarascans controlled both shores of the Balsas River. To the west, the empire included a good part of the current state of Jalisco, and it encompassed in the north the towns of Yuriria, Acámbaro, along with its subjects of Apaseo and Coroneo, which served as border towns with the Chichimecan tribes.[2] Most of this region lay inside the Lerma River watershed, whose main branch flowed through the southern valleys but also had several other tributaries, including the Guanajuato, Silao, Turbio, and La Laja Rivers, to name only the most important.

We know little of the landscape the Spanish encountered upon their arrival in the lands north of the Tarascan empire, on the margins of what is today known as the Bajío of Guanajuato. What has been possible to reconstruct comes from data contributed by the *Relaciones geográficas del siglo XVI*, which contains an eloquent description of the province of Acámbaro in 1570:

> The climate is temperate, more frequently hot than cold, and has a generally dry quality; although there is some humidity in certain parts owing to two large rivers that pass through this province. It is a province with less water than others, because it usually rains two months later than elsewhere. The brunt of the rain comes during the months of June, July, August and September; and the wind is generally easterly, of moderate strength; and it prevails throughout most of the year, except during the rainy season when it moves northward and withdraws somewhat from southern areas.[3]

A contemporary reading of this description suggests that the prevailing regional climate was warm semiarid, or seasonally warm-arid and temperate semiarid, according to the Thornthwaite classification. The altitude of between 1,600 and 1,800 meters above sea level gave it temperate weather, with an average temperature of 20°C. Rainfall was concentrated in the summer, with annual precipitation that varied between 489 and 700 millimeters. The irregularity of the pluvial regime also favored the occurrence of frosts when the rains continued late into the year. Additionally, the concentration of precipitation in a period of four months (June, July, August, and September) produced too much water, exceeding the needs of plants and resulting in evaporation, infiltration into the soil, and above all, runoff.[4]

The conquistadors found the Lerma River basin to be a sparsely occupied region that had not sustained sedentary indigenous populations for many decades, although it had once been home to social groups that developed more-or-less extensive and diverse economies. Archaeological research has uncovered many sites that testify to permanent settlements dating from 800 BC (the Preclassic horizon, as it is known in Mesoamerican chronology), subsisting through the Classic horizon (AD 300 to 650), and arriving at the Postclassic horizon (AD 900 to 1000). Little is understood of the reasons for these populations to have abandoned the region or of the social structures they established in these northern reaches of Mesoamerica. One hypothesis suggests that an intense drought between the tenth and eleventh centuries AD forced native people to abandon the Bajío, long before the arrival of the Spanish.[5] Whatever the case, the Lerma River by the sixteenth century had become the natural boundary between the civilizations of central Mexico and the seminomadic tribes generically known as Chichimecas.[6]

The hydrological topography of the Bajío varied greatly. As map 3.1 shows, the Lerma River was the primary watercourse and carried large volumes of water through its southern extremities. The Lerma ran roughly east to west from its source in the Lerma wetlands in the current state of Mexico to its mouth at Lake Chapala. Along the way, it captured the water of several tributaries as well as seasonal runoff from the mountain ranges to the north and south of its path. Because they originated farther away, the rivers that flowed down from the northern part of the Bajío were longer than the southern tributaries and carried greater volumes of water. Such was the case for the Guanajuato, Silao, Turbio, and La Laja Rivers. The rivers and streams that emptied into the Lerma from its left bank were shorter but included such important tributaries as the Querétaro, Apaseo, and Duero Rivers.

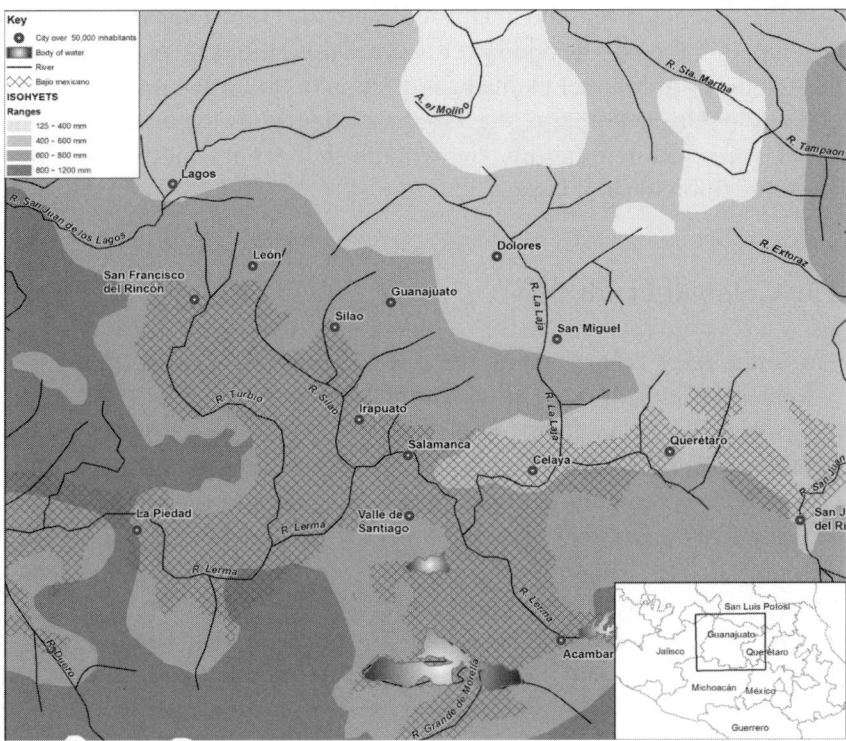

Map 3.1. Historic features of the Lerma watershed.

Also dotting the region were a significant number of permanent and seasonal bodies of water. These included lakes, ponds, and marshes in the upper, middle, and lower parts of the basin. The ponds at the source of the Lerma River and the Lake of Chapala were particularly important; likewise, the Lake of Cuitzeo in the lower part of the basin, with its salty waters, produced rich grasses and an abundance of frogs and fish such as *charales*, *curuengas*, and *mojarras*. Swamps and marshy areas also formed throughout the area—sometimes acquiring significant dimensions—as a result of river flooding during the rainy season, the upwelling of springs, and the lack of drainage as a result of the region's low inclines as well as basaltic dikes that closed off the valleys.

According to contemporary reconstructions, there appear to have been ten major swamps located in the Bajío, the most notable of which occupied the area now covered by the Lake of Yuriria. Another formed at the confluence of Río La Laja with the Apaseo River near Celaya, and yet another along the meandering path of the Silao River close to its confluence with the Guanajuato in the heights of Irapuato. Three large marshy areas developed in the middle and lower basin of the Turbio River between the present towns of Manuel Doblado and Abasolo. Philip W. Powell's history of the Chichimec War also refers to the existence of "the wetlands of the Chichimecs." Moreover, some maps of the eighteenth century still show a marsh near the town of Celaya.

The Colonial Landscape

The water in the Bajío wetlands sustained a profusion of grasses that served as fodder for flourishing herds of livestock in the colonial era. In the case of Cuitzeo, the chroniclers spoke of "an abundance of pasturage throughout the year" and observed that "especially along the edge of the lake there is usually large amounts of green grass."[7] Nonetheless, the same accounts mention that the livestock preferred to browse on mesquite trees rather than the native grasses. This helps to explain why chroniclers singled out Celaya and Apaseo as sources of mesquite groves.[8]

Spanish colonizers' inability to find any precious metals in the region, combined with ongoing indigenous resistance during the first decades of the sixteenth century (in part because livestock caused extensive damage to native crops), made the relatively unpopulated Bajío an attractive site for cattle ranching. The task of developing a livestock trade fell to Catholic missionaries, who, although "less spectacular and slower, proved to be steadier" than independent hacienda owners.[9] By the end of the seventeenth century, the Bajío had emerged as a major agro-pastoral center. Reports of the protoparishes known as *doctrinas* produced by the bishopric of Michoacán tell us quite a bit about these developments (particularly in John Tutino's analysis of them).[10] The first ranching landscape appeared in the northern regions around Guadalajara prior to the development of extensive agricultural economy there in the mid-eighteenth century thanks to the construction of a water management infrastructure.[11]

Nevertheless, the importance of livestock in the initial development of the Bajío and in the formation of the first agricultural landscapes of New Spain is worth examining in this region and elsewhere. So are the efforts to develop irrigated agriculture through the extensive clearing of mesquite forests—a process that was doubtless helped along by livestock—and the construction of a hydraulic infrastructure of dams and canals. One of the first large hydraulic projects that left its mark on the Bajío landscape was the canal and dike that created what is now the Lake of Yuriria. The construction of this artificial reservoir marked a radical transformation of the environment. Taking advantage of the existence of a marsh formed by the flooding of the Lerma River after the rainy season had passed, Friar Diego de Chávez, an Augustinian monk from the province of San Nicolás de Tolentino in Michoacán, "made" the lake by dredging a broad channel to the river and allowing the marsh to flood whenever the river ran high.[12] Little by little, landowners subjected the river currents of the Bajío to hydrological projects that interrupted or diverted their courses. The population growth generated by silver strikes at what became the royal mines in Zacatecas (1543) and Guanajuato (1552) led the second wave of colonization in the region and a vastly expanded demand for wheat flour.

The advances in irrigation were hurried along in no small measure through the process of apportionment (*mercedación*) of land to new cities, villages, and settlements (*congregaciones*). Even so, the regional economy remained centered on livestock until the beginning of the eighteenth century. In other words, ranching was without a doubt the dominant factor in the transformations of the environment and the formation of the first Hispanic landscape in the Bajío. Indeed, cattle themselves played a role in this process. According to some chroniclers, large herds of cattle circulated northward from the Valley of Toluca in search of grass that grew in rich lowlands of the Bajío, and from there moved on to the Lake of Chapala. Along the way, they destroyed the crops of indigenous communities, generating a heated conflict between natives and Spanish colonizers.[13]

Water and the Formation of an Agricultural Landscape

Specialists in the colonial agrarian history of the Bajío and other regions of New Spain agree that the eighteenth century was a period of significant change as a result of three factors: the increasing production of grains to meet the needs of an expanding market; the displacement of ranching activities to less productive lands or to northern frontier areas; and the overall growth of population. The late-colonial economic boom also led to greater investments of capital in hydraulic infrastructure, fences, and granaries. Most authors also agree that the growth in grain production resulted not from technological innovation but rather from placing more land into cultivation, although Van Young does affirm that advances in water management played a critical role.[14] His observations allow us to imagine the impact that an expanding agricultural frontier and an increasingly cultivated landscape must have had on the environment. Not only did landowners clear forests, reclaim land, excavate canals, and shore up levees, they introduced new management technologies for controlling rivers and streams. These they used for irrigation through the construction of dams to create reservoirs, as well as dikes, diversion levees, and sluice gates.

Numerous reasons explain why agriculture in New Spain—particularly north of the densely settled region of central Mexico—was neither intensive nor governed by technological innovation. Most notably, the colony was believed to have an unlimited supply of land. This myth had some basis in fact, at least as far as the Bajío was concerned, because the absence of large, permanent indigenous populations opened a potential space for colonization during the sixteenth and seventeenth centuries. Large cattle estates appeared and eventually grew into vast latifundia but later split into smaller ranchos and haciendas. Fortresses, garrisons, and militarized agricultural colonies (presidios) appeared, as did civilian communities, towns, villages, and cities. The sediment-rich soils ripe for cultivation made the development of alternative technologies seem unnecessary, unlike Europe, where the scarcity of land encouraged a more intensive agricultural regime.

But in environmental terms, we should look for other factors that influenced the practice of extensive agriculture. Water management can be considered one of the most innovative elements of colonial and early-republican agriculture, especially in the Bajío. The pattern of rainfall helps to explain why. Alexander von Humboldt was among the first to notice the region's peculiar weather; he concluded that the altitude of the plain located between the twentieth and thirtieth parallels produced a drop in overall barometric pressure and accelerated water evaporation, while the ascending current of hot air from the plains impeded precipitation.[15] Indeed, Mexico's geographical position and its orography influence the rainfall pattern as a result of prevailing winds, which generally blow landward from the Atlantic and southward from North America. From June to October, the rains in most of the country derive from the humid Atlantic current. Rainstorms arrive from the southeast, ascend along the foothills of the Sierra Madre Oriental, and intensely lash much of eastern Mexico. This is to say that the eastern mountains keep humid winds from reaching the internal plains, but the nearly constant humidity in the east, occasionally augmented by torrential downpours from northers and hurricanes, produces abundant precipitation in that part of the country, where rainfall averages 1,500 millimeters per year.

The Pacific slope, in contrast, has much less average rainfall, and its rains are irregular and torrential. The winds predominantly come from the north, though there are important variations from one part of the region to another. For example, rainfall in the Sonoran Desert usually reaches only 50 millimeters annually, whereas the region of El Salado on the southern plateau receives ten times as much. The central part of the country averages under 1,000 millimeters per year.[16] In the specific case of the Bajío and the mountain ranges that encircle it, variation is also the rule. The Valley of Celaya in the heart of the Bajío registers an average annual rainfall of 700 millimeters, while the neighboring zone of Salamanca, a mere 50 kilometers away, averages 15 percent more, or 800 millimeters per year. So do Guanajuato, Apaseo, and Yuriria. The mountainous part of Pénjamo can receive up to 900 millimeters annually, while in San Felipe, San Luis, and Dolores the average reaches only 500 millimeters. Temporal variations are even greater than geographic ones, as most rainfall is concentrated in a single month. Of the 700 millimeters of average annual rainfall in Celaya, between 150 and 160 millimeters falls in August. That is when Apaseo, Jerécuaro, and Yuriria likewise record the most intense rains, averaging from 170 to 180 millimeters. San Felipe, San Luis, and Dolores have average total rainfalls of only 80 millimeters between June and September.[17]

As Jorge Tamayo explains, the nation's precipitation is distributed in a particularly unfavorable way, because when the rain does fall "it comes in excessive abundance, even to the point of being damaging," whereas the rainwater becomes scarce "when its employment could be considered indispensable."[18] Moreover, much of the rain comes from violent but (usually) short-lived thunderstorms. Because most of Mexico's rivers are short and have shallow basins yet often traverse steep slopes at certain points, the runoff from torrential rains swells watercourses large and small almost overnight. Large volumes of water course through rivers and streams that can dry out a few hours or days later.[19]

The only way for Bajío landowners to increase production was by making better use of these hydrological resources. Because the rainfall pattern was concentrated between the months of June and September, relatively little water flowed in perennial watercourses during the rest of the year. Planters began to make modifications to streams from the very outset of the colonial era in the sixteenth century, but they lacked the means to extend irrigation networks or accommodate the shift from corn and bean cultivation to wheat. That changed in the eighteenth century, when the development of technologies capable of controlling torrential rainfall could no longer wait and landowners began to build more resistant and costly dams and sluice gates, as some historians have demonstrated.[20] Yet the single most important innovation was the adoption of flood farming (*entarquinamiento*) that captured seasonal river flows in fields known as *cajas de agua* (literally, "water boxes"). These fields remained flooded for a few weeks before farmers released the water to plant winter wheat.[21]

Basin Irrigation in the Bajío

Basin irrigation entails the capture of seasonal rainwater runoff (that is, torrential flows) rich in organic materials; the runoff is used to flood agricultural fields for a period of four to six weeks. The water is then drained and the fields prepared for planting. The quantity of water needed to cover each field depends on several factors, the most important of which (naturally) is the total area that the farmer intends to plant. In the Bajío, all of the lands subject to flood farming are called "cajas de agua." Etymologically, *entarquinamiento* comes from *tarquín*, which means to fill with silt (that is, to muddy it). This technique is also known as "pond-making" (*enlagunamiento*), flood farming (*anegamiento*), or siltifying (*entanquinamiento* or *enlame*). These technologies are collectively referred to as *bassin d'irrigatión* or *limonage*, both of which are variants of water harvesting techniques first developed in ancient Egypt. In Mexico, the practice of basin irrigation was concentrated around the Lerma River basin and the region of the Comarca Lagunera located between the current states of Durango and Coahuila.[22]

In Egypt, the practice of basin irrigation first appeared in fields that were already subject to regular flooding by the Nile. Pharaoh Menes formalized the practice when he ordered a levee constructed along the eastern bank of the river. Basin irrigation agriculture became increasingly widespread until the 1820s, when Mohamed Ali, with the help of French engineers, began to promote "modern" irrigation techniques for growing cotton.[23] The essential elements of the Egyptian basin technique included diversion canals running to levees built both parallel and diagonal to the river, which formed shallow reservoirs (*bassins* in French) that completely covered the fields before agriculturalists drained them and planted crops. A series of sluice gates regulated the water that flowed into the canals and from there to the bassins. The levees varied in size, with some measuring up to 6 meters in height and bases as wide as 3.5 meters. The relatively flat terrain, combined with the levees' height and impressive length, permitted the formation of water deposits covering areas as great as 31,500 hectares.[24] The reservoirs were filled during the river's flood phase, known as *axt*, which occurred from early August to the end of November in most parts of the Nile River basin.[25] The process of filling the bassins usually began in the southern provinces during the first half of August, and the last ones were typically emptied in early October. In Lower Egypt, the entire process began and ended a month later. A cycle of flooding the fields and allowing the water to seep into the soils usually stretched over a period of forty-five days. Once the flooded fields had reached their capacity and the Nile had lowered far enough, drainage gates were opened and the water spilled once more into the river channel. If the water remained high, farmers had to leave their fields underwater even if it meant delaying the planting.[26]

In Mexico, references to earthen reservoirs (cajas de agua) first appear in the eighteenth century and pertain exclusively to the Bajío. Two hundred years ago, Humboldt emphasized the role of Mexican basin irrigation in grain production and noticed the similarity to its Egyptian counterpart. In 1900, Karl Kaerger once again compared the cajas de agua to the Egyptian technique. After that, another seven decades transpired before David Brading once again pointed out their existence in his classic work on hacienda owners and rancheros in the Bajío. In 1986, Michael Murphy discussed cajas de agua. Guillermo García Zamacona partially described them in a 1999 publication. Only recently, however, have scholars attempted to document the historical development of basin irrigation and to examine the cajas' benefits to Mexican agriculture—both past and present—in greater detail.[27]

I want to insist on retaining the customary term, *cajas de agua*, because the technique is still in use today. Fields farmed this way varied in size—without ever reaching the dimensions seen in Egypt—and sometimes grew as large as five hundred hectares. The basins (that is, the fields intended for flooding) were created by building embankments of varying lengths, heights, and thickness that formed trapezoidal areas. Water was then introduced via a canal, which in most cases consisted of channels or diversion levees, known as *bocatomas*, situated somewhere upstream. Once the water coursed into the basins, secondary canals and gates located in the fields ensured that the water flowed evenly over the entire surface of the land.

The Bajío's highly seasonal rainfall meant that planters had only one chance to take advantage of the river flows that peaked between June and September, known as "wild waters" (*aguas broncas*). Rivers—some of which were seasonal—maintained their highest flows for no more than a week, so farmers had to act quickly. Some of these waters were considered more desirable than others, however. In the Valley of Celaya, for example, farmers' accumulated experience led them to prize the runoff that surged out of surrounding mountain ranges because it had a particularly high content of organic materials. Whatever its source, water was diverted into canals and deposited in the cajas de agua, where it remained for several weeks if the rains were plentiful. Some of the water was slowly lost to evaporation. More infiltrated into the soil, leaving behind a thin layer of organic sediments (*tarquín*). Once the rains had ended, farmers opened sluice gates to drain the fields and proceeded to sow winter wheat. This technique not only saturated the soils in ways that benefited crops but also provided a series of other environmental services. Flood farming guaranteed the annual renovation of soils as a result of the percolation process, for instance. It also helped to eliminate mineral salts (which dissolved into the water), increased soil fertility through the deposit of highly concentrated organic material, controlled animal and plant pests, and mitigated the need for certain fieldwork—such as aeration—in preparation for planting.[28]

If the advantages of basin irrigation had been recognized for millennia, why did it become systematically employed in the Bajío only during the eighteenth century? Moreover, how important were the cajas de agua for the production of grains in the Bajío? The answers to these questions can be found by investigating the processes of socioeconomic transformation that overtook the Bajío in the late seventeenth and early eighteenth centuries.

After the arrival of the Spanish in the sixteenth century, the Bajío underwent major changes as a result of colonization, forest clearing, the construction of an incipient hydraulic infrastructure, and eventually manufacturing. However, the eighteenth century stands out as the period when the region began to play a major role in New Spain's economy, primarily because of its large-scale production of wheat. Its centuries-old status as a frontier zone gave way to an unmistakably urban network energized by its demographic expansion, increasingly sophisticated agriculture, and growing industrial capacity.[29]

Because other authors have examined the impacts of population growth and the expanding role played by mills in the Bajío's agricultural development, I will merely describe these issues in passing.[30] The population of the bishopric of Michoacán, which included the Bajío, expanded rapidly during the eighteenth century, and the demand for agricultural products naturally grew apace. During the century between 1660 and 1760, the population of Irapuato alone jumped from 1,753 souls to 12,030; that of Celaya expanded from 4,000 to 25,000, and Chamacuero's from 1,000 to 5,500. In the case of Salvatierra-Acámbaro, the growth of the confessional population (which roughly reflects the overall increase in the number of inhabitants) climbed from 2,086 to 17,414 between 1670 and 1760. In San Francisco de Rincón, the number rose from 1,947 to 18,770.[31]

In terms of agricultural production, the statistics also reflect significant gains. In the parish of Silao, 20,558 bushels of corn and 90 *cargas* of wheat were collected as tithes in 1689. These numbers increased to an average of 90,352 bushels of corn between 1751 and 1755, and from there to an average of 166,410 bushels of corn and 1,910 cargas of wheat between 1776 and 1780. In León, the average wheat crop between 1661 and 1665 amounted to 790 cargas, whereas the figure had risen to 2,210 cargas by 1750.[32] In the particular example of the predominantly livestock-raising Jalpa hacienda, administrators reported production of 185.5 cargas of wheat in 1745; forty years later this figure had multiplied to 1,114.5 cargas.[33]

The community of Irapuato, after having been an important livestock producer in the early colonial era, launched into wheat production around 1740. In 1738, the district harvested 24 bushels of wheat, a figure that increased to 369 in 1743 and 453 in 1752. By 1777, it produced more than 2,000 cargas.[34] Farther north of the Lerma River basin but still within the bishopric of Michoacán is the community of San Luis de la Paz. With a drier climate and a more mountainous terrain than Irapuato, the shift from livestock herding took place there as well, albeit later than in the Bajío. Small livestock predominated there until the middle of the eighteenth century; then an impressive expansion of agriculture occurred between 1732 and 1777 and even came to represent over half of the tithes paid to the church (an average of 57.5 percent, to be exact) between 1797 and 1804.[35]

Another development at this time had equally significant implications for the way that water and land functioned as productive resources: wheat displaced corn as the region's most important crop. Encouraged by the economic growth in the mining sector, the haciendas of the Bajío increased production, both by opening new land for agriculture and by planting wheat in land formerly dedicated to corn. As John Tutino has explained, corn—which represented the staple food of the Bajío's lower classes—was relegated to less fertile and typically unirrigated (rain-fed) land. Tutino shows that wheat first arrived in the eastern part of the basin—where by 1785 haciendas were planting three times as much wheat as corn—as commercial landowners began to plant it on their best lands. At that point, corn was planted primarily on less desirable land and the uplands of the more indigenous western side of the basin.[36] An increase in grain production of this magnitude in a region with a predominantly torrential rainfall pattern—but also with significant capital, labor, and available land resources—suggests that the construction of a water management infrastructure lay behind both increased productivity and environmental transformation. From the perspective of natural resource management, basin irrigation clearly played a major role in the expansion of irrigated lands, as well as the increasing production of corn and wheat, the gradual substitution of corn by wheat, the displacement of cattle ranching by agriculture, and hence the conversion of the Bajío into a grain-producing region during the eighteenth century. Three examples of this process will illustrate more about the nature of these developments.

The community of Irapuato is located in the central part of the Bajío and receives water from the Silao and Guanajuato Rivers, whose nearby confluence represents a constant threat of flooding during the rainy season. During the sixteenth and seventeenth centuries, most land in the area was used primarily for pasture, but grain production had gained prominence by the eighteenth century. Fernando Picó has found that Irapuato's expansion of agriculture occurred in two distinct phases. The first lasted from 1715 to the end of the 1730s, when corn production predominated. The second began when wheat displaced corn around 1740 and continued until the crisis of 1785–86.[37]

An eighteenth-century map of Irapuato shows that hacienda owners had established a series of cajas de agua "as is the custom of that place," according to the exegesis that accompanied it. The document explained that the authorities and principal residents sought to put an end to the recurrent floods by opening a new branch of the Silao River. The leading citizens formed a commission to select the best location for the project and settled on a site to the northeast of the village on the lands that belonged to the hacienda owned by George López. The commission further determined that the water would be carried by a canal through the hacienda (as well as two others, named Cubujados and La Soledad) and then empty into a reservoir on land belonging to the Yóstiro hacienda. The canal would be approximately one league in length, four yards deep, and ten or twelve yards wide. A sluice gate set in a stonework base would regulate water flows. Its cost was estimated at eight thousand pesos. The viceroy determined that the project would provide a valuable public service and approved the plan on May 21, 1757.[38] Soon, however, the community's lack of funds became clear, prompting local officials to offer water rights to the haciendas that the canal traversed if the owners agreed to finance the project. This was how the owners of Cuisillos, Cubujados, López, La Soledad, and Yóstiro haciendas acquired the rights to half of the water from the Silao River and began the excavation of the canal and the construction of levees, dams, and cajas de agua. Over time, conflicts arose over the distribution of the water.

The map also identifies specific uses for each river and seasonal watercourse. Listed in order of importance are the Silao, Guanajuato, Derramadero, Amezcua, Manserrua, Carrizal, and Nuevo Rivers, but also listed are the La Caja and Arandas arroyos (an arroyo is a dry creek bed that fills only during the rainy season). The map also clearly distinguished three bridges and, most important, forty dams and earthen levees (*bordos de tierra*) of varying sizes, notably those of Arandas and Las Ánimas. (This map represents only a fraction of the colonial dams and cajas de agua in the area, however.) The most notable example is the stonework San Vicente Dam, commissioned by Vicente Manuel de Zardaneta y Legaspi (the marqués de San Juan de Rayas) for his San Antonio hacienda, which when completed in 1780 measured 3,081 varas (over 2.5 kilometers) in length.[39] In addition, the complementary hydraulic infrastructure (such as the nine sluice gates that managed the flow of water to and from the new dam) must be considered.

David Brading, María Guadalupe Rodríguez, and numerous cartographic documents give us a guide for estimating the agricultural role that cajas de agua played around the city of León. Brading presents a chart that summarizes the area under cultivation and hacienda growth over the course of the eighteenth century. For example, the Cañada de Negros hacienda reported sowing 11 *caballerías*, or more than 470 hectares, of land in 1746. It nearly tripled its productive area to 30 caballerías in the next quarter century, then tripled it again to 94.5 caballerías (4,044 hectares) by 1790. Its value likewise rose a factor of three between 1771 and 1790, although productive improvements such as clearing land and building dams and fences also had something to do with its increased value.[40] However, we can complement Brading's findings by taking a closer look at the topographical and hydrological conditions in the area where Cañada de Negros was located (map 3.2).

In the first place, the hacienda lay at the intersection of two geophysical regions: the Bajío, with its low sloping valleys, deep and silt-rich soils, and torrential water flows; and the more mountainous Los Altos region of Jalisco, which is also characterized by seasonally torrential rainfall but, in contrast, has shallow, rocky soils and a riparian topography defined by innumerable hills and ravines. A close analysis shows that a good portion of the hacienda's territory comprised uncultivated pastures, which undoubtedly served primarily for cattle ranching and some unirrigated crops such as corn, beans, and squash. This topography is precisely what allowed for the use of runoff to fill the dam-and-reservoir system that Brading mentions. Yet the hacienda's only level lands were located in the northeast, far from the dam, which casts doubt on how much of these good flatlands could have been irrigated at all. If we further take into account the fact that the reservoir was relatively small, then factors other than the overall extension of land (or the clearing

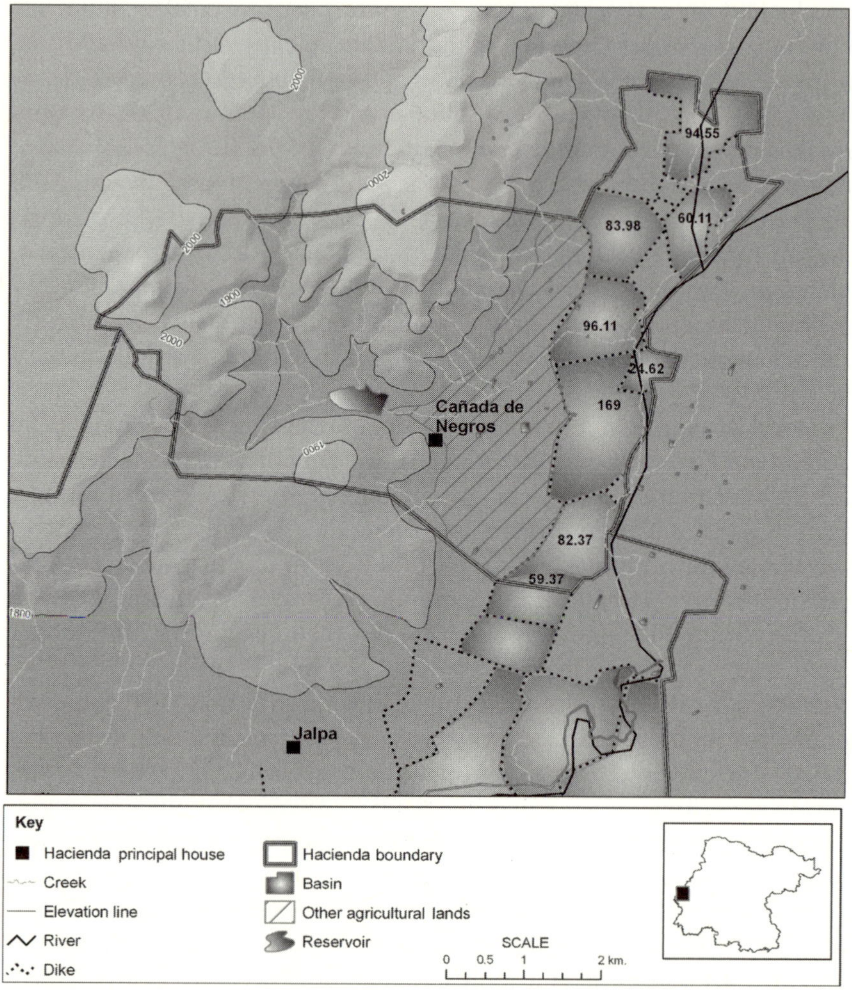

Map 3.2. Cajas de agua (in hectares) on the Cañada de Negros hacienda. The hacienda encompassed 2,956 ha; the cajas de agua, 670.15 ha; and other agricultural lands, including pasturage and nonirrigated cropland, 529.95 ha.

of forests and construction of fences and the dam, as mentioned above) must explain the hacienda's exponential growth in value toward the end of the eighteenth century. The single most likely explanation is the construction of cajas de agua used for flood farming on the flatlands, which allowed the growers to take advantage of the surface waters from the Turbio River as it passed through the hacienda's territory.

Like many of the other rivers that feed the Lerma (also known as the Río Grande), the Turbio is a seasonally torrential river that reaches its largest water volumes between the months of June and September. Before arriving at the Cañada de Negros, significant amounts of water were typically withdrawn by other landowners, to the point that the Turbio's volume diminished almost to nothing by the time it joined the Lerma. The river entered on the north side of Cañada de Negros through a property called Vallado de la Haciendita and crossed nearly the entire hacienda along its eastern edge, enabling the diversion and distribution of its waters through gravity alone, because the land inclined slightly westward. This enabled the growers to contemplate the use of cajas de agua to irrigate the land.

An 1890 map of the Cañada de Negros shows that the hacienda covered an area of 2,862.2 hectares divided as follows: 685.8 hectares of partially irrigated land *de medio riego*; 390.20 hectares of first-class rain-fed land (*temporal*) as well as another 385.6 of second-class temporal; 142.2 hectares of woodland; 1,228 hectares of rangeland (*agostadero ceril*); and 40.4 hectares covered by the reservoir commissioned in the eighteenth century.[41] The Cañada de Negros reservoir simply did not have the capacity or location to irrigate the amount of land that Brading records. Indeed, the reservoir probably could not even irrigate the 390.2 hectares classified as first-class lands in 1890. The greatest increase in irrigated area in fact resulted from vigorous capital investment and manual labor in preparing fields for basin agriculture. This entailed the clearing of forests and the construction of several dozen kilometers of stone and earthen levees and canals.

Like the Cañada de Negros, the neighboring Jalpa hacienda straddled the Bajío and Los Altos de Jalisco, as well as the basin of the Turbio River. Jalpa consequently shared the same topographical and hydrological characteristics, although it possessed considerably more land (map 3.3). The hacienda's origin dated from the year 1542, when Juan de Villaseñor received a royal grant of four *sitios de estancia de ganado mayor* and 8 caballerías, equivalent to 7,365 hectares. By 1714, the territory had grown to five sitios de ganado mayor, plus two others for small livestock, as well as 13.4 caballerías used for crops, or in other words a total of 10,912 hectares. More than 75 percent of the area was located among hills, ravines, and the shallow and rocky soil characteristic of the Los Altos region. Only the quarter of hacienda's total territory located in the valley was suitable for more intensive agriculture.[42]

In terms of productivity, the relationship between livestock and agriculture in 1675 clearly favored the former. The hacienda's herd included 1,348 cattle, 626 horses, 172 foals, 22 donkeys, 43 mule yearlings, 16 full-grown mules, 160 sheep, 13 pigs, and 40 oxen. Only 1,129 bushels of corn and

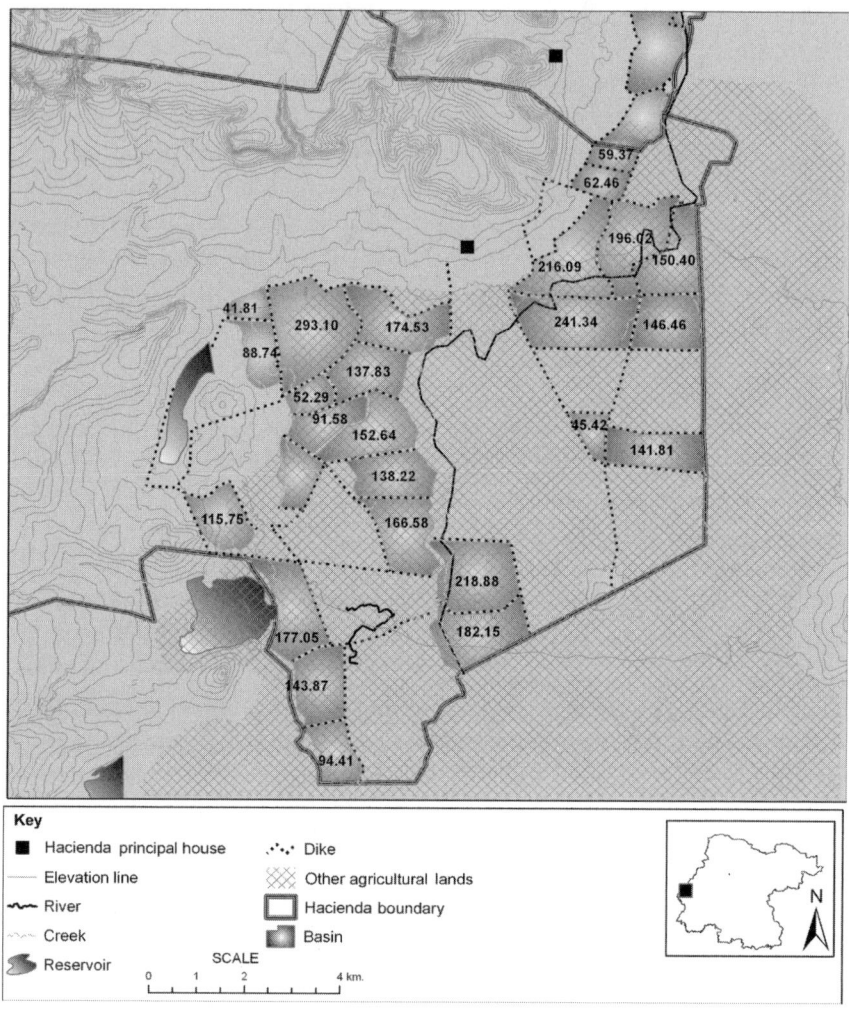

Map 3.3. Cajas de agua (in hectares) on the Jalpa hacienda. The hacienda encompassed 14,077.9 ha and the cajas de agua 3,616.5 ha.

90 of beans were harvested, which undoubtedly went to feed the livestock. By 1710, the number of livestock had increased; the hacienda reported owning a herd of 108 oxen that year, for example. It had also made significant investments in fences, stone walls, and the construction of four dams.[43] The generally poor quality of the hacienda's land meant that raising livestock never ceased to be its principal activity, but it nevertheless succeeded in making significant gains in grain production during the eighteenth century.

Guadalupe Rodríguez notes that Jalpa had partially reinvented itself and become a significant grain producer by 1745, at which point it produced 2,400 bushels of corn, 185.5 cargas (equivalent to 371 bushels) of wheat, and 32.5 bushels of beans. Forty years later, documents indicate that a proportion of the wheat harvest already improved on that figure by a factor of six, amounting to 1,114.5 cargas of wheat.[44]

Without a doubt, the investment in dam and reservoir construction represented a priority for the owners of Jalpa because it allowed them to claim extraordinary volumes of water. What has not been quantifiable—but surely represented an equal or greater investment—was the cost of building levees, canals, and sluice gates for the cajas de agua that allowed far greater access to the Turbio River's seasonal flows and floodwaters. If we examine the location of the reservoirs, the agricultural lands, and the trajectory of the Turbio River shown in an 1875 map of the hacienda, what stands out most clearly is the presence of a system of cajas de agua that the map calls *"vallados"* (referring to an area enclosed by an earthen wall known as a *valla*). As in the case of Cañada de Negros, the waters of the Turbio entered Jalpa on its northeast side and were collected in the San Antonio and Nombre de Dios reservoirs. From there, it flowed into a network of canals and eventually flooded the fields. The total area within the vallados exceeded 3,600 hectares, comprising approximately seventeen individual fields, none of which was smaller than 100 hectares in area. The largest were the vallados of El Conde, with an extension of 293 hectares, El Tecolote at 241, La Purísima at 218, San José at 216, and La Luz at 196.[45] Clearly, the eighteenth-century cajas de agua (or vallados, if you prefer) did far more to transform Jalpa's landscape and, of course, its environment than did the dams and fences that have so often been the object of historians' attention.

The use of cajas de agua in agriculture was limited neither temporally to the colonial era nor geographically to the Bajío. The practice of basin irrigation endured beyond the War of Independence in 1810, the civil wars of the first half of the nineteenth century, and even the Mexican Revolution of 1910 (map 3.4). In the Bajío, archeological evidence points to the continuity of construction of hydraulic infrastructure for the practice of entarquinamiento throughout the nineteenth century. In the zone of Irapuato, for instance, there are several references to cajas de agua in use on the haciendas of El Copal (1854), Serrano (1864), and Cuchicuato (1901). In the northern Mexican region known as Comarca Lagunera, which takes its name from the water that flows from the Nazas and Aguanaval Rivers, basin irrigation is also practiced using so-called water frames (*cuadros de*

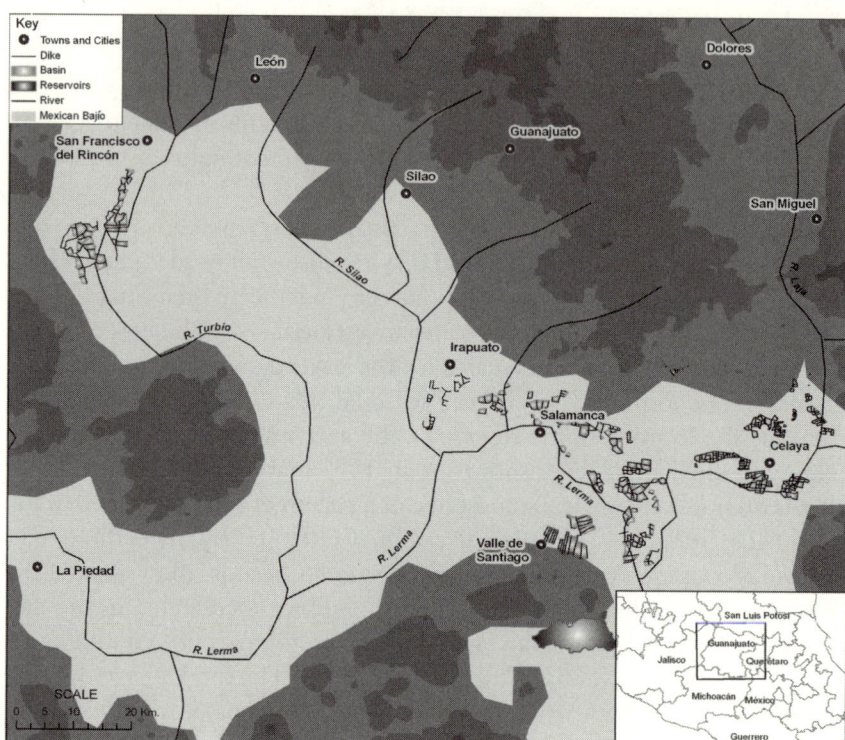

Map 3.4. Cajas de agua in the Bajío by the early twentieth century.

agua). This hydrological technology enabled planters to grow cotton in the mid-nineteenth century; this crop made the Laguna one of Mexico's wealthiest regions by the time of the Mexican Revolution.[46]

Conclusions

Examples of capturing torrential water flows for use in grain production, and the attendant transformation of the Bajío's productive landscape, can be found throughout the central Lerma River basin. Landowners built dams, levees, and cajas de agua in virtually every valley and subbasin in the region, and they continued to do so well beyond the colonial era. We still do not fully understand all the effects that the movement for independence had on the Bajío's productive infrastructure, but it does not appear to have damaged the Bajío beyond repair. We also lack clear records of what happened during the political instability associated with the 1858–61 War of the Reform and the 1864–67 Second Empire, but sources from the period suggest that the agricultural frontier and irrigation projects continued to expand even during that turbulent era. Not until the Porfiriato did a new chapter in rural investment, and of irrigation in particular, begin to be written. During the late nineteenth century and up until the revolution, new hydraulic systems, canals, dams, levees, and sluice gates complemented those built during the colonial period, vastly extending the reach of agriculture and transforming the regional environment.

In this sense, the use of basin irrigation and flood farming both as hydrological technologies and as a means of expanding irrigation frontiers should be seen as a continuous process that extended throughout the eighteenth and nineteenth centuries and even through the first half of the twentieth. Its growth not only displaced a significant amount of previously rain-fed fields known as *tierras de temporal*, which had once been dedicated to the staple crops of corn on which the region's poor subsisted, but also led landowners to clear forestland for wheat production, creating an agricultural landscape that is still recognizable today.

The success of flood farming, or entarquinamiento, in expanding commercial agriculture in the Bajío did not rest solely on its capability to capture torrential water flows as a means of artificial flood control. The technique provided several additional benefits, such as continually fertilizing the soil by depositing organic material on its surface and promoting aerobic and anaerobic processes that further enriched it. Flooding the fields also controlled animal and plant pests. Finally, flood farming washed the soils, desalinized the fields, and promoted groundwater recharge. These are all benefits that modern irrigation techniques (sprinklers, ferti-irrigation, and hydroponics) have difficulty achieving, even today.

Notes

1. Eric Van Young, *La ciudad y el campo en el México del siglo XVIII: La economía rural de la región de Guadalajara, 1675–1820*, trans. Eduardo L. Suárez (Mexico City: Fondo de Cultura Económica, 1989).

2. J. Benedict Warren, *La conquista de Michoacán, 1521–1530* (Morelia: Fimax Publicistas, 1977), 3–4.

3. René Acuña, ed., *Relaciones geográficas del siglo XVI: Michoacán* (Mexico City: UNAM, 1987).

4. Carlos Manuel Castillo, "La economía agrícola en la región del Bajío," *Problemas agrícolas e industriales de México: Anales de la economía agrícola mexicana* 8, no. 3–4 (July–December 1953): 9–12.

5. Efraín Cárdenas García, pers. comm., August 2009.

6. Efraín Cárdenas García, *El Bajío en el clásico* (Zamora: El Colegio de Michoacán, 1999), 93–112; Rosa Brambilia Paz, "La zona septentrional en el Posclásico," in *Historia antigua de México III: El horizonte Posclásico y algunos aspectos intelectuales de las culturas mesoamericanas*, ed. Linda Manzanilla and Leonardo López Lujan, 313–16 (Mexico City: INAH/UNAM/Miguel Ángel Porrúa, 1995); Juan Carlos Ruiz Guadalajara, *Dolores antes de la independencia: Microhistoria del altar de la Patria* (Mexico City: CIESAS, 2004), 1:72–78.

7. Acuña, *Relaciones geográficas del siglo XVI*, 79.

8. Ibid., 58–60.

9. Philip Powell, *La Guerra Chichimeca, 1550–1600* (Mexico City: Fondo de Cultura Económica, 1984), 20.

10. *El obispado de Michoacán en el siglo XVII: Informe inédito de beneficios, pueblos y lenguas*, with an introduction by Ramón López Lara (Morelia, Michoacán: Fimax Publicistas, 1973); John Tutino, *De la insurrección a la revolución en México: Las bases sociales de la violencia agraria, 1750–1940* (Mexico City: Era, 1990).

11. Van Young, *La ciudad*, 204–46.

12. Diego Basalenque, *Historia de la Provincia de San Nicolás de Tolentino de Michoacán, del Orden de N.P.S. Agustín*, with introduction and notes by José Bravo Ugarte (1673; facsimile ed., Mexico City: Jus, 1963), 125–26.

13. Heriberto Moreno García, *Haciendas de tierra y agua en la antigua Ciénega de Chapala* (Zamora: El Colegio de Michoacán, 1989).

14. David A. Brading, *Haciendas y ranchos del Bajío: León, 1700–1860* (Mexico City: Enlace/Grijalbo, 1988); Van Young, *La ciudad*, 231–35, esp. 234; María de los Ángeles Romero Frizzi, "La agricultura en la época colonial," in *La agricultura en tierras mexicanas desde sus orígenes hasta nuestros días*, ed. Teresa Rojas, 210–14 (Mexico City: Consejo Nacional para la Cultura y las Artes, Grijalbo, 1991).

15. Alejandro de Humboldt, *Ensayo político sobre el reino de la Nueva España*, trans. Juan A. Ortega y Medina (Mexico City: Porrúa, 1978), 29.

16. Jorge L. Tamayo, *Geografía general de México* (Mexico City: Talleres Gráficos de la Nación, 1949), 42–43; Brigitte Boehm Schoendube, *Historia ecológica de la Cuenca de Chapala* (Zamora: El Colegio de Michoacán; Guadalajara: Universidad de Guadalajara, 2006), 64–65.

17. Secretaría de Programación y Presupuesto, *Síntesis geográfica de Guanajuato* (Mexico City: Secretaría de Programación y Presupuesto, 1980), 22–23.

18. Jorge L. Tamayo, *El problema fundamental de la agricultura mexicana* (Mexico City: Instituto Mexicano de Investigaciones Económicas, 1964), 47.

19. Jorge L. Tamayo, *Datos para la hidrología de la República Mexicana* (Mexico City: Instituto Panamericano de Geografía e Historia, 1946), 35–37.

20. Van Young, *La ciudad*, 231–35; Brading, *Haciendas y ranchos*; Michael A. Murphy, *Irrigation in the Bajío Region of Colonial México* (Boulder, Colo.: Westview Press, 1986); Claude Morin, *Michoacán en la Nueva España del siglo XVIII: Crecimiento y desigualdad en una economía colonial* (Mexico City: Fondo de Cultura Económica, 1979).

21. Martín Sánchez Rodríguez, *"El mejor de los títulos": Riego, organización social y administración de recursos hidráulicos en el Bajío Mexicano* (Zamora: El Colegio de Michoacán; Guanajuato: Gobierno del Estado de Guanajuato, 2005).

22. For a more thorough description of basin irrigation, its operation, its benefits, and the extension of its practice into different regions of Mexico, consult the following: Herb Eling MacIntosh and Martín Sánchez Rodríguez, "Presas, canales y cajas de agua: La tecnología hidráulica en el Bajío mexicano," in *Antología sobre pequeño riego: Organizaciones autogestivas*, ed. Jacinta Palerm Viqueira and Tomás Martínez Saldaña, vol. 2, 97–132 (Mexico City: Plaza y Valdés and Colegio de Posgraduados, 2000); Sánchez Rodríguez, "*El mejor de los títulos*"; Martín Sánchez Rodríguez, ed., *Entre campos de esmeralda: La agricultura de riego en Michoacán* (Zamora: El Colegio de Michoacán; Morelia: Gobierno del Estado de Michoacán, 2002); Jacinta Palerm Viqueira, ed., *Antología sobre pequeño riego*, vol. 3, *Sistemas de riego no convencionales* (Mexico City: El Colegio de Posgraduados, 2002).

23. William Willcocks, *Egyptian Irrigation* (London: E. and F. N. Spon, 1913), 40, 301–3, and 342; Helen Anne B. Rivlin, *The Agricultural Policy of Muhammad 'Ali in Egypt* (Cambridge, Mass.: Harvard University Press, 1961); J. Barois, *Irrigation in Egypt*, trans. A. M. Miller (Washington, D.C.: Government Printing Office, 1889).

24. Willcocks, *Egyptian Irrigation*, 60.

25. Ibid., 10.

26. Ibid., 38, 46.

27. Karl Kaerger, *Agricultura y civilización en México en 1900*, with an introduction by Roberto Melville (Mexico City: Universidad Autónoma de Chapingo and CIESAS, 1986); Brading, *Haciendas y ranchos*; Murphy, *Irrigation in the Bajío*; Guillermo García Zamacona, "Algunas técnicas agrícolas tradicionales en agricultura moderna: El sistema de cajas en el Bajío," in *Agricultura y sociedad en México: Diversidad, enfoques, estudios de caso*, ed. Alba González Jácome and Silvia del Amo Rodríguez, 249–67 (Mexico City: Universidad Iberoamericana, Plaza y Valdés Editores, and Gestión de Ecosistemas, 1999).

28. Sánchez Rodríguez, "*El mejor de los títulos*," 88–91. See also Martha A. Velásquez, José Luis Pimentel Equihua, and Jacinta Palerm Viqueira, "Entarquinamiento en cajas de agua en el valle zamorano: Una visión agronómica," 261–75, and Elvia López Pacheco, "Buscando la autogestión en las cajas de agua del valle de Coeneo-Huaniqueo," 241–57, both in Sánchez Rodríguez, ed., *Entre campos de esmeralda*.

29. Eric R. Wolf, "The Mexican Bajio in the Eighteenth Century: An Analysis of Cultural Integration," in *Synoptic Studies of Mexican Culture*, ed. Munro S. Edmonson, Glen Fisher, Pedro Carrasco, and Eric R. Wolf, 178–99, Middle American Research Institute Publication 17 (New Orleans: Tulane University, 1957); Brading, *Haciendas y ranchos*.

30. Morin, *Michoacán*.

31. Ibid., 62, 72.

32. Brading, *Haciendas y ranchos*, 61–62, 138.

33. María Guadalupe Rodríguez Gómez, *Jalpa y San Juan de los Otates, dos haciendas en el Bajío colonial* (Salamanca, Guanajuato: El Colegio del Bajío, 1984), 97, 112.

34. Fernando Picó, "Los pequeños y medianos productores agrícolas del bajío en la época del virreinato: Irapuato en los siglos XVII y XVIII," *Relaciones* 72 (1997): 89–137, esp. 95–96.

35. Cecilia Rabell, *Los diezmos de San Luis de la Paz: Economía de una región del bajío en el siglo XVIII* (Mexico City: UNAM, 1986), 67–76.

36. John Tutino, *From Insurrection to Revolution in Mexico: Social Bases of Agrarian Violence, 1750–1940* (Princeton: Princeton University Press, 1989), 64–65. (For the Spanish, see Tutino, *De la insurrección*, 65.) Van Young also mentions the displacement of corn with wheat in the best hacienda lands of the Atemajac Valley, which he attributes to the growth in demand by the city of Guadalajara. Van Young, *La ciudad*, 78.

37. Picó, "Los pequeños y medianos," 95–96.

38. Archivo General de la Nación (Mexico City), Tierras, vol. 1167, exp. 1.

39. Herb Eling McIntosh, Martín Sanchez Rodríguez, and Cristina Martínez García, "Primer informe del reconocimiento arqueológico del sistema de riego de Irapuato y Silao, Guanajuato, basado en mapas de 1792 y 1799," in *Los estudios del agua en la cuenca Lerma Chapala Santiago*, vol. 2, ed. Juan Manuel Durán Juárez, Brigitte Boehm Schoendube, Martín Sánchez Rodríguez, and Alicia Torres Rodríguez, 83–99 (Zamora: El Colegio de Michoacán; Guadalajara: Universidad de Guadalajara, 2005), 87.

40. Brading, *Haciendas y ranchos*, 152.

41. Archivo Histórico del Agua (Mexico City; hereafter AHA), Aguas Nacionales, caja 340, exp. 3632.

42. Rodríguez Gómez, *Jalpa*, 64.

43. Ibid., 71, 86–87.

44. Ibid., 97, 112.

45. AHA, Aprovechamientos Nacionales, exp. 3640, Legajo 01.

46. On the transformation from flood farming to dams in the Laguna, see Mikael D. Wolfe, "Water and Revolution: The Politics, Ecology and Technology of Agrarian Reform in 'La Laguna,' Mexico," PhD diss., University of Chicago, 2009.

CHAPTER FOUR

Nature as Subject and Citizen in the Mexican Botanical Garden, 1787–1829

Rick A. López

The Mexican Royal Botanical Garden, funded by King Charles III in 1787, was no ordinary royal garden. Typically, eighteenth-century European royal gardens mapped their patron's territorial domination. In Versailles, for instance, land nearest the palace emphasized legibility, rationality, and the centrality of regal authority. As visitors ventured farther afield, they moved into spaces that symbolized the territorial reach of the French king's power. Kew Botanical Garden outside London used its layout, architectural structures, and plants to demonstrate England's dominion over tropical resources encircling the globe. Madrid's Royal Botanical Garden similarly asserted the Spanish king's claim over the products found in his territories around the world.[1] The Mexican garden lacks these standard tropes, which gives the false impression that it ignored the organization of power, space, or knowledge. Yet the Mexican garden did comment on such matters. And the particular ways it did so reveal much about the changing (and contested) understandings of nature first within New Spain and then within the Mexican republic.

The garden was part of the Royal Botanical Expedition to New Spain (1787–93), an undertaking that was to signal Spain's reassertion of its colonial might and of its relevance to the Enlightenment. This ambitious expedition had three main prongs: a massive 56.25 hectare garden in the heart of Mexico City to cultivate, propagate, and study plants with economic or scientific potential; a school that would reform the training of doctors and pharmacists in New Spain; and a research apparatus that would scour the colony to identify, analyze, and publicize New Spain's botanical riches. These interdependent arms fell under the supervision of the head of the expedition, the Aragonian doctor Martín Sessé y Lacasta.[2]

Because the Mexican Royal Botanical Garden, unlike most other imperial institutions in New Spain, survived the transition from colony to nation at the turn of the nineteenth century, it offers a revealing point of entry for an analysis of the emergence of Mexico's nationalist ecological imagination(s). Those who have studied the garden or the botanical expedition of which it was a part have tended to focus narrowly on disentangling species identifications, tracing the professionalization of biology and chemistry, or recounting the bureaucratic and personal dramas that unfolded within and around the expedition. This study asks, instead, what the garden can teach us about the evolving relationship between nature and the imagined community during the period when New Spain saw the rise of Creole protonationalism.

The scientists of the expedition claimed to represent the European Enlightenment against rival modes of scientific organization, including those of the Mexican Enlightenment. Sessé and his allies criticized the homegrown Mexican movement as "unscientific" and as inappropriately melding the private sphere (which these agents of the European Enlightenment constructed as feminine, ethnically and locally specific, and immature) with the public sphere (which they constructed as masculine, universal, scientific, and Creole- and European-controlled). The head of the garden, Vicente Cervantes, tellingly, did not care whether Nahuas' and others' ways of naming or using the plants, animals, and natural products of Mexico survived in "plazas or in small gatherings with Indian herbalists and vegetable vendors" so long as they did not intrude upon "places of learning."[3] Rather than blot out other modes of knowing, the expedition sought to impose a hierarchical system of knowledge that exalted and masculinized the system developed by the influential Swedish botanist Carolus Linnaeus (1707–78). It also aspired to centralize botanical knowledge in Mexico City to serve the needs of scientific research and medical practice, while restoring some of the Spanish empire's tarnished glory.

This chapter shows how, from their base in Mexico City, members of the expedition codified their views of the natural world as the new "common sense" that guided how Mexican scientists, the state, and economic and political elites interacted with nature during the late colonial and postindependence eras.

Imperial Claims upon Mexican Nature

The story of the Mexican Royal Botanical Garden begins with Spain's imperial expansion across the New World starting in 1492, which coincided with the Renaissance rediscovery of the botanical and medical knowledge of the ancient Greeks. As it became clear that the flora of the Americas surpassed that known to the ancients, King Philip II appointed his court physician (*protomédico*) Francisco Hernández to carry out a detailed botanical study of New Spain. At the start of 1571, Hernández arrived in Mexico, where he remained until 1577. Following the Crown's instructions, he based his observations on interviews with native people rather than on public texts such as those by Fray Bernardino de Sahagún, Martín de la Cruz, or Juan de Cárdenas.

Hernández encountered a society in which Nahua healers known as *titici* were alive and well. For six years he learned from these titici while practicing medicine at their side in the Hospital of Oaxtepec, where he also benefited from a pre-Conquest botanical garden that remained in use. As the titici taught him the names and medicinal uses of plants, native draftsmen known as *tlacuilos* drew these specimens so that Hernández could compile an authoritative account. His guides also led him and his retinue to nearby fields, woods, and deserts, where villagers taught him about local simples (medicinally useful plants). Hernández was careful to point out that he never took the natives' word for anything; he tested all their claims, along with his own hypotheses, on patients in the hospital.[4]

In his letters to the king, Hernández expressed enthusiasm for the many plants in New Spain that were unknown in Europe and had potential economic value, and he bragged about the detailed studies and illustrations he was preparing from his samples.[5] He also used his time in Mexico to complete his translation of the thirty-seven books of Pliny's *Natural History* into Spanish under the conceit that his work and Pliny's, taken together, would offer the definitive account of global botany.

Hernández characterized the natives as "feeble, timid, and mendacious," as well as lazy and impious, and he criticized the titici for their ignorance of the principles of humorism, which led them to rely on the administration of simples, rather than the supposedly more advanced techniques of bleeding, surgical intervention, and the rebalancing of humors.[6] Despite this, he expressed the highest admiration for the useful knowledge they had accumulated about native plants. "It is a wonder," he wrote, "that among such rude and barbarous people" all the plant names are "adapted ... with such apt skill and wisdom that hearing the name alone is enough to indicate the natural properties that can be known or investigated."[7] Because the Latinate systems (which did not develop uniformity until the Linnaean innovations of the eighteenth century) were proving inadequate even for the plants already known in Europe, he doubted they could accommodate the vast trove of knowledge he was about to introduce. For this reason, upon his return to Spain, he urged integration of the Nahuatl nomenclature system as an ideal language for botanical science.[8]

His work redefined New World nature as a resource that could be studied and exploited through the mechanisms of science and empire, rather than simply the stuff of myth and legend.[9] Spain already enjoyed a special place within Renaissance science thanks to Muslim scholars who had preserved ancient knowledge that had been lost to medieval Christians. To this, Hernández added Spain's exclusive access to Nahua science, which he filtered through European assumptions and, ultimately, through his own position of authority.

The protomédico expected to receive hard-earned royal laurels. Unfortunately for him, the Spanish Crown had become less interested in learning about its resources than in guarding them against English and French encroachment. His European colleagues, moreover, frowned upon Nahuatl as too strange and savage to serve as a real language of learning.[10] The Crown hid away the vast collection of information and images that Hernández hoped to publish, lest they fall into enemy hands. The bits of his findings that eventually become public emerged in abridged and distorted form.[11] Even as Hernández's findings helped shape Enlightenment science on both sides of the Atlantic, his original illustrations and manuscript disappeared into the archives as Spain's imperial power withered.

Enlightened Science, New World Nature, and the Recapturing of Imperial Glory

As England challenged Spain on the global stage during the mid-1700s, Charles III set out to modernize his vast holdings into an overtly extractive empire along the lines of the English, Dutch, and French. To this end, he became an avid patron of botany and cartography, the twin pillars of imperial science. One official revealingly declared that "a dozen naturalists and some chemists scattered in Spain's dominions . . . will offer an incomparably larger utility to the state than a hundred thousand men fighting for the enlargement of the Spanish empire."[12] As Spanish scholars of the time anxiously debated the past, present, and future of the empire, they latched onto Hernández's work as a point of pride as well as a source of information to exploit for future gain.[13] A 1671 fire in the royal palace of El Escorial had consumed Hernández's manuscript and illustrations, but only now did the Crown appreciate the magnitude of this loss. As luck would have it, officials located another draft of part of Hernández's manuscript; however, the text was hard to decipher without its accompanying illustrations.

After failing to find copies of more than a handful of the destroyed illustrations in Europe, Casimiro Gómez Ortega, head of the Royal Botanical Garden in Madrid, contracted three intellectuals in New Spain to comb the Mexican archives. His first two agents were the prominent Creole protonationalists José Antonio Alzate y Ramírez (1738–99) and José Ignacio Bartolache (1739–90). I will leave aside for now these well-studied figures in the history of the Mexican Enlightenment to focus on Gómez's third agent, who became the central figure in the Royal Botanical Expedition: Martín Sessé y Lacasta (1751–1808). Possessed of medical training, an inclination for scientific research, and a taste for adventure, Sessé had served in Gibraltar and Cuba, where he conducted clinical trials to combat the epidemics that were debilitating the imperial army. In 1785, while trying to collect on a 1,000-peso debt he had extended to the Crown, he moved to Mexico City, where he took over a lucrative private medical practice and worked at research hospitals while serving as a physician for the Holy See.[14] The trio failed to find a copy of the manuscript in Mexico, but Sessé convinced Gómez Ortega that he could turn this setback into an opportunity.

Gómez Ortega, from his position as head of the Royal Botanical Garden in Madrid, was a towering figure in European science who fiercely advocated for the Linnaean system in botany and its chemistry counterpart, the Lavoisierian system. When Sessé in 1785 proposed a new expedition to build upon what remained of Hernández's manuscripts, Gómez Ortega embraced the idea. The two men won the backing of the New Spanish viceroy and the intendant of the Indies, and then presented their plan directly to the Crown, which agreed in 1787 to offer royal backing and, of equal importance, the support of the royal treasury.

The initial justification for the expedition was to re-create Hernández's lost illustrations and to fact-check and round out his text.[15] But Gómez Ortega and Sessé were hatching larger plans. Sessé from the start proposed that the expedition establish a botanical garden and a school of botany in Mexico City. As he envisioned it, the expedition, the garden, and the school would work in tandem to conduct original research, reform the medical-training-and-licensing establishment in New Spain, and free the colony and the mother country from reliance on medicinal imports from imperial rivals.[16]

Sessé and Gómez Ortega shared a desire to impose a top-down structure in which the metropolis was to serve as the hub for the imposition of scientific uniformity across the empire. Recently launched colonial expeditions, such as those by Hipólito Ruiz to Peru and Chile (1777–88), José Celestino Mútis to Nueva Granada (1783–1808), and Juan de Cuéllar to the Philippines (1786–97), had unfolded in the absence of any central coordination and without any plan about what to do with the information that was gathered. To lend order to these expeditions, Gómez Ortega created the Royal Office of Botany (Oficina Botánica). The most pressing task of the new royal institution, however, was to design from the ground up the Royal Botanical Expedition to New Spain, which was to use the latest scientific theories and methods to rationalize the Crown's exploitation of New World nature while ensuring Spain's future as a pillar of Enlightenment science and as an empire to be reckoned with.[17]

By the middle of 1787, Sessé and Gómez Ortega had recruited much of their expeditionary team, most importantly the Spaniard Vicente Cervantes (1755–1829) as *catedrático* (chaired professor or director) of the garden and the botany school. Cervantes was born in 1755 in Extremadura. He had shown intellectual talent as a youth, but his family could not afford university, so he entered an apprenticeship in a Madrid drug dispensary. At the end of each workday he studied botany in the company of friends who attended university. Though an autodidact, Cervantes successfully petitioned to take the certification exam, which he passed brilliantly. This attracted the attention of Gómez Ortega, who helped him win a professorship in the Royal School of Botany and a position as botanist to the General Hospital of Madrid.[18] Gómez Ortega then appointed his rising protégé to assist Sessé in the creation of a Mexican garden and school following the model of the Madrid garden and school. Cervantes arrived in Mexico in 1787 to begin his collaboration with Sessé.

Though they drew inspiration from Hernández's earlier work, Sessé and Cervantes intentionally departed in crucial ways from their sixteenth-century predecessor. Hernández had relied openly upon native knowledge and even applauded the Nahua nomenclature system. The new expedition sought to steer clear of such tendencies. Moreover, whereas Hernández had been instructed by the Crown not to interfere with local licensing or medical practices, Sessé won a royal mandate to intervene precisely in these matters.[19] Sessé thought of himself and his team of nine as living in a new scientific era that would build upon past knowledge while constructing a framework for new knowledge that would render that past superfluous. Sessé noted that "no sooner was America discovered and incorporated into the vast territory of the Crown of Castile than Phillip II sent to this continent his illustrious physician Dr. don Francisco Hernández to identify, discover, and collect its medicinal products," which, he claimed, the Crown immediately recognized as more valuable even than precious metals. "Hernández," he argued, "did his best to fulfill the good intentions of the monarch, but the lack of scientific principles, which were not known until 200 years after his death, made it such that his descriptions, despite being no worse than those of the wisest Greeks and Romans who preceded him, and superior to those of almost any other botanist since Pliny, turned out to be so incomplete and defective" that they could offer no real guidance to the present. He credited the sixteenth-century physician for having extracted, tested, and refined indigenous knowledge, and admits that Hernández's native informants "certainly knew more useful plants than those described by Dioscorides." Had Hernández's finding been fully exploited in Europe in the 1500s, they would have added greatly "to the riches of the old continent."[20] But that did not happen, and in Sessé's time it was too late. Enlightenment science, in Sessé's view, provided such powerful tools that there no longer was any need to mine native knowledge; any effort in that direction would bring science up to date with the sixteenth century, but it had little use, he argued, in his own era at the turn of the nineteenth century.[21]

As to Hernández's respect for Nahua science and Alzate's argument that Nahua nomenclature and locally based experts had the advantage of growing out of prolonged experience in the New World and its flora, Cervantes, after several years in Mexico, countered that the expedition's research in the garden and its clinical trials in the hospitals provided more than enough experience and a sufficiently "long period of observation." Because of this, he considered that the Royal Expedition to New Spain would surpass the work of Hernández, the Nahua system, and local expertise; its modern theory and scientific methods, moreover, would set it apart from *all* the recent expeditions that had been carried out by Spain, which he claimed emphasized taxonomy and description at the expense of experimentation and research.[22]

Sessé and Cervantes also set themselves against what they viewed as the backward practices of the Creole medical establishment. In this conflict, the expedition benefited from the weight of the Crown. At the start of the expedition, the two men requested the conversion of several busy general hospitals into research facilities, with fewer beds. When the Mexican medical board balked, the Crown overruled it to grant Sessé the control he sought.[23] This was but the start of a tempestuous power struggle between the expedition on one side and university professors and local scientific society members, intellectuals, and officials on the other.

Lest we accept Sessé and Cervantes's claim (which has been taken for granted by their later biographers) that they were stamping out the backwardness of mestizo and Creole doctors and of Nahua herbalists in favor of science and rationality, we should keep in mind that the methods they described as cutting edge were firmly rooted in humoric theory and other medical assumptions that, today, we might view as less advanced than those of the herbalists they disdained. While they engaged in medical and scientific arguments, therefore, we must not lose sight of the degree to which these were embedded within larger issues of power and authority.

The privileging of metropolitan structures of knowledge was clear in the methods employed by Sessé and his Creole acolyte José Mariano Mociño (the latter a product of the expedition's school and Sessé's eventual successor, who would became a leading scientific figure in Spain). Rather than seek out native guides, as Hernández had done, Sessé and Mociño drew on native knowledge indirectly, through European texts that filtered and reordered that knowledge. They combined this indirect knowledge with their own direct scanning of the ground. Methodically, they gathered specimens that they then examined, drew, and compared to the published record. They conducted clinical trials, identified "unknown" plants, and assigned Latinate names, following the Linnaean system, to everything they found. Their method subjected the flora, land, and people of New Spain to a structure of knowledge that subordinated the colony to the metropolis, while claiming the fruits of that knowledge and the territory of New Spain in the name of European science mediated through the Spanish empire.

Between 1787 and 1803 they traversed four thousand miles, collecting seeds, live plants, and dried herbarium samples. They produced thousands of field botanical illustrations, reorganized medical education and practice in New Spain, and created Mexico's first Enlightenment-influenced botanical garden as well as its first natural history museum. By implanting its institutions and approaches within the colony and training the next generation of Mexican doctors, scientists, and pharmacists, the expedition helped change how nature was viewed within New Spain and how it was understood in relation to the imagined community. Previous efforts, such as those by Hernández, Bernardino de Sahagún, or Juan de Cárdenas, had confined themselves to the immediate environs of the Valley of Mexico. By venturing across the breadth and width of New Spain, this expedition was the first effort to define New Spain not by the political structures that held it together, by the peoples it ruled over, or even by terrain and property lines but by the plants and animals that lived in and wandered about or spread their seed across its soils.

Ironically, even as the expedition subordinated the colony to the metropolis, it also nurtured possibilities for conceiving of New Spain as a distinct imagined community rooted in nature (or natural resources) and centered in Mexico City. The expedition collected, named, illustrated, and otherwise claimed plants as abstract, floating specimens, the knowledge of which was mediated by centralized "scientific" authorities in Madrid and other European centers. The expedition's scientists rarely noted the specific site of collection, nor did they comment on the plant's range or habitat. All that they conveyed, generally, was that the plants were from the Crown's colony of New Spain. By abstracting plants out of their local places and out of webs of local knowledge they reconfigured space and knowledge as beholden to, controlled, and defined by centralized imperial agents in the Royal Botanical Garden of Mexico and, through it, the Madrid garden and the European scientific network of which it was part.

Space, Science, and Nature in the Royal Botanical Garden of Mexico

Much of the material collected or produced by the expedition, such as living and dried plants, seeds, and botanical illustrations, was exported to Madrid to serve the needs of the metropolis and perhaps later to be re-exported to the colonies as part of imperial projects of "improvement."[24] However, the expedition did leave the colony a legacy in the form of the Mexican Royal Botanical Garden and the Royal School of Botany, both officially headed by Sessé but in practice managed by Vicente Cervantes.[25]

After two years of planning, the expedition inaugurated the garden on May 1, 1788, with an audience that included the regent of the Audiencia, a representative of the viceroy, attendees from the university, and members of Mexico's scientific societies, along with other distinguished persons of the capital. After an orchestral performance, Sessé delivered a lecture on innovations in the study of botany and medicine. Then followed a fireworks display by the most renowned pyrotechnician in the colony, Joaquín Gavilán. This incorporated explosions in the form of recently discovered Mexican plants and, reportedly, even a male papaya tree shooting flares of pollen to fertilize two nearby female papayas.[26] The attention to flower structures and pollination emphasized the primacy of the Linnaean system, which categorized plants based on their reproductive features. It also spoke to the ambition of the expedition to spread, like the male plant, its pollen of scientific knowledge so as to fertilize the feminized colonies so that they, through the rationalized exploitation of their natural resources, could nourish the empire.

During the event, the rector of the university was required to affirm the investiture of Sessé and Cervantes as his equals. This eradicated jurisdictional claims by members of the university who had tried to exert supervision over the expedition, and it publicly affirmed the expedition's right to have a say in local practices. The event coincided with royal orders that all medical students had to take courses in the botanical school and that the head of the garden was in charge of administering the medical licensing examination. The next day saw the inaugural class session in botany.[27] The first group of students (which included José Mariano Mociño, who, as the star pupil, later ascended to leadership of the expedition) comprised young doctors eager to learn European science and, above all, to understand the practices and theories that had developed around the illustrious Linnaeus.

The garden's inauguration marked an important triumph for Sessé. He had initially petitioned to locate the garden and its botanical school near the viceregal palace and university, the centers of learning and political power in the heart of Mexico City. When his request for this venue got stuck in bureaucratic red tape, he began to worry that plans for the garden would grind to a halt unless he found an expedient alternative. He turned to a zone known as the Potrero de Atlampa (Atlampa Pasture), halfway between the central square (the Zócalo) and Chapultepec, at the edge of which a prominent architect named don Ignacio Castera offered to rent one of his homes. The large 56.25 hectare plot beside the new Salto del Agua fountain was located within easy access to the Royal Indian Hospital (Hospital Real de los Indios), which, Sessé suggested, might prove useful for clinical trials. Mexico City finalized the transfer of the proposed garden lot to the Crown in 1788 in time for the garden's official inauguration. The land was poor, but that did not worry Sessé. Only later did he learn that he had overestimated the ability of science to bend the site to its will.

In the meantime, he had to figure out how to make scientific research profitable. The Crown agreed to cover all of the expedition's salary and travel expenses but expected the garden and the school, like other royal institutions, to devise ways to become self-sustaining. Sessé felt confident that his efforts would benefit the Spanish empire in countless ways over the long term, but he had to figure out how to make scientific research profitable in the short term. Barring the discovery and rapid development of a wonder drug equal to Peru's cinchona (used to treat malaria), Sessé had to look for sources outside of science. The problem was that most forms of rent, from lottery to other fees and taxes that Sessé was able to imagine, including even licensing fees for pharmacists, already were claimed by other agencies. Among these other claimants was the Mexican medical licensing board (*protomedicato*), which Sessé already was at odds with and did not wish to antagonize further. He even pursued a plan to build a bullfighting ring to fund the garden, until he found himself unable to get the start-up capital and then discovered, in any case, that the receipts would generate inadequate revenue. He secured patronage from among the New Spanish elite, but this was not enough to cover more than a fraction of his expenses.[28]

Though royal institutions were expected to become self-sustaining, Sessé convinced the Crown to support the garden while he continued his search for a secure form of funding. No one predicted at that time that the garden and school would remain permanent dependencies of the state treasury. While the scientists waited in vain for more funding, Cervantes set up a provisional garden of twenty thousand square yards surrounded by a picket fence in the patio of the school's temporary quarters, rented from don Ignacio Castera. They planted seeds imported from Europe and Nueva Granada (modern-day Colombia, Ecuador, Panama, and Venezuela) and, soon after, transplanted specimens that the botanists had begun to gather from the New Spanish countryside.[29]

Anxious about finances, the Crown demanded a detailed estimate of what the garden would cost to run during that first year. Officials were shocked by the price tag and grew more so in August 1789 when Sessé presented plans for the permanent school and garden. The first set of site plans and elevations, now filed in the Archive of the Indies in Seville, was ambitious. It called for the construction of a grand new structure complete with a teaching hall, an herbarium, a natural history museum (*gavinete*), living quarters for Cervantes and the staff, and a research library open to the public, plus an enclosed teaching garden with colonnades and an elaborate irrigation system. Beyond this edifice lay the massive main garden, which was to be encircled by an irrigation ditch and a masonry wall and accessed by a bridge leading into the main gate. The grounds were to be divided into thirty-four beds, each almost as large as a city block. The recently refurbished Chapultepec aqueduct on the south end of the garden was to provide water, channeled through the canal around the perimeter and then into a grid of pipes that extended along paths throughout the garden. Water from the aqueduct was to be supplemented in the dry season by a large pond in the southwestern corner of the garden, and individual beds also would enjoy their own water reserves held in fountains. This was an expensive proposal. With no clear source of income other than royal funding, which the Crown was loath to guarantee, Sessé and Cervantes had to figure out ways to scale back their plans.[30]

These financial and bureaucratic difficulties, however, faded in the face of a more pressing dilemma. It turned out that seasonal flooding and poor drainage not only prevented students from attending class or working on their experiments but also rendered the entire Potrero de Atlampa unsuitable for gardening. The expedition and garden were supposed to organize the study and exploitation of natural resources while affirming man's domination over the natural world. They presupposed that modern science offered all the tools necessary to eliminate the peculiarity of place as a problem, and to discipline nature so that it would conform to Enlightenment schemes of abstract rationality. Nature presented unanticipated resistance.

As the rest of the expedition forged ahead with fieldwork, clinical trials of simples, training a new generation of Mexican doctors and pharmacists, and assembling an herbarium, the garden continued to suffer from flooding. Extended periods of submersion suffocated plant roots and rotted seeds that had been acquired at great expense from Spain, other parts of Latin America, and the far reaches of New Spain. It was with good reason that the common name for Atlampa was "El Sapo," an allusion to the frogs that enjoyed its soggy conditions. The flooding problem was compounded by the lack of the sandy loam that every gardener cherishes. Instead, the ground was heavy clay that remained so soggy after the water receded that people could not walk on it without their feet getting sucked under, much less till or plant it. When it finally dried, it turned hard as stone, until it flooded again.[31]

Confident that modern science must triumph over nature and frustrated at the loss of the seeds and plants, Cervantes desperately commissioned raised beds filled with hundreds of canoe loads of manure and high-quality soil from San Agustín de las Cuevas. When the raised beds failed to solve the problem, he commissioned two thousand flowerpots that he elevated atop racks, but these would not accommodate the trees that Sessé had begun sending to the garden from the countryside. In the meantime, Cervantes hired plowmen to prepare the large lot that he still hoped might serve as the main garden.[32]

Try as he might, Cervantes and the site planner, Miguel Constanzó, could not bend the land to their will. Cervantes had expected to use the garden to propagate plants plucked from all across New Spain, whose ecological zones range from temperate forest to desert, from coastal to alpine, and from rainforest to chaparral. However, nothing but the hardiest of cultivars (which generally were useless for the expedition) could survive these conditions. Cervantes and Sessé at last accepted failure and resolved that the only solution was to abandon the Potrero de Atlampa for higher ground.[33]

During their search for a better location, Cervantes and Sessé eyed the site of Moctezuma's pre-Conquest garden on Chapultepec Hill. In 1784 the Viceroyalty of New Spain had begun construction of a sprawling structure on the old Aztec site. That building, today known as Chapultepec Castle, sat incomplete in 1790, with the government unable to finish it or find a private buyer. Sessé argued that the grounds surrounding the abandoned facility would serve as an ideal garden. The structure could be finished to fill the needs of the garden and the school, and the many levels of elevation and microclimates of Chapultepec would be ideal for cultivating the wide variety of plants gathered by the botanists during their fieldwork. The request fell on deaf ears.[34]

Imperial Science over Mexican Nature

By 1790 the expedition had begun to earn a place within Mexico City society. Its members had initiated reform of the study and licensing of pharmacy, medicine, and surgery and already had identified hundreds of medicinally useful plants that needed to be further studied and propagated in the garden. But the garden stalled, stuck in the mud of Atlampa.[35] Thanks to clay pots, it had managed to grow many plants, some of which had even been transferred to Spain, but it had not yet managed to come into full physical form.

At last, in 1791, Cervantes and Sessé won an alternative space, not in the unfinished castle atop Chapultepec but in the even more highly prized viceregal palace on the Zócalo. Though the space was a fraction of the size of the Atlampa site and lacked the climatic advantages of Chapultepec, its location within the royal residence eliminated the overhead that would come from constructing or refurbishing and then maintaining a separate building, and the existing arcade could be converted into a greenhouse and into a series of teaching laboratories for medicine, surgery, and pharmacy. The patio could hold only one thousand species, but these would be sufficient for the school's needs, provided that they were supplemented with plants collected daily from the countryside and by shuttling plants and students back and forth between the viceregal palace and Chapultepec (which Sessé continued to propose as a site for extensive beds). The downtown location, moreover, allowed ready access to housing for staff and to teaching hospitals for Cervantes and the students. It also afforded public prominence and an opportunity to edify the subjects of Spain's Enlightened absolutism.

By bringing the garden into the center of colonial authority in New Spain, the expedition, the viceroy, and the Crown expected that the union of absolutist politics and Enlightenment science could overcome the obstacles set forth by nature so that they could get on with the business of reorganizing space, knowledge, and power. As it turns out, the new site did not solve their problems. The structures could be remodeled to meet the school's teaching and research needs, but the land presented a more daunting challenge.[36]

The soil was an accumulation of clay mixed with rubble from old Aztec temples. Scarred by his battle against Atlampa, Cervantes insisted to the viceroy that altering the height of the beds or refilling them with better soil would not be enough: the entire site had to be excavated to a depth of one meter so that the clay and rubble could be discarded and replaced with fertile, well-draining soil from Tlalpan.[37]

The garden, like the broader expedition of which it was part, sought to erase the indigenous foundations upon which it was built. By harnessing Enlightenment science to imperial power, Cervantes literally and metaphorically sought to replace the historical accumulation with fertile soil amenable to the appropriation, reorganization, and propagation of Mexican nature. This massive earthmoving was accomplished by Indian laborers working under the supervision of head gardener Jacinto López—who, like Cervantes, had trained at the Royal Botanical Garden in Madrid, where he then worked for fourteen years before coming to Mexico. They removed the clay and rubble with canoes floating in the Aztec canals that still served at that time as Mexico City's roads.[38]

The earthmoving project was not as visually striking or awe-inspiring as the terraces of Versailles or the massive glass greenhouses of Kew or Madrid, but it was no less an achievement. It called for hundreds of canoe loads of good-quality soil; production of large quantities of locally manufactured bricks, tiles, lime, and tools; and a massive procurement of native labor.[39] Whereas the garden of Versailles made the visitor acutely aware of the engineering feats that made possible the transformation of the land, the Mexican garden minimized outward evidence of the effort. The power to eliminate this evidence while making invisible the indigenous knowledge from which it derived was, in fact, central to the garden's Enlightenment mission.

Cervantes and Sessé traded visual awe for a studied admiration for the ability of men of science, backed by the Crown and New Spanish supporters, to make the land look pretty much as it did before but to have it be "better" than it was before. That, after all, was what they hoped to do with New Spain: keep it looking as it did, growing what it grew, and producing what it produced, but to do so "better," under scientific supervision, and for the benefit of the empire. And they hoped to do so in a central location where they could concentrate products from the vast, diverse spaces of New Spain into one centralized hub, whose organization appeared as though it were subject only to the abstract rule of Enlightenment rationality.

The Imagined Community of Plants

Unlike European gardens, the Mexican Royal Botanical Garden did not re-create the dominated territory in cartographic miniature. Nonetheless, it did demonstrate the capacity of centralized scientific authority to know, control, and reorganize the natural resources within its dominion. It then put this capacity on display at the doorstep of the viceroy, thereby affirming the Crown's prerogative, and capacity, to use Enlightenment rationality to claim and exploit the resources of its domain.

Shortly after moving to the viceregal palace, the expedition finally gained the right to claim part of Chapultepec Hill for garden beds. In practice, the two parts of the garden, one in the downtown viceregal palace and the other atop Chapultepec Hill, worked in tandem. As Sessé explained, the beds in Chapultepec grew "the plants from across the diverse climes of this America" while the viceregal site offered students and teachers a space for research and pedagogy, and to the public it offered "the Capital its most beautiful spectacle, visible from its principal square."[40]

Sessé also insisted that the Mexican Royal Botanical Garden in the viceregal palace serve as a public and pedagogical space, which "gente decente" could enter; others would be able to admire its collection from beyond a short barrier. More broadly, the garden was to model how to bring the urbanity idealized by Hispanic culture together with the rural space that constituted most of New Spain. Order could be imposed upon the unruly city by importing nature from the countryside, but only after that nature had been domesticated and rationally regimented by Enlightenment science and centralizing political absolutism. By bringing nature into the city in an organized manner, therefore, the garden could serve as a model for how to organize space within the city, as well as a model for how to tend nature beyond the city, treating the expanse of New Spain as a massive garden that could be subjected to top-down regimentation.

Their struggles with the soil in Atlampa and then on the Zócalo had forced the viceroy and the members of the expedition to accept certain limitations regarding the power of modern science to obliterate the importance of place, space, and soil. The Crown, however, would have none of it. In 1792, after Sessé and Cervantes had relocated the garden from Atlampa to the hills of Chapultepec and the downtown viceregal palace, the king's minister fired off an irate missive excoriating the members of the expedition for failing to bend the Potrero de Atlampa to their will and impugning the abilities and integrity of Sessé, Cervantes, and even Constanzó. He ordered them to move the garden back to the Potrero de Atlampa and make it work.

Sessé responded with an extended review of the Chapultepec site, examining its woods, ditches, and slopes and its ready access to water, with which to naturalize plants from such diverse climes as deserts, forests, marshes, and alpine regions, as well as those imported from Africa and Europe. All this would be impossible in Atlampa but could be accomplished in Chapultepec, he explained, while avoiding the high cost of irrigation. To make his case as forcefully as possible he reminded the minister that only by cultivating this wide variety of species could the garden find ways to propagate and exploit the medical and commercial potential of New Spanish plants for the benefit of the empire.[41] In the end, Sessé prevailed through a deft combination of argument, obstinacy, and mobilization of alliances in Madrid and Mexico City.

The layout of the garden, first in Atlampa and then in the viceregal palace and Chapultepec, suggests a union of symbolism and practicality even as Sessé and Cervantes learned to adapt their methods to the limitations of nature. This was designed to be a working garden, every aspect of whose design speaks to the practice of science and to the study, centralization, organization, and transfer of New Spain's natural products. Because of this, the garden's plan resembled a nursery except that, rather than growing seedlings to be transferred elsewhere for maturation, as does a nursery, it propagated frames of understanding so as to nurture *new knowledge*. The frames of understanding, meantime, were to be imported from the metropolis into New Spain by way of the garden and the school so that they might take root and prosper in colonized soil. These new frames of understanding were to be used in pursuit of new knowledge that then could be transplanted from New Spain to the metropolis, where it might be cultivated to full maturity through its incorporation into European science.

The garden reduced New Spain to a collection of individual plants that could be moved, sorted, and otherwise traded, observed, or manipulated in isolation from their local places of origin. Yet it also bestowed two commonalities upon the plants that moved through its beds. First, they all possessed potential economic and scientific value that could serve both the colony centered in Mexico City and the empire based in Madrid. Second, except for the handful of plants explicitly treated as imports, they all came from somewhere within the territory claimed by New Spain. Consistent with this, as it collected the plants for the garden from across the colony, the expedition did not concern itself with their potential value to local people. Nor did it record the exact location where they were collected or concern itself about the range of particular plants. All that mattered was that they had potential economic and scientific value, and that they came from New Spain.

From Empire to Republic

Whereas the garden initially had been thrust upon the colony by the Crown, it ceased to be viewed as an imperial imposition by the end of the century. Many of the scientists and doctors trained within its walls (it trained between ten and seventy-eight per term) had risen to positions of authority within the colony by the time of the independence wars (1810–21). Creole patriots, rather than rejecting the garden along with other imperial institutions, incorporated it into their own modernizing vision for Mexico. This support was evident in 1800 when the new viceroy tried to cut the Crown's costs by forcing Cervantes and López back to Spain on the grounds that they already had exceeded their original six-year contract.[42] A decade earlier, such a plan might have won the consent of the Mexican establishment, but by 1800 Cervantes had become sufficiently embedded within local society that locals backed him, rather than the viceroy, and he was allowed to stay provisionally and, after 1803, permanently.

Sessé returned to Madrid in 1803 to present the expedition's findings but made clear that he, too, planned to return to Mexico to serve as the permanent director of the garden, with Cervantes as the *catedrático* of the school and López as the head gardener. The commitment to Mexico on the part of Cervantes and Sessé likely owes as much to personal circumstances as to a dedication to the garden and the school. Cervantes's lucrative pharmaceutical business and his work as an esteemed researcher and physician at the San Andrés teaching hospital afforded him wealth and prestige that would have been difficult to reproduce in or transfer to Spain. He also had a family in Mexico. Sessé, similarly, benefited economically from a lucrative medical practice in New Spain and had married into a Mexican family. His wife, María Guadalupe Morales, and two children, Alexandro and Martina, along with his wife's sister Josefa Morales, joined him on his return to Spain, and they all anticipated a return to Mexico. As it turned out, Sessé died in Spain before he could present his findings. His Creole protégé José Mariano Mociño succeeded him as head of the expedition while winning fame in Europe as a professor and then vice president of Spain's prestigious Royal Academy of Medicine in Madrid and director of the Royal Museum of Natural History (Real Gabinete de Historia Natural).[43]

The garden and school in Mexico, meanwhile, entered into a productive routine. Now that the Linnaean methodology was firmly established in the colony, Cervantes could comfortably delegate intellectual reproduction to his Creole protégés, whom he invited back to the school to help teach the next generation of Mexican scientists and doctors. Together they cured the sick and expounded upon the virtues of the rubber tree, the *árbol de las manitas* (hand tree), and the many other plants described and illustrated by the expedition and at that time under cultivation in the garden.[44]

When the celebrated German naturalist Alexander von Humboldt visited in 1803, he was appalled at Mexico's social inequalities but stood in admiration of the scientific accomplishment of this colonial capital. "No city of the new continent, without even excepting those of the United States," he wrote, "can display such great and solid scientific establishments as the capital of Mexico." He pointed particularly to the School of Mines, the Mexican Royal Botanical Garden, and the Academy of Painting and Sculpture. He praised Cervantes, Sessé, and Mociño, along with the Creole botanical illustrator Atanasio Echeverría, "whose works will bear a comparison with the most perfect productions of the kind in Europe." Humboldt estimated that, on account of their efforts, the "new philosophies" of botany and chemistry were embraced more in Mexico than in much of Spain.[45]

The uprising that Miguel Hidalgo announced with his September 1810 *Grito de Dolores*, along with Napoleon's invasion of the Iberian peninsula, growing Creole discontent with the Bourbon reforms, and the development of a grassroots popular movement, set off the chain of events that culminated in Mexican independence. More irksome to Cervantes than the general impact of the war and the occasional mistreatment that his workers and students suffered during their collecting expeditions was the callous treatment he and the garden received from Spanish government officials and the troops that commandeered his space, dismantled many of the garden's structures, destroyed the irrigation system and beds, and uncaringly trampled to death the plants that had been acquired with such difficulty.[46]

To these setbacks was added the death of López in 1813 during a wartime epidemic and the Spanish government's reluctance to authorize a replacement. With no head gardener, the laborers had little supervision and the garden lost hundreds more "interesting plants that were vital to the lessons" in botany, medicine, and surgery.[47]

Cervantes's relationship with the Spanish forces deteriorated further when a military official named Velasco tried to strip him and the garden of their functions. Velasco readily admitted that Spanish troops had severely damaged the garden and destroyed its "choice plants." But rather than make amends, he argued that on account of this damage and of the garden's inability even to "provide the medicinal herbs most frequently necessary for the public," it no longer merited the rank of "botanical garden" and was not worth its expense. Moreover, given the need to mobilize against the rebels, the government could no longer afford the luxury of scientific research. By his estimation, the students who daily drained resources through their studies at the garden and its school would better serve the empire by joining the army and fighting in the front lines. He proposed that Cervantes, who already had exceeded his initial six-year commission, should be replaced by a certain Diego Martín, who could fill his role at 16 percent of Cervantes's current salary.[48]

Outraged, Cervantes retorted that even Napoleon, during his aggression against Spain, had seen fit to continue to fund science and learning, yet the Spanish Crown was proving unwilling to do so. He refuted Velasco's presumption that the garden was nothing more than a source for medicinal herbs. It was not some dispensary but a center of scientific learning and research. Though its collections had suffered at the hands of the Spanish troops, it still contained seven hundred species that, together with two hundred to three hundred plants brought in routinely from the countryside by assistants, enabled the sixteen to twenty students to conduct botanical and chemical research while studying medicine, pharmacy, and surgery, thus providing a benefit to society that he insisted was no less valuable than service in the battlefield. Cervantes personally chafed at Velasco's suggestion that his role as catedrático of the school and garden could be filled by Diego Martín, whom Cervantes insisted was but a poor, illiterate Indian who knew only enough to grow chiles, tomatoes, garlic, and onions on his chinampa. Cervantes convinced the *fiscal* to reject Velasco's proposal but, in exchange, accepted further cuts to his already strained budget.[49]

In 1817 the garden began to show signs of recovery. Students and workers resumed their collecting and research expeditions, and Cervantes even managed to coordinate a complicated exchange of live plants with agents in Havana, Cuba. The following year, he finally secured a successor for López. The new head gardener was the Italian Juan Lazari, who previously had headed the private garden of don Manuel Tolsá, the recently deceased head of Mexico City's famous institute for art, the Academy of San Carlos, where the expedition's botanical illustrators had studied.

By June 1821, just months before the August 1821 conclusion of the conflict, Cervantes, who by this time had managed to set the garden on the road to recovery, already had ceased referring to the garden as the Royal Botanical Garden, preferring to call it simply the Botanical Garden. The documents do not reveal whether this was because Cervantes pragmatically responded to a shift in the political tide or whether it was because the royal forces had alienated him with their disregard for his scientific endeavors (or, more conjecturally, that his treatment by Spanish forces was the result of sympathies Cervantes may have had for the rebels). Whatever may have prompted the subtle name change, Cervantes had tilted in favor of local society and perhaps even the republic.[50]

When Mexico gained independence in 1821, the new government expelled most agents of the Spanish Crown, including Vicente Cervantes's colleague Fausto de Elhuyar, head of the Mexican School of Mines. Cervantes, by contrast, was invited to stay. He escaped the general anti-Spanish sentiments of the independence era because of his services to the common good of Mexicans. Moreover, despite the economic and political crises it faced, the postindependence government continued to fund his scientific and educational work almost until his death on July 26, 1829, and to support the garden until the middle of the century.[51] The new state, with its ambition to assert dominion over the territory previously defined as New Spain, seems to have recognized the utility of the garden for claiming and ordering natural resources from across the new republic and placing these resources at the real and symbolic disposal of central authorities.

During Spanish imperial rule, the garden had devoted its energies to gathering in Mexico City knowledge and natural resources from across New Spain, claiming plants through the authority of the central government, and studying these in pursuit of "pure science." This trend became more pronounced after Sessé's and Mociño's departure in 1803 and continued under the independent republic.

The garden suffered the postindependence economic downturn along with other state institutions but soon found a champion in Lucas Alamán (1792–1853), the conservative Mexican politician, scientist, and former student of Cervantes.[52] Together, Cervantes and Alamán drafted plans to reestablish the declining garden on firmer footings and to combine it with a museum of Mexican antiquities and natural history. They aspired to create an institution that would "rival or surpass establishments of [their] kind in Europe." After the fall of Agustín de Iturbide's brief Mexican empire (May 1822–March 1823), they won state approval to move the garden to the cemetery of the Hospital de los Indios (two blocks south of the modern-day Palacio de Bellas Artes) and even began the removal of the buried bodies and demolition of the chapels and walls as well as work on an irrigation system, quarters for workers, paths for public viewing, and amendment of the soil.

One of their most intriguing objectives was to compile an encyclopedic collection of seeds from all Mexican plants. This move from the imperial practice of sampling New Spanish plants for medicinal and imperial uses to the nationalist goal of assembling a comprehensive collection of native plants, however unattainable, points to a new way of bestowing Mexican citizenship upon the natural world. It meant claiming all of nature, rather than select products, as a national resource, and it used plants to define the extension of the imagined community and as validation of the new state's territorial claims.

Plans to move the garden's operations from the presidential palace to the hospital stalled, and the goal of creating a microcosm of Mexico's native plants never got past the planning stage. Nevertheless, initiatives continued to emanate from the cadre Cervantes had trained. By 1831, the combined botanical garden and museum of antiquities and natural history, along with its school of botany, was under the charge of Cervantes's successor and former student, Miguel Bustamante, a native of Guanajuato. A congressional decree of that year ordered Bustamante and the heads of Mexico City's other scientific organizations to form themselves into a council to promote scientific development and advise the government.[53] This council persisted, working through the Colegio de Minería, at least into the 1850s, when the combined botanical garden and museum of natural history and antiquities was headed by the botanists and zoologist Pío Bustamante, whose father, Benigno Bustamante y Septién of Querétaro, had been a student of Cervantes.

Through its continued patronage of the garden and the school, the independent government defined itself in terms of its commitment to science and rationality as it laid the foundations for a new order. The garden that had been created to serve the needs of empire now offered its services to the new elite, who used it to assert dominion over the spaces and resources claimed by the state. To the end of his days, Cervantes continued to emphasize the study of plants and nature as "pure" botanical science and never made any mention of agriculture or the other practical uses beyond medicine.[54] As such, the garden, from the time of its founding until the mid-nineteenth century, served abstract principles related to empire and then nationhood rather than providing technical services for the new economic elite to engage in export-led monocrop agriculture, as happened in other Latin American countries.

No doubt to Cervantes's postmortem chagrin, the survival of the gardens after midcentury in the presidential palace and on Chapultepec Hill came to rest upon their conversion into pleasure grounds for the rulers of the state and as symbols, rather than tools, of scientific modernity. Under Emperor Maximilian during the French Intervention (1864–67), the space on Chapultepec Hill completed its transition into a pleasure garden, and it continued as such under the rule of Porfirio Díaz (1876–80, 1884–1911). The main garden in the presidential palace remained in operation until the 1940s but never regained its scientific purpose or stature. In the 1880s the Department of Development revived the scientific objectives of the defunct botanical garden. Not content merely to pick up where the garden had left off, the Department of Development placed its resources, including its acceleration of research and plant exchanges, at the disposal of private enterprise, applied agricultural science, and forest management.

Conclusions:
Imperial Science and the Nation's Nature

The turmoil in Spain unleashed by Napoleon's invasion contributed to the death of Mociño. In cruel irony, the expedition's botanical illustrations met the same fate as had Hernández's when, after Mociño's unanticipated death, they became lost. The expedition's manuscripts, meantime, sank into relative obscurity until they were recovered in the 1880s by the Mexican government under Porfirio Díaz. Though the Mexican government lacked Sessé's and Mociño's images, it brought their text to light in a celebrated 1887 publication.[55] Díaz's advisors, known as *científicos* on account of their exuberant faith in the transformative power of science and political order, embraced a nationalist push for top-down modernity. Similar to the way Spanish officials of the eighteenth century had looked to Hernández's sixteenth-century manuscripts as they sought to modernize the empire and claim a space for Spain within the Enlightenment, Porfirian reformers of the nineteenth century looked to the eighteenth-century Royal Botanical Expedition as a guide for their effort to modernize and claim a place for Mexico within the global economic, industrial, and scientific revolutions of their own day.

During this era when modernity was being imported from the United States and Europe and nationalists became anxious about foreigners' claims to Mexico's natural resources, the Porfirian government's celebration of Sessé's and Mociño's manuscripts established patrimonial claims to the diverse products of nature that occurred on Mexican soil. The state was particularly interested in those plants whose economic value might be unlocked and exploited through science and top-down management and those that, following the precedent set by Alamán's use of the expedition, could be exploited as nativist symbols.

Part of the appeal of the expedition went beyond the information found in its texts. To Porfirian científicos, much of the expedition's appeal came from the manner in which it had used science to enshrine hierarchies of knowledge, of social rank, and of unequal access to the products of nature. In this regard, the científicos also found a use for the expedition's practice of rendering these resources "alien" to the very people who were most intimately familiar with them by renaming plants and removing them from their local cultural contexts, devaluing alternative ways of knowing nature, and challenging local communities' claims to what elites now declared as "national" resources.

The Porfirian revival of interest in Mexico's native flora, and in the systematic transfer of control over this flora to elites, was derailed by the Revolution of 1910. Nationalist botany was picked up again in the 1930s, but this time with populist objectives as botanists, searching for new natural resources to exploit for the good of the public, rummaged through the expedition's surviving herbarium samples.[56] In 1981 the Hunt Institute in Pittsburgh quietly placed a winning bid on two thousand botanical illustrations from Barcelona that turned out to be the long-lost illustrations from the Sessé and Mociño Royal Botanical Expedition to New Spain. Since that time there has been a renaissance of international interest in Mexican botany, a reshuffling of scientific names, and discovery of many new plants, along with growing tensions among peasants, nationalists, and transnational pharmaceutical and agricultural companies. This has raised the stakes in relation to the question of whether Mexico can lay intellectual claim to its botanical resources. Across this multilayered history, there has been a consistent view of Mexican nature as a massive working garden that elites have sought to control and exploit but upon which competing levels of society, and now transnational corporations, each continue to assert their own claims.

Notes

1. Chandra Mukerji, *Territorial Ambitions and the Gardens of Versailles* (New York: Cambridge University Press, 1997); Richard Drayton, *Nature's Government: Science, Imperial Britain, and the "Improvement" of the World* (New Haven, Conn.: Yale University Press, 2000); F. J. Puerto Sarmiento and A. González Bueno, "Política cientifica y expediciones botánicas en el program colonial español ilustrado," in *Mundialización de la ciencia y cultura nacional*, ed. A. Lafuente, A. Elena, and M. L. Ortega, 331–39 (Madrid: Doce Calles, 1993); and Keith Thomas, *Man and the Natural World: Changing Attitudes in England, 1500–1800* (New York: Oxford University Press, 1983).

2. In my forthcoming book I discuss the rich literature on science in Spain and the New World, and about the botanical expedition in particular. Because of space limitations, this chapter narrowly limits citations to the most relevant texts.

3. Cervantes, quoted in Roberto Moreno, ed., *Linneo en México: Las controversias sobre el sistema binario sexual, 1788–1798* (Mexico City: UNAM, 1989), xiii.

4. King Philip II, Instructions to Dr. Francisco Hernández, 11 January 1570, in *The Mexican Treasury: The Writings of Dr. Francisco Hernández*, ed. Simon Varey, trans. Rafael Chabrán, Cynthia L. Chamberlin, and Simon Varey (Stanford: Stanford University Press, 2000) (hereafter *MT*), 46–47; Jesús Bustamante, "The Natural History of New Spain," in *MT*, 34–36. Hernández drew on them for his separate study of the antiquities of New Spain; see *MT*, 65.

5. Francisco Hernández to King Philip II, November/December 1571; and Hernández to King Philip II, 30 April 1572, in *MT*, 48–50.

6. Hernández, "The Antiquities of New Spain," in *MT*, 72 and 77.

7. Hernández, *Antigüedades de la Nueva España*, quoted in Bustamante, "Natural History," 36.

8. Xavier Lozoya, *Plantas y luces en México: La expedición científica a Nueva España (1787–1803)* (Barcelona: Serbal, 1984), 11–12; Bernardo Ortíz Montellano, *Aztec Medicine, Health, and Nutrition* (New Brunswick, N.J.: Rutgers University Press, 1990), 25–26; Rafael Chabrán and Simon Varey, "The Hernández Texts," in *MT*, 3.

9. Though James Scott locates the transition from "nature" to "natural resources" within Europe, evidence from Mexico suggests that the demystification and atomization of the parts of nature, and the resulting reduction of it from a whole into a collection of economically exploitable parts, advanced more rapidly in the colonies. James Scott, *Seeing Like a State: How Certain Schemes to Improve the Human Condition Have Failed* (New Haven, Conn.: Yale University Press, 1998), 13.

10. Londa Schiebinger, *Plants and Empire: Colonial Bioprospecting in the Atlantic World* (Cambridge, Mass.: Harvard University Press, 2004), 195–223.

11. The most important was the work of the Italian Nardo Antonio Recchi, which appeared in 1615 in Mexico and subsequently in 1651 in Rome. In Europe, leading works that drew upon Hernández's work, usually indirectly through Recchi, included Fabio Colonna's *Minus cognitarum stirpium* (1616), Johannes Faber of Bamberg's *Animalia Mexicana descriptionibus scholisque expostia* (1628), Robert Lovell's *Pambotanologia* (1659), Henry Stubb's *The Indian Nectar* (1662), John Ray's *Historia plantarum* (1686), Hans Sloane's *Natural History of Jamaica* (1701–25), and James Newton's *Enchyridion* (1752, written c. 1689). In Mexico, other drafts of Hernández's work—drafts that no longer exist—were reproduced in all or in part in such works as Gregorio

López's *El tesoro de medicinas* (written in the 1580s and published in 1727), Juan Barrios, *Verdadera medicina, cirugía, y astrología* (published in 1607), and Francisco Ximénez's *Quatro Libros: De la naturaleza, y virtudes de las plantas, y animales* (published in 1615). See Chabrán and Varey, "Hernández Texts," in *MT*, xvii–xix and 6–9; and David Freedberg, "The Doctor's Dilemmas," in *The Eye of the Lynx* (Chicago: University of Chicago Press, 2002), 275–304.

12. Quoted in Schiebinger, *Plants and Empire*, 7–8.

13. Bustamante, "Natural History," 26.

14. Numerous studies have traced out the early details of his personal and professional life. See, for example, José Maldonado Polo, "La Expedición Botánica a Nueva España, 1786–1803: El Jardín Botánico y la Cátedra de Botánica," *Historia Mexicana* 50, no. 1 (July–September 2000): 5–56, esp. 13–15; and Harold William Rickett, "The Royal Botanical Expedition to New Spain (1788–1820) as described in documents at the AGN (Mexico)," *Chronica Botanica* (Waltham, Mass.: Chronica Botanica Company, 1947), 2:6.

15. 23 November 1787, folio 323, vol. 138, Reales Cédulas, Archivo General de la Nación, Mexico City (hereafter AGN). My thanks to Rachel Meketon and Chris Wisniewski for their help managing the documents.

16. Sessé, "Expediente sobre los efectos de las plantas medicinales en los enfermos de Hospital San Andres," Mexico, 1800, WMS.Amer.44, Microfilm #F2634, Wellcome Library, London (hereafter WL).

17. Antonio González Bueno, "Scientific Knowledge and Power in the Illustrated Spain: Toward the Commercial Supremacy through the Medicinal Botany," *Antilia* 1, no. 2 (1995), http://www.ucm.es/info/antilia/revista/vol1-en/arten1-2.htm; and R. Rodríguez Nozal, "La Oficina Botánica (1788–1835): Una institución dedicada al estudio de la flora americana," *Asclepio* 47, no. 2 (1995): 169–83.

18. José García Ramos, "Elogio histórico del farmacéutico Don Vicente Cervantes, Catedrático que fue de la Botánica," *Boletín de la Sociedad de Geografía y Estadística de la República Mexicana*, 2nd series, vol. 1 (1869): 753–65; and Rickett, "Royal Botanical Expedition," 7.

19. King Philip II, Instructions to Dr. Francisco Hernández, 11 January 1570, in *MT*, 46–47.

20. Sessé, "Experiencias clinicas y terapeuticas de la Real Expedicion Botanica a la Nueva-España," Mexico, 1802, WMS.Amer.43, Microfilm #F2633, WL. Also see Sessé to Gómez Ortega, draft of a letter, 3 July 1785, V, 1, 1, 3; and "Relación de plantas enviadas al Real Jardín Botánico de Madrid," 1788–1791, V, 2, 6, 2, Archivo Histórico del Real Jardín Botánico, Madrid, Spain (hereafter RJB).

21. Sessé, "Experiencias clinicas."

22. Ibid. For documents related to the debates over the value of the Nahua system, see Moreno, *Linneo en México*. On Alzate and the protonationalist debate over Creole expertise, see Moreno, *Linneo en México*; and Jorge Cañizares-Esguerra, "Postcolonialism *avante la lettre*? Travelers and Clerics in Eighteenth-Century Colonial Spanish America," in *After Spanish Rule: Postcolonial Predicaments of the Americas*, ed. Mark Thurner and Andrés Guerrero, 89–110 (Durham, N.C.: Duke University Press, 2003).

23. Sessé, "Experiencias clinicas."

24. On improvement as a justification for European imperialism, see Drayton, *Nature's Government*.

25. 23 November 1787, folio 323, vol. 138, Reales Cédulas, AGN; Juan Carlos Arias Divito, *Las expediciones científicas españolas durante el siglo XVIII* (Madrid: Ediciones Cultura Hispanica, 1968), 67; Letter, 26 April 1788, in ibid., 72.

26. Ricardo Ramírez, introduction to *Flora Mexicana*, 2nd ed., by Martín Sessé y Lacasta and Joseph Mariano Mociño (Mexico City: Secretaria de Fomento, 1894), v; Rickett, "Royal Botanical Expedition," 5; and Ramos, "Elogio histórico," 758–59.

27. Ramírez, introduction to *Flora Mexicana*, v; and Patricia Aceves, "La difusión de la química de Lavoisier en el real Jardín Botánico de México y en el Real Seminario de Minería (1788–1810)," *Quipu* 7, no. 1 (January–February 1990): 5–35.

28. Rickett, "Royal Botanical Expedition," 9–10.

29. Ibid., 9–11; and Arias Divito, *Las expediciones*, 90–91.

30. Annotated site plan, "Mexico City—Botanical Garden," p. 1, microfilmed documents, William L. Clements Library, University of Michigan (hereafter WCL); "Plano del edificio del Jardín Botánico que se proyectaba hacer en México," "Plano del Terreno destinado para Jardín Botánico," "Fachada principal que mira á oriente y corte que mira al mediodía á lo largo del edificio sobre la línea a.b.," and "Plano, elevación y perfil de una casa para habitación del Catedrático de Botánica, que debe construirse en el terreno destinado para el jardín botánico de esta Capital de Nueva España," MP-MEXICO, 416, 417, 418, and 419, Archivo General de las Indias, Seville, Spain (hereafter AGI); D. Ignacio Castera, "Plano iconográfico de la ciudad de México," 1794, G4414.M6 1794.A3, Vault, Geography and Map Division, Library of Congress, Washington, D.C.; Rickett, "Royal Botanical Expedition," 9–12; Arias Divito, *Las expediciones*, 93–96.

31. The loss of these plants and seeds was a major blow and would take five years and much effort and cost to replace. On the effort that went into acquiring the seeds, see Sessé to Gómez Ortega, 26 April 1786 and 28 October 1787, V, 1, 1, 7 and 12, RJB.

32. Rickett, "Royal Botanical Expedition," 13–14.

33. Ibid., 13–16.

34. Folders 69–70, vol. 464, Historia, AGN; Rickett, "Royal Botanical Expedition," 16.

35. Sessé to Pedro Acuña y Malvar, Ministro de Gracia y Justicia, 9 January 1794, V, 1, 4, 30, RJB.

36. Ramos, "Elogio histórico," 761; Rickett, "Royal Botanical Expedition," 16; Vicente Cervantes to Viceroy Revillagigedo, 1 October 1791, pp. 1–8, folder 8, vol. 464, Historia, AGN.

37. Vicente Cervantes to Viceroy Revillagigedo, 1 October 1791, pp. 1–8, folder 8, vol. 464, Historia, AGN.

38. Ibid.; Bonilla to Viceroy Revillagigedo, 28 January and 25 February 1791, pp. 1–2, folder III, vol. 464, Historia, AGN; illegible name to Viceroy Revillagigedo, 31 December 1790, p. 3, folder III, vol. 464, Historia, AGN; Rickett, "Royal Botanical Expedition," 16 and 58; illegible name to Viceroy Félix Berenguer de Marquina, 6 April 1801, folder III, vol. 464, Historia, AGN; Sessé to Viceroy Félix Berenguer de Marquina, folder III, vol. 464, Historia, AGN; and illegible name to Viceroy Félix Berenguer de Marquina, 16 January 1803, folder III, vol. 464, Historia, AGN.

39. Mascarós to Viceroy Revillagigedo, 2 October 1791, folder 8, vol. 464, Historia, AGN.

40. Sessé quoted in Rickett, "Royal Botanical Expedition," 19.

41. Ibid., 18.

42. Ibid., 59.
43. Ibid., 66; "María Guadalupe Morales," Arribadas, 441, N.305, AGI; Miguel A. Puig-Samper and Sandra Rebok, "El reconocimiento oficial de Alexander von Humboldt en España," *Humboldt im Netz: International Review for Humboldtian Studies* 5, no. 8 (2004): 1–13.
44. Ramos, "Elogio histórico," 761.
45. Alexander von Humboldt, *Political Essay on the Kingdom of New Spain*, trans. John Black (London: Longman, Hurst, Rees, Orme, and Brown, 1811), 1:134–217.
46. Cervantes to Viceroy Francisco Javier Venegas de Saavedra, 2 April 1811, folder 8, vol. 462, Historia, AGN; Rickett, "Royal Botanical Expedition," 57 and 67.
47. Cervantes quoted in Rickett, "Royal Botanical Expedition," 67.
48. Velasco quoted and translated in ibid., 68.
49. Ibid.
50. Various, April 1818–July 1818, folder 17, vol. 466, Historia, AGN; Rickett, "Royal Botanical Expedition," 69; and Cervantes to unknown, 14 June 1821; Cervantes to unknown, 16 June 1821; Secretary of the Viceroy, 1821; Jose Maldonado to unknown, 21 June 1820; Mariano Lopes to unknown, 18 September 1820; and various, folders 31–32, vol. 466, Historia, AGN.
51. Ramos, "Elogio histórico," 753–76; "Expedición al Virreinato de Nueva España, Tras las huellas de F. Hernández," *Historia del botánico, Real Jardín Botánico*, http://www.rjb.csic.es/historia_nuevaespana.php.
52. Lucas Alamán, "Épocas de los principales sucesos de mi vida," 28 August 1843, series I, box 1, exp. 236, p. 5, Lucas Alamán Papers, 1598–1853, Benson Latin American Collection, University of Texas Libraries, University of Texas at Austin; William Bullock, *Six Months Residence and Travels in Mexico; Containing Remarks on the Present State of New Spain* (London: Murray, 1825), 183–86.
53. Lizardi, report, caja 10, exp. 2, 1821; Cervantes to exmo. Señor, 27 August 1823, caja 49, exp. 16/1, 1823; correspondence of Lucas Alamán, various, 1823–29; congressional decree, 21 November 1831, caja 411, exp. 1; all from Gobernación "sin sección," AGN, Mexico City. Thanks to Eric Van Young for directing me to the correspondence of Lucas Alamán. On Bustamante, see caja 118, exp. 6–9, 1829, Gobernación "sin sección," AGN, Mexico City; and Nicolás León, *Botánico-Mexico: Catálogo, biográfico y crítico de autores y escritos referents a vegetales de México y sus aplicaciones desdes la conquista hasta el presente* (Mexico City: Secretaría de Fomento, 1895), 78. On the location of the Hospital de los Indios (Naturales), see Andrés Romero-Huesca and Julio Ramírez Bolla, "La atención médica en el Hospital Real de Naturales," *Cirugía y cirujanos* 7, no. 6 (November–December 2003): 496–505, esp. 497.
54. Rickett, "Royal Botanical Expedition," 61.
55. Martín Sessé y Lacasta and Joseph Mariano Moçiño, *Plantae Novae Hispaniae* (Mexico City: Escalante, 1887); and Martín Sessé y Lacasta and Joseph Mariano Moçiño, *Flora Mexicana* (Mexico City: I Escalante, 1887).
56. In this regard, it is similar to the shift that McCook has identified between late nineteenth-century practice and that of the populist 1930s in the Spanish Caribbean. Stuart McCook, *States of Nature: Science, Agriculture, and Environment in the Spanish Caribbean, 1760–1940* (Austin: University of Texas Press, 2002).

CHAPTER FIVE

Besieged Forests at Century's End

Industry, Speculation, and Dispossession in Tlaxcala's La Malintzin Woodlands, 1860–1910

José Juan Juárez Flores

Mexican forests are in a precarious condition today. Forests and jungles cover somewhere between 55 and 67 million hectares, or around 28 percent of Mexico's total landmass, and they are disappearing at an alarming rate. About 600,000 hectares of woodland vanish every year, giving Mexico the dubious distinction of ranking as the world's fifth most rapidly deforesting nation—and second in Latin America. One recent study shows that the total area covered by forests has fallen by 16.3 million hectares in the past thirty-five years; others indicate that 13 million hectares of temperate forests and 4 million of tropical forests disappeared between 2000 and 2005 alone. Still others suggest that the net loss of forests occurred at a rate of 300,000–490,000 hectares per year between 1998 and 2003. While data varies, the trend does not: deforestation in Mexico has reached an average annual rate of 22 percent in the past decades, considered to be one of the highest in the world.[1]

A representative case of deforestation occurred on the foothills of the volcano known as La Malintzin, on the central Mexican plateau. The combination of complex social, cultural, political, economic, and of course environmental processes surrounding this ecological system, rich in forest and hydrological resources, has resulted in massive environmental degradation. The different uses of natural resources over the past century and a half have given rise to many forms of environmental decomposition.

Incessant deforestation and consequent erosion have degraded this immense structure's function to such an extent that it may have altered many local microclimates.

From the "Idyll of the Volcanoes" to Ecocide

In the central highlands of Mexico, particularly in the Puebla-Tlaxcala Valley, the sadness that now consumes La Malintzin is not due to her unconsummated love of El Citlaltepec, or "Mountain of the Star" (as Orizaba Peak is sometimes known), nor is it due to the indifference of Popocatépetl, "La Mujer Blanca." Much less does her sadness originate from spite toward the mountain range governed by El Tenzo (her unrequited lover) or because of the dissatisfaction that Cuatlapanga, a tiny hill nestled in one of her great skirts, has perhaps struck out on her own.[2] Alone in the Poblano-Tlaxcalteca Valley, distant from the pretenses of these towering buttes, the sad silhouette of what was once called the "goddess of the blue skirts," or Matlalcuéyetl, succumbs under the predatory momentum of those who only yesterday venerated her vestment of forests and now dedicate themselves to ripping them apart.

At an altitude of 4,460 meters above sea level, La Malintzin (or La Malinche) is the fifth highest peak in the country and holds a transcendental importance for the Puebla-Tlaxcala region. Its abiotic composition of "water and soil" and biological composition of "plants and animals" constitute a vital ecological system for the provision of forest and hydrological resources. The moisture captured by La Malintzin's forests generates immense amounts of water, which is possibly the peak's most important resource, and on which the whole state of Tlaxcala and 80 percent of the municipalities in Puebla depend. Nonetheless, this crucial ecological service has been damaged by the destruction of the forest cover as a result of inexorable historical processes. The mountain's importance to the indigenous people of Tlaxcala was recorded in the chronicles and early historical accounts of the colony, which register the rituals of "a culture intimately connected to water."[3] The millennial relationship that native communities forged with the mountain constituted an essential element of their existence and helped to shape their worldviews. Yet this link was ruptured in the second half of the sixteenth century with the imposition of the colonial economic system that followed in the wake of the Conquest, which resulted among other things in the intensive exploitation of natural resources.

By the second half of the nineteenth century, wood-dependent industrial processes had begun to contribute to the destruction of the forest. Adding to the traditional uses of wood as material for construction and fuel (as both firewood and charcoal), new industrial technologies used forest products to make turpentine (primarily for urban lighting) and railroad ties. The siege on the forests was brutal. Observers in the 1930s already noted a "rapid acceleration" in the rate of deforestation.[4] In 1991 some estimates suggested that of the 75,000 hectares of forest believed to have covered the mountain at the time of European contact, only 15,000 remained.[5]

Population growth has been another factor putting pressure on forest ecosystems on the mountain's lower reaches. The twenty-six settlements that a report from the 1940s listed in the foothills of the volcano have grown in population at a startling rate. For example, the village of Santa Isabel Xiloxoxtla had a total population of 694 inhabitants in 1880; by 1995, that number had grown to 3,171, during which time the population density increased from 178 to 813 inhabitants per square kilometer.[6] Population pressure has contributed, among other problems, to the decline of biodiversity. The uncontrollable felling of trees, destruction of new growth, improper use of soils for agriculture, intentional burning of vegetation to open up pasture lands, and irrational use of the grasslike broom root have all damaged the landscape and continually threatened to nibble away at the edges of the forests that still cover some parts of La Malintzin.

The loss of forest cover accelerated during the latter part of the nineteenth century, when the Liberal state adopted a model of political economy that promised to increase the circulation of goods and domestic production, both of which depended on access to publicly owned resources that had been seized and placed under private ownership.[7] Two factors decisively influenced these transformations: the centrality of wood and water for industrial production; and federal policies that intervened in local affairs and appropriated public funds by imposing taxes on consumption. Indeed, the process of modern industrialization demanded that natural resources be put into circulation to promote economic activity, thereby producing a transformation in the basic functions of these resources. Textile factories and paper mills were founded, oftentimes over the objections of towns and villages concerned about the water required to drive electrical turbines and the wood needed for paper pulp. The assault on these resources was as intense as it was extensive and caused substantial changes in the landscape and its ecology. Furthermore, social relations were profoundly marked by the intensification of conflicts over the appropriation of water and forest resources. The assault on communal property transformed social relationships, economic activity, and the rural landscape of the towns located at the foot of the mountain and in the valleys of the volcanoes that border the central Mexican plain.[8] At the same time, the growing demand for wood by critical industries signaled a transition away from its traditional use as a material for the construction of rural houses and fuel in the forms of firewood and charcoal. The forests were being subjected to new forms of exploitation by the lumber, chicle (chewing gum base), paper, and turpentine industries,[9] as well as by railroad companies for railroad ties. The industrialization of the forest was also encouraged by growing demand in the more developed nations for raw materials. As Mexico became integrated into the world economy, local production was incorporated into national and international markets.

The expanding forestry industry in particular placed the woodlands at the center of disputes over ownership and acquisition. From the mid-nineteenth century on, a series of laws and liberalizing regulations bypassed the legal and institutional obstacles that had previously protected the forests from privatization. Disentailment (the forced privatization of collectively owned land) enabled outsiders to seize forests that until then had usually been held as village commons or as the assets of municipal corporations. Disentailment also opened the way for the central government to intercede in the ownership of allegedly "public" land. The legal instrument for this intrusion into community property rights was the fiscal system applied after the restoration of the republic in 1867, intended to free up production and circulation to bolster consumption.[10] Moreover, logging projects on formerly collective lands generated new streams of tax revenue for the federal government.

From Abuse to Confiscation in the Foothills of La Malintzin

Despite the original etymological meaning of the name Matlalcuéyetl, "the lady of the blue skirts" (which alluded to the mountain's thick and extensive forest covering), the image that La Malintzin projects today is one of a mountain with great scars and deforested spaces that attest to the forest's alarming retreat. Species of trees such as pines, oaks, and fir still cover some areas between 2,800 and 4,000 meters of altitude, while other tree species such as sacred fir and ocote pine can be found in the intermediate regions. Yet a richer diversity of species existed during the late nineteenth century. Historical accounts refer to ocote pine, jacolote pine, sacred fir, oak, Mexican white pine, Montezuma cypress, and madrone—trees that produce different qualities of wood for different uses.

According to these sources, the ocote pine "produces much resin which is used for pitch and oil of turpentine"; the jacolote pine, "less resinous," produces a wood that is "softer and of better complexion, easy for the woodworker to work with." The oyamel pine "does not produce resin," and its wood "is less compact and longer lasting"; of the oak, "two classes are known, the encino and the live oak, both . . . of the same consistency, equally wide and of solid wood"; finally, the Mexican white pine, a member of the cedar family, produced a denser wood "of a light purple color and a lustrous consistency." These woods' particular qualities suited them to specific industries such as carpentry. In contrast, the Montezuma cypress, "many branched," with its "fibrous, strong smelling, resinous" wood, and the madrone, with its "porous, soft wood of poor consistency," were generally used as fuel.[11]

The forest did not legally belong to the native communities; rather, it was classified as public municipal property (*bienes patrimoniales*) owned by the city of Tlaxcala. The town council had rented these lands to indigenous communities ever since the colonial era. Rather than payment in cash, these "Indian Republics" agreed to meet certain obligations, such as providing wood products including logs, boards, beams, and stakes for urban public works projects as well as shoring up the banks of the Zahuapan River when it needed periodic reinforcement against floods. The indigenous people had once owned modest parcels of communal land where they could plant corn, wheat, and agave, the last of which they used in ceremonies to honor their patron saints. Although some statistics from the early nineteenth century indicate that these lands were lost to village headmen and colonial officials charged with managing indigenous properties, this is an issue that historians have yet to study in detail.

The use of trees for wood, firewood, and charcoal—along with the increasing commodification of forest products generally—produced considerable income for native communities, but it also sparked incessant disputes among them. The growing trade in forest goods also led to changes in the nature of landownership, particularly after the liberal Law of the 25th of June of 1856 (the "Ley Lerdo") decreed the disentailment and eventual privatization of collectively owned lands, including Tlaxcala's municipal property. Native communities understood that disentailment represented an unprecedented opportunity to (re)acquire their patrimony.[12] Although the Tlaxcala city council tried to oppose the requirement to disentail its lands, the municipality acquiesced toward the end of the nineteenth century and transferred ownership and administration of the woods to the village councils of communities on La Malintzin.[13]

Another factor enabling native communities to use liberal policies to obtain forestlands was that Mexico's process of industrialization demanded vast quantities of wood. Native people had supplied forest products to Creoles, Spaniards, and other outsiders for a very long time, as we have seen. Ever since the colonial era, the Matlalcuéyetl had supplied charcoal and firewood for markets in the city of Puebla and other population centers, as well as "to the factories that needed this fuel."[14] As late as 1865, wood was used "mostly as fuel" in the state of Puebla.[15] After the 1860s, however, it increasingly went to supply the needs of industries such as railroads, which required large quantities of railroad ties.[16] Great extensions of ocote pine were tapped to produce resin needed to make turpentine, which was used to light city streets.[17] The state government took steps to ensure the flow of resources from the forest, regardless of who owned it. A government circular issued by the State of Tlaxcala in 1860 announced that "the public" should be "granted access" to the forests "to cut wood and extract turpentine," which the government promised to purchase at a set fee.[18]

The Uses of Wood

The restructuring of property rights after 1870—despite the Tlaxcala city council's continued opposition to the privatization of its assets—eventually transferred ownership to village councils, thus returning the land to its original owners. The mountain communities created "forest commissions" to oversee the woods. They also imposed their own taxes on the harvesting of wood, firewood, and grasses. The municipal council of Huamantla, for example, published a fee schedule in January 1877 that set the prices that "any citizen" would pay to fell wood or use pasturelands. It set down the following charges:

For removal of wood:
- For each load of split firewood $0.02
- For each load of pieces of wood hauled by donkey $0.03
- For each load of pieces of wood hauled by mule $0.04
- For each cut beam that remains in the municipality $0.06
- For each cut beam that is taken out of the municipality $0.12
- For each dozen planks of 2.5 rods in length $0.25
- For each dozen logs (*morillos*) $0.25
- For each dozen trim pieces (*cintas*) $0.25

For grasses:
- For each head of large livestock per day $0.02
- For each 100 head of small livestock per day $0.25

This document also stipulated that the payments be made to the city treasurer, who would then issue a receipt without which forest wardens would not grant the buyer access to the woods. Anyone found extracting wood without the necessary paperwork would be subject to a fine or, for subsequent offenses, the confiscation of the illegal merchandise. These provisions made an exception for "the residents of the neighborhoods," who could use the woods without paying any fee "as long as they remove no more than two loads of wood for their own use." The forest commission reviewed the petitions "of individuals that enjoy this privilege" and made recommendations as to whether the municipal president should grant the licenses.[19]

This administrative process opened a path for *aguardiente* distilleries, glass factories, and iron foundries to encroach on natural resources at a time when Puebla and Tlaxcala were emerging as major centers of a rapidly developing textile industry. Furthermore, in 1875, Juan Martinez Zorrilla founded a glass factory in the town of Apizaco that used a muffle furnace capable of firing eight crucibles at a time.[20] By 1885, it had expanded to three furnaces operated by ninety employees and consumed 15,000 cords of wood per year, or approximately 42 cords per day. Historical sources also mention the existence of another glass factory operating at this time in Chiautempan with one oven, fifty employees, and a consumption of 8,000 cords of wood annually.[21] By the end of the 1830s, an iron foundry had also appeared on the Panzacola hacienda. Once the difficulties of the early industrial age had been overcome by around 1885, the foundry operated with two smelters and employed fifty-five workers, consuming 18,000 bushels of fuel per year.[22] Another iron foundry existed in the town of Apetatitlán. Additionally, the industries that had traditionally consumed great quantities of wood still existed: forty-four earthenware pottery kilns, two aguardiente distilleries in Huamantla that produced 1,500 barrels annually, and various other factories in several of the towns surrounding La Malinche.[23]

The consumption of wood had grown to the point that timber extraction had become an industry. In fact, this is the term used to describe the felling of timber, firewood splitting, and charcoal manufacture by 1890, not only around La Malinche but also in other Tlaxcalan regions such as the Sierra Nevada and in the Tlaxco forest.[24] Commerce on this scale attracted the attention of federal, state, and municipal taxing bodies. In an arrangement typical of liberal policies of the 1880s, the Mexican state levied both a federal tax on consumption and a state tax on circulation of forest products.[25] This arrangement remained in force in the 1890s as well. A decree issued by the state government on February 19, 1894, imposed a 2 percent state sales tax on the sale of forest products on top of the corresponding federal levy.[26]

Municipal governments also depended on tax receipts from the forest trade. The 1893 municipal account of Altzayanca reflects income from logging permits, for example. The people who cut wood included a certain Juan Pérez, who had a tequila distillery and a mescal liquor plant, as well as shopkeeper Jesús Pardo, each of whom paid 6 pesos per year, or 50 cents per month, "for usage rights" to the woods.[27] In 1909, the municipal council of Teolocholco received 392 pesos for the sale of 112 cords of firewood: among others, Nicanor Flores from the barrio of Quilehtla paid 60 pesos, and Lucio Gutiérrez from the town of Ayometla paid 52, both at a price of 3.50 pesos per cord. Additionally, 45 pesos were charged for the sale of "treetops," that is, boughs, to twenty-one residents of the same town.[28]

The consumption and sale of firewood had also reached staggering proportions. In 1905 alone, nearly 500,000 tons of wood were extracted from Tlaxcalan forests, with a value of slightly more than 2.5 million pesos.[29] But the wood was not the only resource that industrial interests appropriated or that fiscal powers taxed. Another forestry product that emerged as important commodity was broom root, known in Mexico as *zacatón*.

Broom Root Extraction

Mexican broom root is a species of wild palm (*Epicampes macroura*) that grows widely on La Malintzin. The severe erosion that currently afflicts the mountain is caused not only by the deforestation of its foothills but also by the overexploitation of this plant, among whose primary benefits was its ability to fix the soil. Around the time period of our study, broom root grew "perfectly" in the intermediate altitudes of the mountain. It was a "very useful" plant used as fodder for cattle, as roofing material for huts and granaries, and as brooms, brushes, and scouring pads. The international trade in broom root had become a "very promising" business by the 1880s, and exports continually grew to meet the "great demand" of American and European markets.[30]

The industry's remarkable development spelled trouble for the forest ecosystem of La Malintzin. The rental agreements that the municipal council of Huamantla issued for its woodlands in 1883 guaranteed that private interests received precedence over all other users and reflect the increasingly severe pressures on the forests. These agreements stipulated that "every entrepreneur or worker" was permitted to collect broom root, "which is currently of vital [economic] importance" to the region. The agreements indicated the area and qualities of the land where collectors could work and instructed them to "immediately cover the hole left by [removing] the root by planting ocote pine seeds therein and by no means leaving that place uncovered." It also prohibited collectors from removing the root if it threatened nearby trees. The municipal council appointed forest wardens to enforce these ambitious provisions and charged them with ensuring that only authorized individuals worked in the woods and that the collectors fulfilled their contractual obligations, as well as managing the overall quality of work. Finally, the municipality charged 1.25 pesos per month for each collector.[31]

Merchants rushed to collect broom root as quickly as possible and could not be bothered with details such as covering up the little holes left behind or protecting the lives of exposed saplings in the middle of an immense forest. Indeed, entrepreneurs reaped rewards for increasing the scale of their operations. For example, Rafael Rivera Díaz obtained a license in January 1883 allowing him to employ between twenty and two hundred workers to gather broom root from the woods owned by the municipality of Huamantla. Instead of paying the 1.25 peso fee for each worker that the rental agreement required, the municipality charged him 50 cents and allowed him to send workers into the woods as long as he pleased. He was assigned lands that bordered the towns of Tetlanohcan and Contla and was promised the right to work in the San Juan Valley if he chose to hire more men.[32]

Broom root extraction had become the "most productive" industry in the district of Huamantla and produced its "only export product," according to a statement dated in November 1891.[33] The municipality of Chiautempan was another important location of broom root removal on La Malintzin. Yet only half a dozen workshops were involved in the industry, fumigating, washing, combing, pressing, and packaging the root. Some of these workshops processed as much as six thousand kilograms per month and employed up to one hundred workers. The manufacture of brooms, scrub brushes, and brushes was the primary business, though some broom root was exported for processing outside of Tlaxcala or even outside the country. Small amounts arrived in Hamburg, Le Havre, and Antwerp, where workshops apparently owned by a French consortium processed it.[34] In 1906, the increasing prosperity and technological sophistication of the broom root industry led Carlos and Ramón Maldonado to install steam-powered machinery in Junguito (Huamantla district) to extract the fiber from wild palm.³⁵ That the removal of broom root, along with logging and the extraction of firewood, had degraded the woods on La Malintzin was already clear. Nevertheless, other industries such as turpentine producers appeared in this period and further industrialized the woods in both intensive and extensive ways that squandered the mountain's soils and compromised its biodiversity.

Turpentine Production

International demand for products derived from the sticky resin secreted by most species of pines and other conifers expanded rapidly in the second half of the nineteenth century. Pine resin was used to make rosins needed for the manufacture of paper and soap, and it could be distilled into turpentine used in industrial products such as varnishes, paints, and enamels. The demand for resin also propelled the expansion of the pine-tapping industry on a global scale. In the case of Spain, for example, resin industrialization led in 1898 to the establishment of the Spanish Resin Union (L.U.R.E.).³⁶ Scholars still have little information about the early history of resin production in Mexico, perhaps because they have yet to recognize its role in Porfirian industrial development. However, some of the processes that initially drove the industrialization of pine resin have been identified; most notable among these is the increasingly widespread use of turpentine for urban lighting to replace earlier sources of illumination, such as the burning of lard, oils, and tallows.³⁷ The heir of a large and little-understood rural tradition, pine-tree tapping began its slow modernization around the middle of the nineteenth century. In the 1870s, the first successful entrepreneurial ventures appeared, based primarily on the sale and consumption of turpentine.³⁸

In the urban areas of Puebla and Tlaxcala, turpentine was used for public lighting until sometime between 1898 and 1900, when electric lighting became commonplace.³⁹ Even then, it was still used in some neighborhoods and suburbs to light the streets at night. Although an international market for turpentine evolved in this period, the specific industries or businesses that were the primary consumers is still unclear. As a consequence, whether the turpentine refinery built in 1887 in the foothills town of Zitlaltepec was intended to produce turpentine for lighting or to meet the needs of some other domestic or international business cannot be established.

Although only partial, the data on this company allow us to fill in some details about its operation generally, and its use of natural resources in particular. The turpentine refinery apparently was founded by the Agricultural Society of the city of Zitlaltepec in mountains that had once belonged to the estate of San Bernardino. We do not know the year in which the work began, but by May 1887 there were already shipments of barrels of turpentine to Puebla for a Mr. Sacramento Reyes and a payment of nearly two hundred pesos received from one Cristóbal Báez, who appears to have been from Ixtenco. One of the refinery's most influential figures was Antonio Sánchez de la Vega, who from 1888 to 1890 was its treasurer. In 1896, he also served as the municipal president of Zitlaltepec. Like the Spanish L.U.R.E. plants, the Zitlaltepec turpentine refinery received resin from pine trees throughout the region, and while the use of resin to make turpentine distinguished it from other businesses in the area, it also made integral use of forest resources by participating in the lumber and broom root trade.

An examination of the company's daily income and expense reports allows for a partial reconstruction of its operation. Records have been located from 1887 to 1890 but unfortunately are not complete and cover a total of eleven months during these four years; nevertheless, the following charts demonstrate that these records can tell us something of the company's business and the intensity of forest exploitation in the region. The data in figure 5.1 suggest that broom root production, rather than turpentine, was the mainstay of the factory's business. Figure 5.1 compares the total income generated by these products even though some months saw no sales of specific products; for example, the records reflect sales of broom root in four months, compared to nine for turpentine.

These incomplete data show that the company made nearly 1,000 pesos from broom root compared to 600 from turpentine. The records for October 1889 also show that 572 bushels and 17 pounds of broom root were sold to Bernardo Vara at the price of 1.62 pesos per bushel.[40] In contrast, turpentine was sold in barrels, each of which contained between four and five measures (*arrobas*) of turpentine. According to the records, each arroba was worth 12, 14, 16, 18, or 20 reales, depending on its quality.[41] The 625 pesos that appear in the chart were for the sale of 338 arrobas and 19 pounds of turpentine in 77 barrels, 68 of which were sold in Puebla and 9 in Huamantla. Sacramento Reyes, in Puebla, was the most significant buyer, at 527.30 pesos; the remaining 100 pesos were divided among Gregorio Covarrubias, a woman named Luz Pérez (both of Huamantla), Sito Mora of Puebla, and Jesús Rosas of the Carpinteros neighborhood of Puebla.

By comparison, income from the sale of pine pitch and firewood was more modest. The revenue from pitch sales can be deceptive if we compare

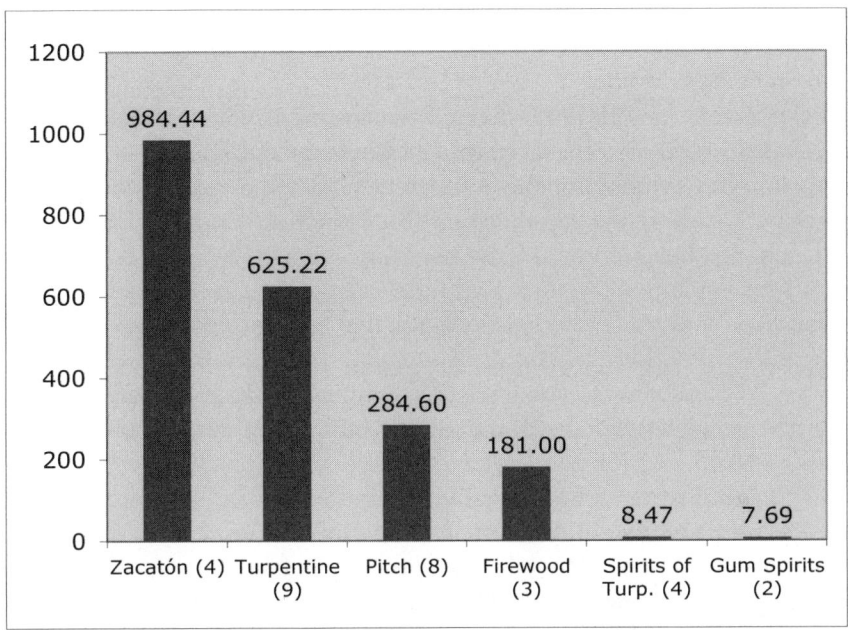

Figure 5.1. Total income (pesos) of the Esquilmos del Monte turpentine factory, 1887–90 (selected months). *Source*: Archivo Municipal de Huamantla, Sección Tesorería, cortes de caja (monthly accounting statements) de la Fábrica de Aguarrás, May–June 1887, February–March 1888, March–June and October–November 1889, and April 1890. The numbers in parentheses correspond to the number of months in which sales were made.

it to the more highly refined forms of turpentine: spirits of turpentine (*trementina*) and gum spirits (*cola*). A closer examination suggests that price was closely correlated with the different levels of quality of these resin products. According to the records, glue was worth 6 reales per arroba, turpentine 4 reales, and pine pitch between 1.5 and 2.5 reales. In other words, pine pitch was a product of lesser quality and value. Spirits of turpentine fetched 8.50 reales per arroba. The 7.70 pesos for glue derived from the sale of two barrels containing little more than five arrobas each, so the price per barrel was 3.8 to 3.9 pesos, while the 285 pesos for tar were for the sale of 1,264 bushels to various buyers, among whom was Sixto Mora, who also bought (standard) turpentine. This allows us to see that spirits of turpentine and gum spirits were by-products of the company's overall distilling activity, which centered on producing standard-quality turpentine in industrial quantities. The 181 pesos earned from firewood collection derived from the sale of approximately 190 cords (*medidas*) at a price of 7 to 8 reales in May 1889. The clients included the Hacienda de Tamariz and one Cándido Mier.

The records include other miscellaneous sources of income as well. The company earned 13.14 pesos from a category labeled "harvests of firewood and charcoal," which probably refers to fees that the company charged villagers who collected scrap pieces of wood such as branches, leaves, and fallen wood (but not tree trunks) used for firewood or to make charcoal. What the 43.68 pesos that derived from "damages to the woods" refers to is less clear, but some references suggest that communities may have paid fees in lieu of performing certain services such as maintaining paths in the woods. The category may also refer to fines the company assessed on people who collected wood without authorization.[42] Among those listed as causing "damage," for example, are "some people from Ixtenco."

The examination of the company's expenses allows us to envision some aspects of the work and the workers in the turpentine refinery. The men who harvested broom root (the *zacatoneros*) faced several risks, because the plant was a common refuge for snakes that thrived in the woods. Additionally, they needed to have sufficient strength to pull up the plant with its root intact. As a consequence, the 9.5 reales (1.18 pesos) that the zacatoneros received for each bushel of broom root harvested placed them among the refinery's best-paid workers.[43] The expenditures of 493 pesos reflected in figure 5.2 were payments for a total harvest of 417 bushels and 14.5 pounds of broom root. The intensity of this harvest, like the overall income it produced, indicates that the broom root industry had placed the forest under significant pressure, to the point that it threatened to compromise the forest's integrity and overall biodiversity.

The work of turpentine workers (*trementineros*) consisted of tapping trees to induce the secretion of oleoresin, which they collected in pails or simply chipped off the bark. The work was more difficult during the rainy season and cold spells, when the workers had to tap a larger number of trees to collect their accustomed amount of resin. Unfortunately, the accounts do not record how many workers there were or how much each received of the total 340 pesos in wage expenses. Elsewhere, the 20.38 peso expense for firewood used as fuel for the refinery paid for 71.5 individual "tasks" (*tareas*). Under the rubric of "other," one peso is registered as having been paid in March 1889 for four small barrels (known as *castañas*)[44] filled with turpentine that the refinery apparently purchased. Another peso's worth of expenses in October of the same year went to pay for additional tree tapping; perhaps the rainy season caused the trees to produce less resin than

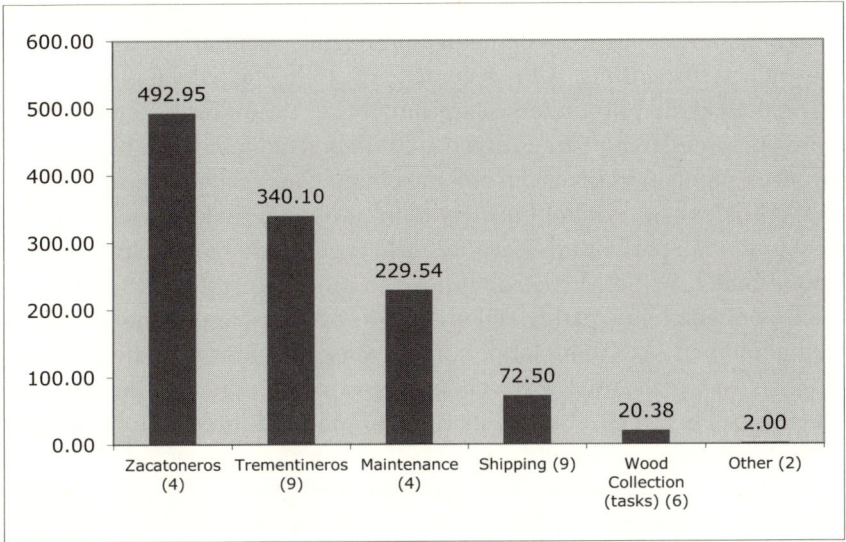

Figure 5.2. Operating expenses (pesos) of the turpentine factory, 1887–90 (selected months). *Source*: Archivo Municipal de Huamantla, Sección Tesorería, cortes de caja (monthly accounting statements) de la Fábrica de Aguarrás, May–June 1887, February–March 1888, March–June and October–November 1889, and April 1890. The numbers in parentheses correspond to the number of months in which payments were made.

the factory needed in order to produce its projected quantity of turpentine. The shipping charges for barrels of turpentine cost one peso per unit for transport to the city of Puebla and fifty cents for transport to the city of Huamantla.

Some brief references to the expenses incurred in the maintenance of the refinery can give us further insight. The largest investment was made in the maintenance of the still used to refine resin into turpentine: a single entry lists repair costs of 180 pesos. Additional sums went to repair and fabricate the still's boiler header and for a vat that cost nearly 20 pesos. Minor expenses included the purchase of utensils for collecting resin (buckets, cans, barrels), loads of shingles, and payments to a mason who helped to repair the refinery building.

The refinery appears to have been profitable, or nearly so, until Sunday, November 22, 1891. On that day, someone set fire to the Agricultural Society's turpentine refinery in Zitlaltepec, destroying most of its machinery and damaging the still. No one was arrested. Even so, the disappearance of the refinery did not spell an end for resin tapping in the La Malintzin forests, as increasing global demand continued to fuel the market for products such as turpentine.[45] On December 14, 1898, the federal government decreed a law that promoted establishment of "an industry to extract the resins contained in pine trees, and the distillation of these resins by the most advanced procedures known in order to obtain chemically pure products."[46] In 1904, Mexico's leading business journal opined that the industry had great potential, particularly because resin production had declined in the United States, while Mexico abounded in pine forests that secreted oleoresins of outstanding quality and quantity. The newspaper concluded that there remained "vast room for development of this industry."[47] In 1911, the proposal was made that resin collecting work be done for the Spanish L.U.R.E.[48] These expectations soon led to renewed interest in the potential of La Malintzin's forests.

In May 1913, a letter from the Mexican Agricultural and Colonization Company asked the government of Tlaxcala to determine whether the turpentine industry continued to operate within the state, and if so to describe the procedures it employed and its overall importance to the regional economy. In June of that same year, the government responded that "San Francisco Tetlanohcan is the only town that engages in this industry. Resin is extracted from ocote pine trees and . . . once distilled, is used to produce turpentine." The report concluded that the industry had "little significance" for the regional economy and produced only a modest amount of turpentine.[49] Despite the experience of the refinery in Zitlaltepec, early twentieth-century resin tapping in the Malintzin forests had once again become a domestic industry that used rudimentary technology to satisfy the requirements of nearby companies such as pharmacies in the city of Puebla.[50] Nevertheless, the intensive and extensive use of the forests, as well as the conflicts it provoked, left a mark on rural society that endured for years to come.

Social Conflict and the Deterioration of the Woods

The industrial utilization of wood, broom root, and pine resins for making turpentine and other industrial products devastated La Malintzin's forest cover. The clearing and girdling of trees and the destruction of grasses by cattle and controlled burning of pastures (*chamusquinas*) all had intensified significantly by the middle of the nineteenth century.[51] In April 1861, some of the residents of Huamantla denounced this destruction, accusing the Báez and Brito haciendas of having seized, cleared, and sold the wood from several plots of land usurped from the villages. They also suggested that the aguardiente factory that Antonio Lozano had built on the San Juan sugar plantation should be removed "because the forest is being destroyed." The villagers further complained that cattle owned by outsiders who had rented communal land had damaged their woods.[52] Conflicts over boundaries and the disputes over forest ownership made this situation even more acute. In that same year of 1861, the residents of Huamantla blamed the neighboring community of Tetlanohcan for causing significant "destruction" to the forests between the two villages.[53]

Accusations of the clearing and destruction of La Malintzin forests began to appear as early as the 1750s, when urban growth first created a demand for the volcano's forest products for firewood and charcoal. Around the middle of the nineteenth century, however, the Liberal economic policies that promoted the circulation of goods as a means of encouraging production and generating wealth heralded a period of intense deforestation. An edict issued by the Tlaxcala state government on September 26, 1860, is a case in point. It recognized that the government had received "repeated reports of the destruction of the woods" and that the forests continued to suffer from misuse even though laws had been enacted that prohibited "clearing the forest, cutting the trees or their bark [to girdle them], or setting fire to the forest [understory] or to pastures." Yet the edict also reminded local authorities that despite these provisions "the public should in no way be deprived" of access to the woods or its products. The circular dealt with this apparent contradiction by proposing a tax on anyone working in the forest, which it promoted as a means to "reconcile" the use of natural resources with their "conservation." The law also gave municipal councils a strong incentive for complying with the new regulations and collecting the tax, because any fees for the use of municipal forests went directly into municipal coffers.[54]

The edict concluded that these new provisions were necessary to ensure the "conservation and replenishment of the forests, which are necessary not only for the supply of firewood, wood, etc., but also because their influence on the atmosphere guarantees abundant rain."[55] The reference to environmental issues had less to do with officials' preoccupations about the ecological consequences of resource exhaustion than with their desire to justify seizing control of the forests by imposing tax levies that made it difficult or impossible for most people to use them legally. In other words, the new regulations implied that preservation of public goods was only possible by restricting access to them: only those with the ability to pay or the capacity to invest in business ventures were in a position to enjoy the benefits of "free" access to the woods.

The dispute over the appropriation of forest commons had always generated unease in the communities around La Malintzin, but the Liberal legislation that demanded their privatization while restricting access seemed to threaten traditional usufruct rights and set off an intense social conflict. In 1862, several residents of the town of Huamantla insisted that "all farmers and ranchers" should henceforth lose the right to let their goats graze in the woods and that only recognized (and putatively indigenous) members of the community (*los hijos del pueblo*) should continue to enjoy grazing rights. A village council would determine how many animals each villager was allowed. The villagers insisted that "it is not fair that the good of one individual should bring harm to the entire population"; they demanded that "the state of affairs" they were denouncing "be returned to their primordial state, before the illegitimate ambitions of a few individual mercenaries changed them."[56] The city council of Huamantla, for its part, challenged Tlaxcala's claim to the forestlands between the two towns. In a bid to press its claim to the disputed woods, the council of Huamantla granted licenses to various people from the town of Tetlanohcan "to work in the woods" in 1864. The Tetlanochtecos agreed to pay a fee of three pesos, which the township intended to use "to prosecute the lawsuit against the city council of Tlaxcala."[57] The expanding industrial uses of the woods, particularly the collection of oleoresins for turpentine, exacerbated these social conflicts, as in the case of a controversy that broke out between Huamantla and Tetlanohcan in 1864.

That year, Albino Flores and Felipe de la Cruz Cuapio were among the thirty residents of Tetlanohcan who requested permits from the municipal council of Huamantla to tap pine trees for resin. Each of these men believed he had obtained the rights to the same trees, which led to mutual recriminations and lawsuits. On June 7, 1864, Flores appeared before the fourth court of Huamantla and testified that a municipal official had ordered Cruz to stop tapping trees "in the woods that belong to this city" but that Cruz had ignored him because he had a license from the municipal government. Two weeks later, Cruz appeared in court to testify that Flores and Andrés Bernardino "were felling trees that belonged to the woods of this town" without the council's permission; he also disputed Flores's earlier accusation and stated that he, Felipe de la Cruz, "had been using the trees to harvest resin" legally and in accordance with the license that the municipal president had issued.[58] Having heard the complaint from both sides, the judge ruled "that each had committed abuse" and that Felipe de la Cruz had taken advantage of Huamantla villagers by logging the land in exchange for making token payments to the holders of religious *cargos* (the community political/religious hierarchy).⁵⁹ The judge believed that it was preferable to allow as many people as possible to make a living in the woods rather than reserve them for a handful of entrepreneurs. To mitigate the damage "to the common good," he decided to cancel all of the permits given to the residents of Tetlanohcan, since they "intend to tap an excessively large number of trees to extract resin without planting even one, so there is no doubt that we will see the complete destruction of the forest in a very short time and all the public goods of the city of Huamantla reduced to the most appalling misery—and all without compensation, because even though we are told that a tax for extraction of resin has been paid, the municipal treasury has no record whatsoever of having received a single peso for the aforementioned privilege."⁶⁰

Friction over the profits from resin tapping, unclear ownership of the woods, and imprecise boundaries all contributed to the growing tensions between Huamantla and Tetlanohcan. In early July, Lucas Antonio and José Longinos, two sons of Felipe de la Cruz, made a formal complaint to a Huamantla official that Andrés Martín, Andrés Bernardino, and José Albino Rodríguez (or perhaps Albino Flores) continued to damage the forest by "removing resin from its trees," apparently with permits issued by the Tlaxcala city council.⁶¹ A month later, the same official notified the mayor (*presidente municipal*) of Huamantla that these men had been discovered in Huamantla's woods, along with "a great number of others" from the town of Tetlanohcan. The troop had equipped itself with hatchets and resin collecting pails and intended to parcel out the woods among themselves. This provocation was confirmed the following day, August 6, by Juan Climaco of Chiautempan when he testified that the three men intended to divide the woods that "they claim to own."⁶² The antagonism between the two communities could no longer be contained. The first outbreaks of violence occurred in January 1865.

The prefect and military commander of the Huamantla district reported that the brothers Manuel Victoriano and José María González and two others, all of whom were residents of Huamantla, had gone into the woods to tap pine trees on January 19, when Andrés and Albino Flores, Mariano Cuapio, and a certain Martín, surrounded them. Other outraged residents of Tetlanohcan soon appeared as well and demanded to know what the men were up to. The two sides shouted insults at each other until the villagers from Tetlanohcan finally killed Manuel Victoriano, then disappeared with his body. That same afternoon, troops unsuccessfully searched the woods for the assailants. The military commander observed that "these events will no doubt lead to very dire consequences" unless the authorities somehow managed to overcome villagers' objections and fix a clear boundary line between the two communities.⁶³

This incident radicalized each side's position. In early March, the village council of Huamantla presented a writ to Tlaxcala city officials asserting the "right of possession of the woods" that the residents of Tetlanohcan had rented. As the ultimate owner of the forests, the Tlaxcalan city council responded that Huamantla could not impede the residents of Tetlanohcan from using the woods they occupied as tenants.⁶⁴ Even so, violence broke out again on May 5, as José Ceferino Tlahchi, a resident of Tetlanohcan who worked as a woodcutter for Leandro Rodríguez, headed into the forests with sixteen companions. The workers were walking along the edge of Tello Canyon when they came across some Huamantlans tapping pine trees, at which point the Tetlanohcans confronted and chased away the Huamantlans. When they passed through that area a bit later, forest wardens from Huamantla demanded to see the Tetlanohcans' permit to extract resin, prompting two of their party—including the one who carried the permit—to flee. The wardens surrounded the party and began thrashing them with dead branches and eventually carried Ceferino off to Huamantla as a prisoner.⁶⁵ Four months later, in September, another confrontation occurred in which the victim, the tree tapper José Eulogio of Tetlanohcan, accused the three Huamantlans of chasing him away from his work site. According to his testimony, his assailants said that if they succeeded in catching him, they "surely would have killed or injured" him.⁶⁶

Not until 1869 did the municipal councils of Tlaxcala and Huamantla reach an agreement about jurisdiction and boundaries in the woods. Yet these initiatives failed to stem social conflicts because the daily life of people who lived in the vicinity of the volcano depended directly on forest resources. Indeed, in March 1871 two residents of Huamantla along with three oxen were captured after having been caught in the act of harvesting resin from the trees in the woods of Tzompantepec without a permit; they were also named the prime suspects in a series of fires that municipal authorities could not put out because of strong winds. Officials suspected the two men had set the fires to conceal an illegal resin tapping operation. Moreover, authorities reported that the two men behaved suspiciously: not only did they linger in the vicinity of the forest fire, they "did not even try to contain it or warn anyone."⁶⁷

Tlaxcala's state government attempted to limit intentional and unintentional damage to the forests, which it declared to be the result of the "unthinking manner" in which they were used. By 1879, the government was worried that overexploitation had brought both public and privately owned forests "to the point of exhaustion." It published circulars and required municipal councils to include conservationist provisions in their ordinances.[68] Yet these restrictions on the use of forest resources may have increased the price of forest products. Wood became more scarce, for example. In this light, we can interpret the increasingly common theft of firewood as a resource conflict that pitted the villagers against the administrators of some estates. In 1861 the hacienda of Xaltelulco, of the district of San Pablo del Monte, for instance, complained to the warden that it had been the victim of "many harmful individuals who do not respect any authority" and consistently gathered firewood on hacienda lands without permission. The theft of firewood continued to generate complaints thirty years later in San Pablo del Monte: in 1891, Gabriel Copaltomatl and Félix Potrero received fines of eight and six pesos, respectively, for gathering dead branches and sticks from the San José Buenavista hacienda and the Guadalupe rancho, both owned by Pablo Sánchez.[69]

The social malaise was directed primarily against the factories and businesses that had deprived villagers of access to wood and pastures for their cattle. No doubt such social tensions also motivated the "malefactor" or malefactors who burned down the turpentine refinery in 1891. The restriction on collecting dead branches and sticks also contributed to an increasingly serious reduction in the number of oak seedlings (which were not covered by conservationist edicts), as wood gatherers began to use them instead of deadwood.[70] By the end of the nineteenth century, conflicts over access to resources had set many of the communities of La Malintzin against each other and heightened their antagonism toward haciendas.[71]

In light of native communities' resistance to the encroachment on their property and appropriation of their most prized resources, official discourse and policy moved away from the wholesale promotion of resource extraction (combined with a conservationist policy that sought to limit access to the woods via the imposition of usage fees) and settled instead on a formula that promised to rationalize resource use and hence "resolve the problem of forest conservation and systematic utilization."[72] In the face of local resistance, the state rephrased its interventionist policy and emphasized instead the better and more rational utilization of forest resources. Starting with the criminalization of traditional practices and uses of the woods defined as "archaic," "backwards," or "irrational," state agencies used a new, rationalizing discourse to override the objections of rural people and consummate the dispossession of communal rights to the forests. These efforts were driven by the conservationist ideology of scientists such as Miguel Ángel de Quevedo and the state's distillation of his ideas in the Forest Code of 1926.[73]

Mexican forests exist in deplorable conditions today. Nearly a million hectares of forest disappear every year, and mangroves are destroyed at an equally disturbing rate. Yet institutional mechanisms to put an end to overexploitation are weak, pliant, and beholden to corporate interests in the energy sector, as well as tourism interests, agroindustry, and ranching. The ineffectiveness of these legal measures puts the significance of this 1926 code, and succeeding laws, very much into question.

Notes

1. "Árboles de la vida," *La Jornada*, 11 March 2008.
2. Stanislaw Iwaniszewski, "Y las montañas tienen género: Apuntes para el análisis de los sitios rituales en la Iztaccihuatl y el Popocatepetl," in *La montaña en el paisaje ritual*, ed. Johanna Broda, Stanislaw Iwaniszewski, and Arturo Montero, 113–47 (Mexico City: CONACULTA-ANAH, UNAM, BUAP, 2001), 117–18.
3. Ismael Arturo Montero García, "Matlalcuéye: Su culto y adoratorio prehispánico," 74, and Gabriel Espinoza P., "El eco del agua: El pasado lacustre de Tlaxcala," 68; both in *Coloquio sobre la historia de Tlaxcala* (Tlaxcala: Gobierno del Estado de Tlaxcala, 1998).
4. Ángel García Cook, ed., *Tlaxcala: Textos de su historia* (Tlaxcala: CNCA, 1991), 11:35.
5. *Ecología del Estado de Tlaxcala: Región Malinche* (Mexico City: INEA, 1995), 52.
6. Alba González Jácome, *Cultura y agricultura: Transformaciones en el agro mexicano* (Mexico City: Universidad Iberoamericana, 2003), 81.
7. Marcello Carmagnani, *Estado y mercado: La economía pública del liberalismo mexicano, 1850–1911* (Mexico City: Fondo de Cultura Económica and El Colegio de México, 1994), 19.
8. Alejandro Tortolero Villaseñor, "Tierra, agua y bosques en Chalco (1890–1925): La innovación tecnológica y sus repercusiones en un medio rural," in *Agricultura mexicana: crecimiento e innovaciones*, ed. Margarita Menegus and Alejandro Tortolero, 174–233 (Mexico City: Instituto de Investigaciones Dr. José María Luis Mora, 1999); Mario Camarena Ocampo, "Fábricas, naturaleza y sociedad en San Ángel (1850–1910)," in *Tierra, agua y bosques: Historia y medio ambiente en el México central*, ed. Alejandro Tortolero Villaseñor, 317–41 (Mexico City: Centre Français d'Études Mexicaines et Centraméricaines and Instituto de Investigaciones Dr. José María Luis Mora, 1996). See also Mario Trujillo Bolio, "Producción fabril y medio ambiente en las inmediaciones del Valle de México 1850–1880," 343–60, and Rodolfo Huerta González, "Transformación del paisaje, recursos naturales e industrialización: El caso de la fábrica de San Rafael, estado de México, 1890–1934," 283–315, both in Tortolero Villaseñor, *Tierra, agua y bosques*.
9. On mahogany, see Jan de Vos, "La contienda por la selva Lacandona: Un episodio dramático en la conformación de la frontera sur, 1859–1895," *Historias* 16 (1987): 73–98. On chicle, see Herman W. Konrad, "Capitalismo y trabajo en los bosques de las tierras bajas tropicales mexicanas: El caso de la industria del chicle," *Historia Mexicana* 143 (1987): 465–505, esp. 468; see also Martha Patricia Ponce Jiménez, *La montaña*

chiclera, Campeche: Vida cotidiana y trabajo (1900–1950) (Mexico City: Cuadernos de la Casa Chata, 1990). On paper, see Huerta González, "Transformación." Turpentine began to be used for illumination and industrial purposes in the 1860s. See José Juan Juárez Flores, "Alumbrado público en Puebla y Tlaxcala y deterioro ambiental en los bosques de La Malintzi, 1820–1870," *Historia Crítica* 30 (2005): 13–38.

10. Marcello Carmagnani, "El liberalismo, los impuestos internos y el Estado federal mexicano, 1857–1911," *Historia Mexicana* 151 (1989): 471–96; Carmagnani, *Estado y mercado*, 266.

11. García Cook, *Tlaxcala*, 11:461–71.

12. José Juan Juárez Flores, "Las finanzas municipales y la desamortización de los bienes corporativos en la ciudad de Tlaxcala: El caso de los montes de La Malintzin (1856–1870)," in *Agricultura y fiscalidad en la historia regional mexicana*, ed. Alejandro Tortolero Villaseñor, 123–48 (Mexico City: Universidad Autónoma Metropolitana–Iztapalapa, 2007).

13. Archivo Histórico del Estado de Tlaxcala (hereafter AHET), Fondo Siglo XIX (hereafter XIX), C116/1869, exp. 2, and C125/1875, exp. 2.

14. García Cook, *Tlaxcala*, 11:211.

15. Ibid., 11:468.

16. AHET, Fondo Ayuntamiento (hereafter AHET-A), C110/1866, exp. "mayo."

17. Juárez Flores, "Alumbrado público"; and José Juan Juárez Flores, "Malintzin Matlalcuéyetl: Bosques, alumbrado público y conflicto social en la desarticulación de un entono ecológico (Puebla-Tlaxcala, 1760–1870)," master's thesis, Department of History, Universidad Autónoma Metropolitana–Iztapalapa, 2007.

18. "Circular del 26 de septiembre de 1860," AHET-XIX, Sección Hacienda, C122/1860, exp. 14; see also García Cook, *Tlaxcala*, 11:392–93.

19. "Dictamen de la Comisión del monte sobre el proyecto de Reglamento para el cobro por el uso del monte por extracción de maderas y uso de los pastos, presentado al P. Ayuntamiento de Huamantla," Archivo Municipal de Huamantla (hereafter AMH), Sección Presidencia (P), Año 1877.

20. García Cook, *Tlaxcala*, 12:177–79.

21. Ibid., 14:18–22.

22. Ibid. One bushel (*arroba*) equals 11.5 kilograms.

23. Alfonso Luis Velasco, *Geografía y estadística del Estado de Tlaxcala* (1892; facsimile ed., Tlaxcala: Gobierno del Estado de Tlaxcala, 1998), 69–70, 82–84, 98–99.

24. Ibid., 118.

25. Carmagnani, *Estado y mercado*, 269.

26. Ricardo Rendón Garcini, *El Prosperato: Tlaxcala de 1885 a 1911* (Mexico City: Siglo XXI Editores and Universidad Iberoamericana, 1993), 144.

27. "Cuenta municipal de Altzayanga, 1893," AMH, Sección Tesorería, Año 1893.

28. "Libro de cuentas de Ingreso y Egreso que manifiesta el P. Ayuntamiento de esta cabecera [de Teolocholco] del año de 1909," AHET, Fondo Revolución, Régimen Obregonista (hereafter AHET-R-O), C2/1910, exp. 83.

29. Rendón, *El Prosperato*, 144.

30. Velasco, *Geografía y estadística*, 26–27, 78, 118.

31. "Una copia de las bases bajo las cuales se concede la licencia para extraer raíz de zacatón del monte de este municipio [de Huamantla,]" AMH-P, Año 1883.

32. Ibid. Other contracts signed in Huamantla included those made by Plutarco

Montiel in 1895 and by Castor de la Fuente in 1899 and 1903–4. See Rendón, *El Prosperato*, 253n40.

33. AHET-XIX, C14/1891, fojas sueltas.

34. Rendón, *El Prosperato*, 252–53.

35. René Cuéllar Bernal, *Tlaxcala a través de los siglos* (Mexico City: B. Costa Amic, 1968), 240.

36. Rafael Uriarte Ayo, *La Unión Resinera Española (1898–1936)* (Madrid: Fundación Empresa Pública, 1996).

37. Juárez Flores, "Alumbrado público."

38. Identification of these ventures is according to the characteristics indicated by Uriarte (in the case of the Spanish resin industry) in *La Unión Resinera*, 6–7.

39. The proposal to convert to electrical wiring in Puebla was made in 1883. See *Memoria Urbana de Puebla* (CD-ROM), "Expedientes de Alumbrado" (Mexico City: CONACYT, 1998).

40. From October to December 1891, the gross bushel (arroba) of broom root cost 1.62 pesos in Zacatelco, a town situated on the other side of the mountain from Zitlaltepec, also on La Malintzin. See AHET-XIX, C14/1891, fojas sueltas.

41. One peso equals eight reales.

42. This was in accordance with Article 8 of "Dictamen de la Comisión del monte sobre el proyecto de Reglamento para el cobro por el uso del monte por extracción de maderas y uso de los pastos, presentado al P. Ayuntamiento de Huamantla [1876]," AMH-P, Año 1877.

43. This data, collected for March–June 1889, should be considered with reservations; in the month of October in 1889 and 1891, the bushel (arroba) was sold at 1.62 pesos, as indicated in the text. Nonetheless, it is indicative of the relevance of the broom root industry.

44. A castaña was a container similar in form to half of a barrel; the flat part was made to fit the back of an ox for transport, and it had a capacity of approximately thirty liters.

45. Uriarte, *La Unión Resinera*, 17.

46. "Relativo a la industria de extraer los jugos resinosos que contienen los pinos," AHET-R-O, C150/1913, exp. 12.

47. *El Economista Mexicano* (Mexico City), 23 July 1904, cited in Lucia Martínez Moctezuma, "Máquinas, naturaleza y sociedad en el distrito de Chalco, estado de México a fines del siglo XIX," in Tortolero Villaseñor, *Tierra, agua y bosques*, 267.

48. Uriarte, *La Unión Resinera*, 40.

49. "Relativo a la industria de extraer los jugos resinosos que contienen los pinos," AHET-R-O, C150/1913, exp. 12.

50. The Medina drugstore, an establishment with a long tradition in the city of Puebla, was a habitual buyer of oil of turpentine and other resin products for the making of creams or ointments generally known as "ocotzotl" because they were derived from pine pitch (ocote). Information given by Mr. Esteban Rodríguez (born 1932), of the town of San Francisco Tetlanohcan, who "learned from his grandparents and whose occupation was that of resin gatherer." Interview, 4 April 2004.

51. See "Circular del 26 de septiembre de 1860," AHET-XIX, Sección Hacienda, C122/1860, exp. 14.

Tlaxcala's La Malintzin Woodlands · 123

52. "Sobre los hechos que denuncian varios vecinos principales del pueblo," AMH-P, Año 1862. Elsewhere I have shown how the *ocotes* harmed the health of trees; see Juárez Flores, "Alumbrado público."
53. AHET-A, Cabildos, Legajo 2, Sesión del 21 de junio de 1861, p. 40.
54. "Circular del 26 de septiembre de 1860," AHET-XIX, Sección Hacienda, C122/1860, exp. 14.
55. Ibid.
56. "Sobre los hechos que denuncian varios vecinos principales del pueblo," AMH-P, Año 1862.
57. "Partes relativos al monte de esta municipalidad. Año de 1864," AMH, SP-ST, C14/64, exp. 59, pp. 14–16.
58. Ibid., exp. 59, p. 7.
59. "El Ayuntamiento de Huamantla requiere al de esta capital para que haga efectivo el pago de los perjuicios causados por Felipe de la Cruz del monte de la Malintzin perteneciente a la primera corporación," AHET-A, C107/64, exp. "julio," p. 1; "Partes relativos al monte de esta municipalidad. Año de 1864," AMH, SP-ST, C14/64, exp. 59, pp. 4–5.
60. "El ayuntamiento de Huamantla requiere al de esta capital [Tlaxcala] para que haga efectivo el pago de los perjuicios causados por Felipe de la Cruz del monte de La Malintzin perteneciente a la primera corporación," AHET-A, C107/64, exp. "julio," pp. 5–7.
61. "Partes relativos al monte de esta municipalidad. Año de 1864," AMH, SP-ST, C14/64, exp. 59, pp. 14–16.
62. Ibid.
63. AHET-A, Cabildos, Legajo 4, Sesión del 24 de enero de 1865, p. 82fte.–82vta.; "Comunicaciones . . . ," C108/65, exp. "enero," p. 12.
64. AHET-A, Cabildos, Legajo 4, Sesión del 2 de marzo de 1865, p. 87vta.
65. "Acto y declaración que se le toma al enfermo José Ceferino Tlahchi, casado con María Juana, Tetlanohcan, mayo 8 de 1865," AHET-A, C109/65, exp. "mayo," p. "s."
66. AHET-A, "Comunicaciones . . . ," C109/65, exp. "septiembre," p. 2.
67. "Sobre daños hechos en el monte de esa municipalidad [de Huamantla]," AMH, Serie "Daños al monte," Año 1871.
68. See this point in "Circular del 28 de octubre de 1879 sobre la conservación de montes," AMH, Sección Montes, C73; and in Article 8 of "Dictamen de la Comisión del monte sobre el proyecto de Reglamento para el cobro por el uso del monte por extracción de maderas y uso de los pastos, presentado al P. Ayuntamiento de Huamantla [1876]," AMH-P, Año 1877.
69. AHET-A, C103/1861, exp. "enero"; Fondo siglo XIX, C11/1891, exp. "julio," fojas sueltas.
70. According to Pablo Sánchez, whose firewood was stolen in 1891; see AHET-XIX, C11/1891, exp. "julio," fojas sueltas.
71. Rendón, *El Prosperato*, 144–47.
72. "Relativo a la industria de extraer los jugos resinosos que contienen los pinos," June 1913, AHET-R-O, C150/1913, exp. 12.
73. See also Christopher R. Boyer, "Revolución y paternalismo ecológico: Miguel Ángel de Quevedo y la política forestal en México, 1926–1940," *Historia Mexicana* 57, no. 1 (July–September 2007): 91–138.

CHAPTER SIX

Water and Revolution in Morelos, 1850–1915

Alejandro Tortolero Villaseñor

In 1892, the French prospector and consular official Louis Lejeune announced that Mexico had made significant strides toward modernity thanks to the political stability and economic efficiency that had become the hallmark of the 1876–1911 Porfiriato, as the regime of President Porfirio Díaz was known. Lejeune judged that Mexico had actually carried out two revolutions under Díaz: a political one that produced social peace; and an economic one based on railroad construction, which had made mining and other industries more profitable than ever before. Now Lejeune felt that the nation needed a third revolution, one that would overhaul its hydrological regime and bring wealth and development to the countryside. The problem, from his perspective, was not the lack of water but rather the absence of a sophisticated irrigation system. The construction of such a system, he believed, had the potential to generate even greater profits than industry or mining.[1] Other experts agreed. As one observer put it in 1906, "the true difficulty—we could even say the only difficulty—is the *shortage of irrigation*."[2]

No one doubted that only a tiny proportion of Mexico's vast territory of nearly two million square kilometers was used as cropland. Raoul Bigot wrote in 1907 that farmers cultivated a mere 6.1 percent of Mexico's total surface; 87 percent of these croplands (5.3 percent of the national territory) consisted of rain-fed (*temporal*) fields, meaning that a mere 13 percent of cropland (1.55 million hectares, or 0.6 percent of the surface of Mexico) was irrigated. He further estimated that pastures of one sort or another accounted for a quarter of the nation's surface area and forests for another

9 percent, leaving him to conclude that the remaining 60.4 percent of Mexico was unused altogether.[3] These figures do not seem unreasonable when we consider that studies conducted as late as 1921 concurred that only 6 percent of the land in Mexico was used for agriculture.[4] Not all of this unproductive terrain was suitable for farming, of course, but most experts agreed that up to 40 percent of the national territory had some potential use if the agricultural frontier could somehow be extended. They emphasized the potential benefits of irrigation and the application of modern agricultural technology on medium-sized and large properties. Andrés Molina Enríquez, one of the most passionate critics of the Porfirian regime, considered the issue of irrigation so critical that he named it one of the five *grandes problemas* (along with credit, population, politics, and the unequal distribution of property) that confronted the nation in 1909.[5]

The state of Morelos had become the nation's primary sugar producer by the time Molina Enríquez made these observations. Indeed, Morelos plantations held some of the most productive sugar lands in the world even though they lay just a short distance southwest of Mexico City's far cooler climate. Yet the state's continued development depended on landowners' ability to expand irrigation networks and, crucially, to gain access to enough water for their thirsty crops. Three idiosyncratic features of the state's geography made this hydrological puzzle all the more complicated. In the first place, the state had an enviable abundance of water, for two major watersheds lay within its border: most of the state, including the Valley of Cuernavaca, lay within the Cuautla/Amilpas watershed, which itself forms part of the larger Balsas River basin; a far smaller proportion lay within the Nexapa River basin on the eastern border of the state of Puebla. Second, Morelos had some of the most intensively irrigated land in the country by the eve of the 1910 revolution because sugar planters had built numerous irrigation projects to take advantage of all this water.[6] Finally, landowners' monopolization of the state's abundant water resources helped to make Morelos a focal point of the revolutionary upheaval.

In this chapter, I suggest that access to water—and the management of water resources in particular—constituted the primary challenge for the continued growth of Morelos sugar haciendas. This hypothesis is not new, yet it merits reconsideration because historians appear to have lost sight of the fundamental role water played in the state's environmental and social history. While a generation of observers during the Porfirian and revolutionary eras had suggested that social conflict over water limited the possibilities for agricultural growth and helped to spark the Zapatista revolt in 1910,[7] subsequent analysts rejected this "hydraulic thesis" and suggested that access to land was the key bottleneck to increased sugar production and therefore lay at the core of the state's agrarian problem.[8] Most historians have adopted what amounts to a simplified version of Domingo Diez's argument that sugar planters' adoption of centrifuges around 1880 increased the efficiency of the refining process and touched off a "radical change" to social relations because it led sugar planters to seek new land to put into production, which inevitably came at the expense of peasant communities: "The *pueblos* [peasant communities]," Diez wrote in 1918, "were obliged to yield their land and water," and villagers' communal property "gradually diminished, in some cases disappearing altogether." This "social imbalance" intensified until the revolution broke out in 1910.[9]

Historians have tended to seize on the agrarian component of Diez's observations. Their explanations for revolutionary upheaval in Morelos routinely subordinate the issue of water to the supposedly more important question of land. After all, that analytic simplification squared with the Zapatistas' revolutionary slogan, "Land and Liberty!" Yet there can be no doubt water was a crucial source of wealth disparities and social tension. For example, the Ministry of Development assessed a hectare of irrigated land in Morelos at 877 pesos on the eve of the revolution, whereas rain-fed land was assessed at a mere 100 pesos.[10] Moreover, cadastral figures show that sugar haciendas planted only cane on the small proportion of their property that could be irrigated.[11] Water, not land, therefore represented the fundamental bottleneck to sugar production; as landowner Manuel Araoz observed as early as 1914, sugar haciendas occupied 60 percent of all land in the state, yet only 15 percent of the state's territory was irrigated.[12]

We must reconsider the hydraulic thesis if we hope to understand the challenges that sugar planters (and, indirectly, the pueblos) confronted in Morelos. While it might be an overstatement to say that water was the source of the Zapatista uprising, a comprehensive view of the Morelos countryside on the eve of the revolution must include an analysis of both land and water.

Land or Water? Dueling Interpretations

Domingo Diez was one of the most renowned experts on development in Morelos during the late nineteenth century. He left his mark both as a civil servant and as a civil engineer responsible for some of the state's most important irrigation projects. He disclosed his ideas in an impressive number of publications, most of which dealt with questions of land and water.

Many of them also addressed the ideas of Fernando González Roa and José Covarrubias, two thinkers who had proposed diametric strategies of promoting rural development: the conservative solution, which argued in favor of concentrating control over land and water in the hands of hacienda owners; and the liberal solution, which favored the formation of small property and irrigation cooperatives.[13] These divergent models implied very different uses of the land, insofar as González's "conservative" approach entailed extensive agriculture, whereas Covarrubias's "liberal" approach favored an intensive land-use model.[14]

According to Diez, the economic situation in Morelos on the eve of revolution favored the conservative (extensive) pattern, in which more and more land and water came under the control of haciendas at the expense of rural communities. In his words, "the haciendas increased their territories and gradually the hacienda owners took over *ejidos* [village common lands] through purchase or coercion. Little by little, the land held by pueblos was reduced to a minimum, and in some cases, even their core land grant [*fundo legal*] came under attack, causing some pueblos to disappear. The ruins of many of them still remain among the fields."[15]

This is where the historiographic simplification begins. The agrarian argument rests on the idea that haciendas deprived peasant communities of their lands. To be sure, Diez's analysis is much more sophisticated and never dissociates land from the water needed to make it productive. Indeed, he constantly refers to what he calls an "agrarian feudalism" in Morelos, characterized by hacienda owners' monopolization of lands and water together.[16] In other work, he insisted that "water is immensely important: in Morelos, it has become a true and imperious need, since water and land is controlled by a single class. This situation has resulted in one of the most forceful and bloody demonstrations of social unrest," that is, the uprising led by Emiliano Zapata.[17]

Diez's arguments echo the works of another author who has influenced historians' understanding of the Morelos sugar industry: Felipe Ruiz de Velasco, an agronomist who had studied at the Agricultural Institute of Gembloux, Belgium.[18] It was a striking choice for an aspiring agronomist. In the early twentieth century, the Institut National Agronomique in Paris was the most influential such institution in the world, and Mexico's own Escuela Nacional de Agricultura had adopted its curriculum. Yet Ruiz preferred a course of study based on the intellectual tradition of the Netherlands, which was the world leader in hydrological engineering.[19] His decision to become an agronomist was no coincidence, either. His father, Tomás Ruiz de Velasco, was the administrator of the hacienda of Zacatepec, a middling estate situated in a valley with significant amounts of standing water—so much, in fact, that the hacienda could increase its productive land only by draining the wetlands. Ruiz's father had encouraged him to pursue a course of study specifically centered on drainage and land reclamation, leading the young man to study under J. M. Leclerc, the Belgian government's chief expert on wetlands drainage, and M. Weemaels, who had directed reclamation projects in the Bravante.[20]

Ruiz went further than Diez in his analysis and argued that water had greater importance in Morelos than land itself. He insisted that "the wealth in this state does not reside in the land for which so much blood has been spilled" but rather was "a consequence of the water that bathes them." He pointed out that only 10,000 of Morelos's 491,115 total hectares were irrigated, and even that modest number faced an uncertain future if rapid deforestation changed the regional climate and diminished river flows.[21] Ruiz soon became a key spokesman for water's role in the Morelos sugar boom, and his diaries provide many examples of such advocacy. For example, he described don Pedro Lamadrid's properties in El Higuerón and Tlaquiltenango as rich cattle lands but suggested that they would be worth a veritable fortune if properly irrigated.[22] In a 1925 history of Morelos's sugar economy, he repeatedly pointed to instances in which the magic of irrigation transformed land in southern Morelos formerly considered worthless into first-class property valued at 2,000 to 3,500 pesos per hectare in 1910.[23]

Nevertheless, the hydraulic thesis first articulated by Diez and Ruiz remains virtually unknown, while the agrarian thesis has become well established in the literature thanks to the work of Gildardo Magaña, Jesus Sotelo Inclán, John Womack Jr., Roberto Melville, Guillermo de la Peña, and William Bluestein.[24] Magaña (who was one of Zapata's chief generals and inherited the movement after Zapata's assassination) emphasized the plundering of communal lands in his 1934 reflection on Zapatismo. It located the movement's roots in agrarian dispossession, noting that "the pueblo of San Pedro was absorbed by the Hacienda de Hospital; Cuachichinola by the hacienda of the same name; Sayula by the San Vicente [hacienda], and many others in the same way, culminating the infamies in the case of Tequesquitengo," the latter of which was the most famous example of a hacienda owner's assault on a peasant community.[25] Womack concurred and explained that while there were 118 towns in Morelos in 1876, they had diminished to 105 in 1887, and by another 5 in 1910, their land having been expropriated by haciendas. Womack observed, "Through the 1890's and after the turn of the century villages continued to disintegrate. By 1909 only a hundred were registered. Hidden darkly in the fields of high, green cane, the ruins of places like Acatlipa, Cuauchichinola, Sayula, and Ahuehuepan rotted into the earth."[26] He also echoed Magaña's reference to Tequesquitengo, where the owner of the San José Vista Hermosa hacienda retaliated against villagers' complaints of dispossession by directing water into a nearby lake that flooded the entire pueblo. Soon, only the church spire remained above water to remind other communities about the cost of tangling with the hacienda owner. Womack also found that many of the pueblos still intact when the revolution erupted in 1910 were mired in land disputes with haciendas—so much so that land titles were nearly sacred texts for the Zapatistas.[27] Guillermo de la Peña extended this line of reasoning still further, pointing out that the 1880–1910 period was characterized by a sudden expansion of the world sugar market, which led to new growth for Morelos refineries and, correspondingly, a strong incentive for haciendas to acquire more land. Lowland villages lost practically all their property, and haciendas had begun to put pressure on highland communities by the first decade of the twentieth century.[28]

The agrarian thesis, and its emphasis on land expropriation, has all but eclipsed the insights of the hydraulic thesis. As late as 1996, historian Horacio Crespo observed that the role of water had yet to be understood in the history of Morelos, though he disagreed with the idea that water constituted an obstacle to the growth of haciendas.[29] In light of these arguments, we must ask whether water was in fact an obstacle to the growth of Morelos's sugar industry. The answer is not clear. Even the specialists recognize that we do not have enough information about hacienda irrigation.[30] Nevertheless, an analysis of archival data and the secondary literature will allow us to begin to address the issue.

Water in Morelos

The state of Morelos has an area of approximately 5,000 square kilometers divided into two regions (figure 6.1). Most of the north lies at an elevation of over 1,600 meters, comprising highlands and forests that extend northward through the slopes of Mount Ajusco and into the mountain range of Huitzilac and the massive Popocatépetl volcano. This upland region also includes the communally owned woods granted to fifteen pueblos during the colonial era. The sugar regions lie to the south and are divided into two fertile zones: the Valley of Cuernavaca to the west and the Amilpas watershed to the east. These two regions were home to most haciendas, family

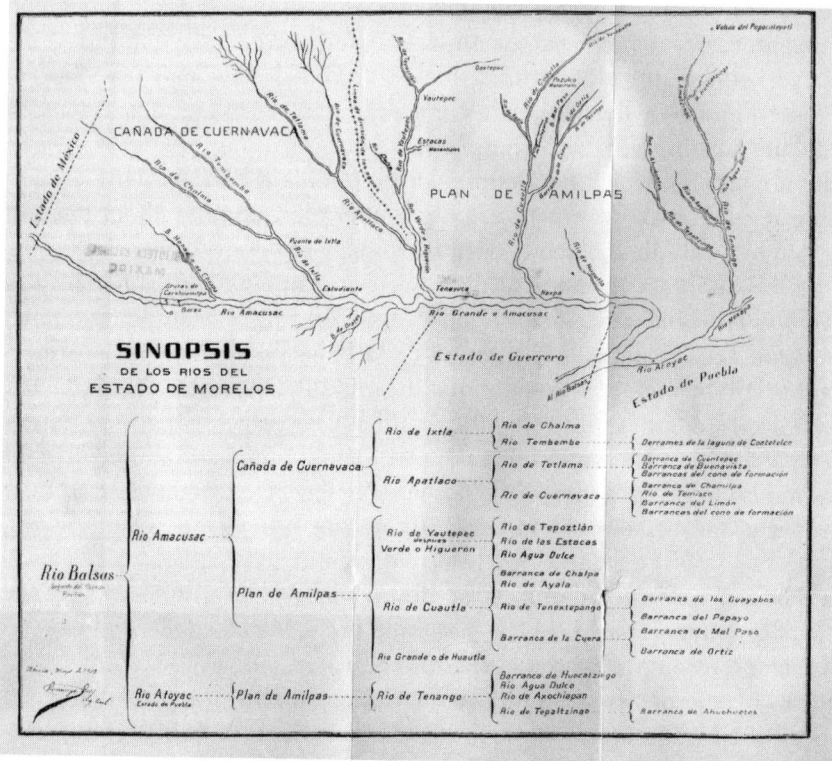

Figure 6.1. Rivers and tributaries of Morelos. *Source*: Domingo Diez, "Observaciones críticas sobre el regadío del Estado de Morelos," presentation to the Escuela Nacional de Ingenieros, 14 May 1919 (Mexico City: Antigua Imprenta de Murguía, 1919).

farms (ranchos), and cities, as well as the springs and other perennial sources of water used for irrigation and the production of hydroelectricity. The main characteristic of these southern hotlands—the so-called *tierra caliente*—is its flatness and lack of rainfall, which made irrigation an absolute necessity.

The expansion of sugarcane in these two regions spurred the development of haciendas during the colonial era. Landowners have sought to acquire the best land for sugar cultivation ever since. In the nineteenth century, hacienda production of sugarcane was transformed into an agro-industry, particularly after the introduction of the steam-powered vacuum pan refining technique in the 1880s, which replaced the earlier system of producing sugar by simply heating cane juice in large kettles until it crystallized. However, the new system also put greater pressure on water resources.

In 1908, seventeen haciendas occupied more than a quarter of the total surface of Morelos and possessed nearly all land suitable for cane cultivation. Statistics produced two years later showed that haciendas occupied 56.34 percent of the total surface of the state, communal properties 25.66 percent, and small private landholdings 18.00 percent.[31] By 1912, the haciendas had expanded farther still and occupied 63 percent of the state.[32] While these haciendas had colonized most of the its surface, we also know that space itself is not undifferentiated or devoid of meaning. Instead, it embodies the pressures and understandings of different social actors. The question is, what kind of lands did the haciendas possess? For this, a clear answer exists: they occupied the best lands, especially those with irrigation.[33]

Irrigation in Morelos depended almost exclusively on rivers, nearly all of which have fast-moving and torrential flows that course through deep banks and often produce seasonal flooding. Its impressive hydrographic system irrigated nearly every hacienda in the state, whose complex water rights are reflected in figures 6.2 and 6.3.

These figures demonstrate that practically all the haciendas of the state drew water from rivers and springs, yet they leave open the question of how water was actually used in sugar agriculture. Sugar haciendas invariably held more than 1,000 hectares; of the total of twenty-seven haciendas historians have studied, seven of them comprised between 1,000 and 2,000 hectares, nine between 2,000 and 5,000 hectares, five between 5,000 and 15,000 hectares, and six were larger still. Hacienda owners therefore owned enough property to plant several different crops (including cane, rice, and corn) and in fact used only a small proportion of their land for sugarcane. In general, the haciendas irrigated less than 10 percent of their lands, except for the San Nicolás and Zacatepec haciendas, which succeeded in irrigating almost 40 percent of their territory. According to the hydraulic engineer and hacienda administrator Felipe Ruiz de Velasco, haciendas in Morelos divided their sugar lands into three zones: those used for *plantillas* (the first cutting of cane); *socas* (the second cutting) and *resocas* (third cutting); and fields that were being fallowed.[34] Apart from these, we must add the lands leased to sharecroppers and those left unworked, such as woods and hillsides. In 1910, the area used for first-cutting sugarcane accounted for no more than 3.47 percent of hacienda lands, and fields dedicated to second and third cutting constituted another 7.08 percent. The remaining land (occupied by sharecroppers, forests, and fallows) constituted 89.45 percent

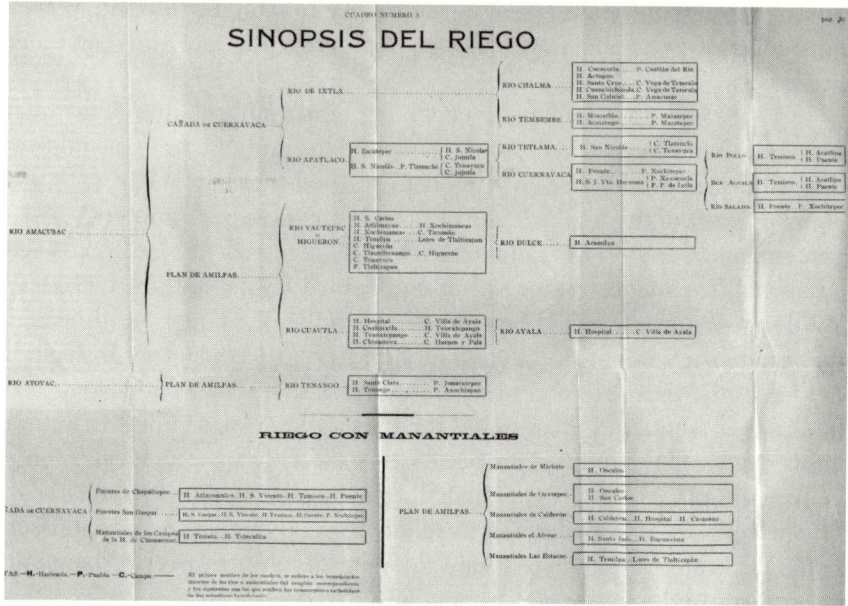

Figure 6.2. Sources of irrigation for principal haciendas. *Source*: Domingo Diez, "Observaciones críticas sobre el regadío del Estado de Morelos," presentation to the Escuela Nacional de Ingenieros, 14 May 1919 (Mexico City: Antigua Imprenta de Murguía, 1919), 30.

of all hacienda acreage.[35] In reality, the haciendas had more lands than they could possibly work, and most of their property was used only when water became available. Wherever water was scarce, haciendas converted land to summer pastures or turned it over to sharecroppers, who grew corn or some other commercial crop. The haciendas' problem, then, was not the lack of land but rather the limited availability of water. The experiences of El Puente and Zacatepec haciendas, both of which left detailed accountings of their irrigation practices, help to illustrate this point.[36]

The two haciendas used their water resources in very different ways. The administrator of El Puente adopted a program of intensive irrigation, whereas Zacatepec followed a careful management plan and tried to conserve water. The explanation for this difference might be traced to the quantity of irrigated acreage that each estate owned. El Puente had an extent of 1,609 hectares, of which only 200 were irrigated. Zacatepec, in contrast, irrigated nearly 40 percent of its property, or 700 of its 1,684 hectares. The fact that Zacatepec had access to 2,200 liters of water per second and El Puente

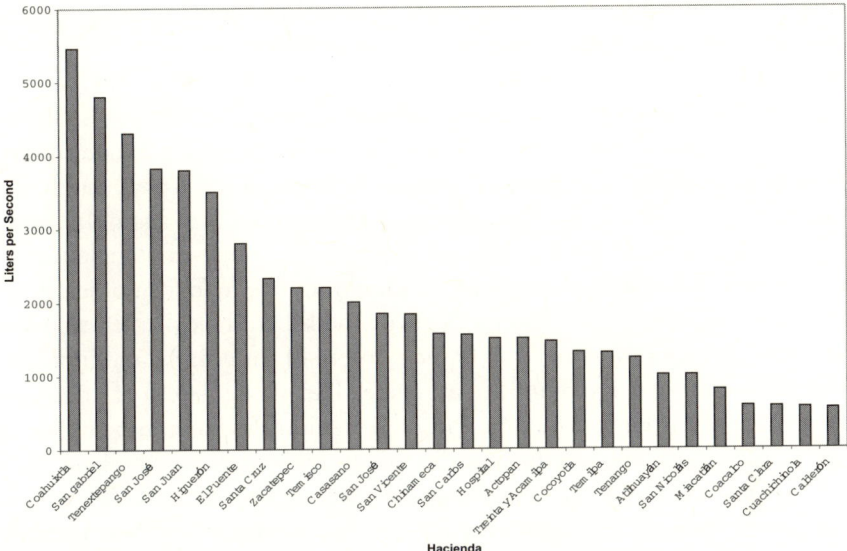

Figure 6.3. Water used in hacienda irrigation canals (liters per second). Data from Domingo Diez, *Bosquejo histórico geográfico de Morelos*, rev. ed. (Cuernavaca: Editorial Tlahuica, 1967), 118.

2,800, raises the question of why El Puente appears to have used its water so inefficiently.

The administrator of El Puente, Ramón Portillo y Gómez, gave a detailed description of the hacienda's production cycle. The process began as soon as local weather conditions permitted cane to be planted. On that day, the seeds received an initial round of irrigation known as the *asentadera*, which the hacienda performed by opening sluice gates that conducted water into two consecutive furrows measuring up to 80 varas (67 meters) in length. After that, irrigation occurred every eight, ten, or twelve days. Weeding began eight days after planting and was repeated four more times, at fifteen-day intervals, usually interspersed with two rounds of irrigation.[37] Fifteen days after the fifth weeding and the two subsequent rounds of irrigation, the furrows between the cane were plowed to maintain their depth. After that, two more rounds of irrigation were necessary, one on the same day as the plowing and another eight days later. In these later rounds of irrigation, water was channeled through three consecutive furrows measuring 120 varas (approximately 100 meters), using a technique known as *riego por mitad*. Fifteen days after the first plowing, workers releveled the field in a process known as *quita de tierra*, using hoes, after which another three rounds of irrigation took place. Twenty days later came the second plowing, which shored up the cane stalks as the plant matured and began to bud. After that point, the field was watered using only an intermediate *apantle* ditch in the middle of the field. When the cane had fully matured, the field was "straightened" by connecting all the irrigation ditches end to end (a technique known as *riego de punta*) and water was allowed to flow until each field had been watered in succession and the process could begin all over again. In some cases, this rotation among fields took place in the course of a single night. Most often, this nonstop irrigation, moving from one field to another, continued for anywhere from eight days to two months, until the planters stopped watering in order to reduce the liquid content of cane stalks and increase the proportion of sucrose in the cane juice. The whole system of irrigation that Ruiz describes was modified in other regions to accommodate local variations in climate, weather conditions, and the quality of the land.[38]

A detailed examination of irrigation practices appears in figure 6.4, which reflects El Puente's records for the fields of San Enrique, San Nicolás, and La Vizcaína for 1883. The data show that the most intensive use of water

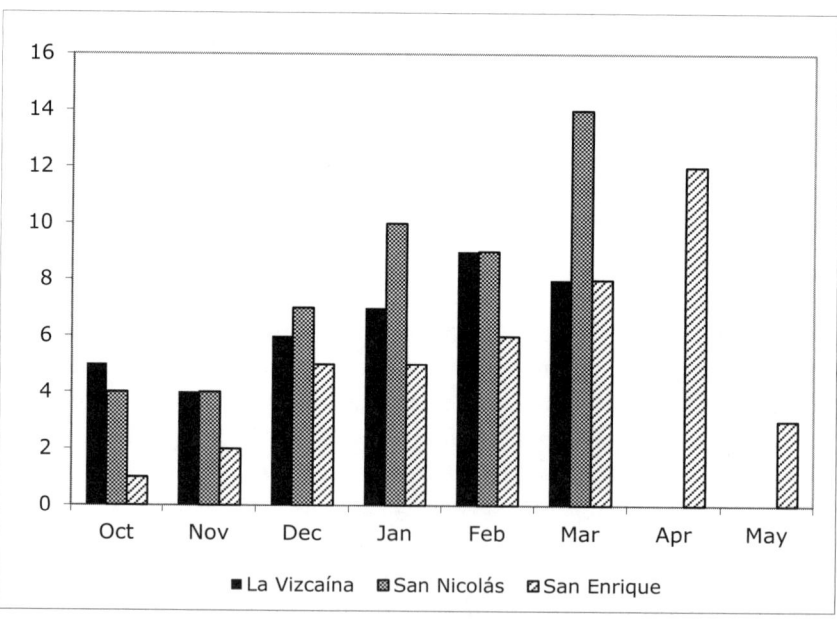

Figure 6.4. Number of days spent irrigating individual fields on Hacienda El Puente, 1883. Data from Ángel Ruiz de Velasco, *Estudios sobre el cultivo de la caña de azúcar* (Cuernavaca: Imprenta del Gobierno del Estado, 1894), 111–15.

occurred during the months of the least rainfall (January through April), both in terms of the number of days spent irrigating and in terms of the type of the irrigation (*por mitad* and *de punta*) carried out. Of a total of thirty-nine days spent irrigating at La Vizcaína, for example, twenty-three (approximately 60 percent) took place in the dry season. The field of San Nicolás was irrigated for a total of forty-nine days, thirty-three (67 percent) of which fell in the dry season, and fully 80 percent (thirty-four of the forty-two total days) of the irrigation in San Enrique took place during the dry season. Clearly, the demand for water in the hacienda of El Puente was concentrated in the driest months. Because most of the water in Morelos's rivers derives from rainfall, haciendas' demand for water clearly peaked at the very time that the rivers were at their smallest volume.

This pattern is further confirmed when we consider the type of irrigation that hacienda owners employed during the dry season. As Ruiz de Velasco pointed out, some forms of irrigation required significantly more water than others: asiento required about 34 cubic meters of water per hectare, while each round of apantle irrigation used about 74 cubic meters per hectare. Irrigation por mitad used 138 cubic meters per hectare, and irrigation de punta demanded about 170 cubic meters. It turns out that the haciendas employed the most water-intensive forms precisely during the dry season. In Santo Tomás, irrigation por mitad began on February 19; the de punta irrigation came in April. In San Enrique, irrigation por mitad began on February 26 and de punta on April 1. In San Nicolás, these two irrigation cycles occurred on February 2 and on February 15, respectively; and in La Vizcaína, on February 3 and on February 12. Thus, El Puente not only irrigated most frequently during the dry season but also practiced the most water-intensive techniques then. It drew the greatest amount of river water during the months with the least rain, when the rivers reached their lowest flows. Both Ramón Portillo and Ángel Ruiz de Velasco attest that this was the general pattern of irrigation throughout Cuautla and Cuernavaca, all of which helps to explain why hacienda owners put so much emphasis on building hydraulic works capable of overcoming water shortages at all times of the year.[39]

The Zacatepec hacienda did not follow this general pattern, thanks in part to the overabundance of water on its land. Ruiz's graduate thesis and the publications of the following twelve years when he served as its administrator provide valuable descriptions about the hacienda's irrigation practices in the late nineteenth and early twentieth centuries.[40] A fervent believer in water conservation, Ruiz analyzed irrigation practices on Zacatepec's eight cane fields and became convinced of the fundamental importance of irrigation in sugar production. After all, between thirteen and sixteen months were required to grow a single crop of sugarcane, whereas the rains did not fall with any regularity except between May and October; therefore, only irrigation could redress the shortfall during the dry season. The data for eight growing seasons suggest that the cane had ripened enough to harvest by sometime around the end of the calendar year. That means that administrators were using de punta irrigation primarily between October and December, while the crop reached maturity. Ruiz always sought to conserve water and typically planted early so that most of the growing season coincided with the rainy season, but even so the hacienda required substantial doses of irrigation because the decisive moment in the growth cycle came during the drier months of late autumn. As a result, sugarcane production in Zacatepec remained entirely dependent on irrigation, which was the only way to provide the quantity of water that the plant required, especially during the dry season from January to April.

A comparison of rain-fed versus irrigated watering of cane fields at Zacatepec and El Puente confirms that irrigation represented the most important source of water in every case except for Zacatepec's San Pedro field, where black clay grounds and flat terrain minimized the need for irrigation and allowed the hacienda to use seasonal rainfall almost exclusively. Everywhere else, irrigation provided at least half of the water used in cane agriculture.

As we have seen, Ruiz tried to convince other haciendas, including El Puente, to conserve water. Almost 90 percent of the water used in El Puente's cane fields of Santo Tomás and San Nicolás came from irrigation (table 6.1). Indeed, El Puente was uncommonly dependent on irrigation, with an average consumption of 17.5 liters per second per hectare, second only to the San Gabriel hacienda (table 6.2). In Ruiz's opinion, such a profligate use of water constituted an enormous waste.

On average, Morelos haciendas used 7.08 liters of water per second to irrigate a single hectare of sugarcane. Although Zacatepec used an above-average quantity, El Puente used more than double the water that most other haciendas did. In any case, practically all the cane-producing haciendas of Morelos relied heavily on irrigation for their economic survival. Ruiz observed that near the end of the Porfiriato, sugar producers collectively used 57,364 liters of water per second to irrigate their lands, which totaled fewer than ten thousand hectares and produced nearly fifty thousand tons of sugar per year.[41]

By 1880, Morelos sugar haciendas such as El Puente had already installed the most advanced milling technology available, so the only way to increase output was to expand the reach of irrigation. But where would the water

Table 6.1 Water used in the haciendas of Zacatepec and El Puente, 1883

Field	Irrigation (mm)	Rainfall (mm)	Percentage of water from irrigation
San Pedro	0.307	0.732	32.71
San Ferrer	0.857	0.876	49.46
San Felipe	0.901	0.735	55.06
Santa Rosa	0.952	0.739	56.27
Santa Gertrudis	1.397	0.781	64.14
San Alejandro (Purísima canal intake)	1.781	0.924	68.37
El Carmen	2.307	1	68.44
San Alejandro (Abillas canal intake)	4.384	0.921	81.84
Santo Tomás (El Puente hacienda)	7.522	1	87.48
San Nicolás (El Puente hacienda)	8.346	1	88.57

Source: Felipe Ruiz de Velasco, *Historia y evoluciones del cultivo de la caña y de la industria azucarera en México hasta el año de 1910* (Mexico City: Editorial Cultura, 1937), 365.

Table 6.2 Water use by Morelos haciendas, 1910 (liters of water per second per hectare)

Hacienda	lps/ha	Hacienda	lps/ha
San Gabriel	19.30	Temilpa	5.37
El Puente	17.50	San Vicente	4.50
Santa Inés	12.84	San Juan	4.13
Cocoyotla	12.62	Hospital	3.47
Couachichinola	11.27	San Carlos	3.32
Cuasasano	11.22	Miacatlan	2.77
Higuerón	11.22	Calderón	2.69
Temisco	11.00	Trenta y Acamilpa	2.47
Zacatepec	10.47	Tenango	2.01
Coauhuxtla	8.41	Oacalco	1.93
Actopam	8.20	Atlihuayan	1.70
Chunameca	7.57	San Nicolás	1.52
San José	7.50	Santa Clara	1.40
Tenextepango	5.89		
Santa Cruz	5.74	Average: 7.08	

Source: Ruiz de Velasco, *Historia y evoluciones*, 364.

come from? Ruiz suggested that hacienda owners could make more effective use of the water they already had by reducing irrigation levels from seven liters per hectare to three. That would allow them to plant cane on another 19,000 hectares and increase production to 262,000 tons of sugar per year. If the landowners did not accept his recommendations for water conservation, Ruiz concluded that the only alternative was to claim the water used by peasant communities and redirect it to sugar haciendas. Such a course of action seemed viable in light of the Díaz administration's decision to offer water concessions to Morelos hacienda owners and allow them access to as much as 79,000 liters per day, enough to cultivate 26,333 hectares and produce 363,213 tons of sugar.[42] The hacienda owners chose the latter course of action. Their endeavor to put greater acreage into production led them to claim more and more water, eventually sparking conflicts over the Higuerón River itself.

The problem was that nearly every perennial source of water had already been claimed by 1910. Hacienda owners therefore concentrated on gaining control of river flows and building hydrological works to domesticate them. This was a difficult proposition because most of Morelos's rivers originate in the mountains and descend quickly into the plantation zone, where the Higuerón River had become a key source of competition and conflict. The Ministry of Development (Fomento) had already granted water rights totaling 22,000 liters per second to private landowners by 1910, even though the river had a capacity of only 13,769 liters per second. In other words, the government had promised landowners 8,231 liters per second more than the river's normal capacity. (Table 6.3 shows these concessions on the eve of the revolution.)

Not every landowner actually exercised his water rights, but even so, the amount of water drawn from the Higuerón sparked social conflicts and competition among concessionaries and demonstrated the extent to which water had become a politicized commodity in Morelos during the waning years of the Porfiriato. Aggravating these tensions was the fact that hacienda owners and rich villagers were the only ones with the means to build irrigation networks capable of capturing river water. As bureaucrats continued to distribute water rights without regard for rivers' actual capacities, new tensions emerged and worried landowners requested still more water rights. In this sense, Morelos exemplifies one of the primary national problems Mexico confronted on the eve of revolution: the dependence on irrigation works to make the commercial land productive. This situation gave rise to a series of social conflicts that traced their origins to the middle of the nineteenth century and helped to trigger the Zapatista revolution in Morelos.

Table 6.3 Higuerón River water rights and flow measurements, 1910–12

Concessions for irrigation	Currently in use (lps)	Currently unused (lps)	Totals (lps)
Pueblo de Tlaltizapán	500		
Sres. Reyna hermanos	3,000		
Sr. Lorenzo Brito	200		
Sr. Valeriano Salceda	3,500		
Sr. León Castresana	500		
Manuel Araoz		3,500	
Eugenio Cañas		3,000	
León Castresana		2,500	
Eugenio Cañas[a]		5,000	
Subtotal	7,700	14,000	21,700
Eugenio Cañas (watermill)		8,000	
All concessions	7,700	22,000	29,700

Flow measurements (date)	Methodology	Totals (lps)
February 17, 1910	Surface floats	18,769
January 25, 1912	Flow meter	13,400
January 26, 1912	Surface floats	16,734
April 2, 1912	Flow meter	14,139
Average flow	Flow meter	13,769
Average flow	All methods	15,761

Source: Domingo Diez, "Observaciones críticas sobre el regadío del Estado de Morelos," presentation to the Escuela Nacional de Ingenieros, 14 May 1919 (Mexico City: Antigua Imprenta de Murguía, 1919), 29.

[a] The Cañas irrigation concession was for Apatlaco River water.

The Politics of Water

Manuel Mendoza Cortina, a rich merchant, owned the hacienda of Coahuistla. According to an 1886 survey, the hacienda encompassed 10,745 hectares divided between cane fields in the low, easily irrigated flatlands, and the pasturelands and rain-fed fields for food crops found on higher, drier land. Coahuistla became one of the top sugar producers in the nineteenth century, a quintessential example of a sugar hacienda that increased its production through the use of agricultural technology and intensive irrigation. In 1881, it produced 200,000 *arrobas* of sugar (about 2.3 million kilograms), more than twice as much as its neighbor, Santa Inés, which produced a mere 90,000 arrobas (or approximately 1 million kilograms).[43] Seven years later, the two haciendas fell into a dispute over the use of water. The owner of Santa Inés, Luis Róvalo, complained that Coahuistla did not legally own the water that it drew from two springs located just inside of Santa Inés's property line. He charged that Mendoza had illegally diverted the water to his own hacienda by destroying the ditches that once directed the water to the Santa Inés cane fields. Róvalo based his claims on the fact that both haciendas had once formed a single property owned by the convent of Santo Domingo and had shared the spring water until the convent divided the land and leased Santa Inés to Pedro del Riesgo in 1785. Róvalo leased the hacienda in 1855, along with its water rights. Liberal reform laws enacted the following year required all religious corporations in Mexico to privatize their property. At that point, he bought the hacienda outright and hired experts to map his new property. The resulting survey by Francisco Berruecos and Pedro Martínez placed the value of the estate at 36,234 pesos and showed that the property included a section known as Barreal, where the springs were located.[44]

Mendoza responded with a lawsuit claiming that the water rights actually belonged to Coahuistla and that until recently he had simply tolerated the neighboring hacienda's use of water from the spring. Now, however, he had decided to revoke this dispensation. He told the judge that he had ordered the destruction of a sluice gate located next to one of the springs so that the water would flow directly to his property and asked the court for an order that recognized the legality of his actions and confirmed his water rights. He assured the court that Coahuistla had always owned the springs in Bárcena and that he had lent the water rights to Santa Inés only because Luis Róvalo and his sons were his friends. Even now, he recognized that his neighbors needed water to raise their cane crop and explained that he had no "objection to letting the hacienda of Santa Inés use excess water, as long as the owners of Coahuistla approve any such use."[45]

Even though Róvalo had shown that Santa Inés had managed the water for the past twenty years and that the springs were on his property, he decided not to fight Mendoza's lawsuit. Apparently, he felt cowed by his powerful neighbor, who was one of the richest hacienda owners in the state and a leading investor in the Morelos Railway. Róvalo knew only too well that political heavyweights had ensured that the railroad would pass through his rival's hacienda, that the president of the Supreme Court paid regular visits to Coahuistla, and that the minister of war arrived by horseback to pay respects to the *hacendado*. Róvalo concluded that he had no alternative but to "look the other way and drop his appeal for justice." He wrote,

"I did not think Morelos should be deprived of the railroad in exchange for giving me back the water I was deprived of. I thought I should make sacrifices.... The dispossession was consummated, legalized through that lawsuit, and I kept quiet."[46] In light of Mendoza's influence, it is hardly surprising to find that the court's decision of May 19, 1879, not only confirmed Mendoza's possession of the springs but also ordered Róvalo to pay court fees.

If Mendoza could use his authority to pressure another landowner into compliance like this, it is easy to imagine how the villagers of San Pedro Apatlaco must have felt when they had run afoul of him a few years earlier.[47] In an 1864 pleading filed with the Supreme Court, the villagers said that they worried they were too weak to fight "such a powerful opponent" but had nonetheless decided to sue Coahuistla for having dispossessed their water.[48] They claimed that Coahuistla's administrator, Gumesindo González, had appropriated their village spring and mistreated them with the "wicked intention of wearing us down to the point that we abandon [our land] voluntarily."[49] The Apatlaco villagers complained they had enjoyed access to the water from time immemorial but that the administrator had cut off their access a few years earlier, then seized 108 of their cattle and herded them into a pen, killing 5. He refused to return the livestock until the townspeople paid a fine of one peso per head.[50]

This episode strengthened the image of Mendoza Cortina as a powerful, nearly feudal landowner whose command of the land, the knife, and the gallows allowed him to decide the fate of entire populations. According to the villagers, Mendoza regarded the people of Apatlaco as "less than beasts," who were born "to suffer and keep quiet."[51] For his own part, Mendoza argued that he rightfully owned the water and that if his administrator had appropriated the spring, "he has done well, and that no one, other than me, has the authority to make claims on it."[52] The villagers responded that they had always used the water with the express permission of the previous landowner, which constituted a de facto legal obligation. "If the owner of a property has assumed an obligation by an express or tacit agreement, he cannot renounce it no matter how much he defends his property, because, in a manner of speaking, he has relinquished his rights."[53] In the opinion of the Apatlaco villagers, their use of the spring water to irrigate their fields since time immemorial had placed an obligation on the hacienda that even Mendoza needed to respect.

The problem according to Mendoza was that he could not recognize Apatlaco's rights because the community had never established its legal status before the court. Was it a village with a commons? A town? An unincorporated settlement (*congregación*)? Or merely a group of individuals taking legal action on their own account? Arguments of this sort were often used by hacienda owners after the disentailment law of 1856 (which called for the privatization of all collective lands) to challenge the legal status of pueblos. From a strictly juridical standpoint, hacienda owners often suggested that the law had deprived the communities of their legal status as colonial-era "corporations" and reduced them to a mere collection of individuals with no legal standing before the court. The villagers of Apatlaco responded that the hacienda owner intended to absorb their lands, but he could not simply wave his hand and make the thick walls of their church disappear: in their opinion, those walls testified to the pueblo's ancient heritage and collective identity. By the same token, they charged that the hacienda had no right to expropriate the land and water upon which the community had depended for so many generations.[54] Finally, they pointed out that they had contributed to the hacienda's wealth with their own sweat: "[O]ur ancestors and we ourselves have worked for the hacienda and helped to make profits of at least ten thousand pesos a year; or more than a million over the course of a century. Should we who have helped that enterprise earn a million pesos be repaid with such ingratitude?"[55] If they were no longer a legal entity because their lands had been privatized with the stroke of a pen, it was nonetheless obvious to them that their community still had its own authorities, still owned property extending eight leagues in circumference that formed "the village commons," and hence still had a sufficiently clear legal character to allow the trial to proceed.[56] Unfortunately, the documents are silent about its final outcome.

Another episode related to the haciendas of San Vicente, Chiconcuac, and Dolores also illustrates the pivotal role that water played in Morelos's sugar economy. The three haciendas formed a single agro-industrial complex that had expanded to more than 6,416 hectares by the end of the nineteenth century, yet only 735 were irrigated.[57] Nevertheless, the fact that a mere 11 percent of the hacienda lands had access to irrigation suggests that water concessions were not uniformly large during the Porfiriato. Indeed, Dolores was the only one with access to a reliable source of water in the mid-nineteenth century. Dolores lacked good agricultural lands or its own sugar mill, and indeed lacked any "importance as a sugar cane producing hacienda," according to its administrator, but did have "a good quantity of its own water, more than enough to be profitably used on the lands of San Vicente."[58] In other words, Dolores's primary value resided more in its access to water.[59] Even then, irrigation problems proved so intractable that Miguel Ajuría, an aspiring cane planter who intended to buy the properties, committed suicide when he discovered that he had vastly overestimated the amount of water he could use.[60]

Ajuría had rented the haciendas and planted a great quantity of cane in anticipation of raising the funds to complete his purchase, but he soon recognized the dimensions of his miscalculation. The greatest problems occurred in La Vega cane field, which in May 1852 suffered from a complete "lack of irrigation." The fields known as Barba and Tomatal were scarcely any better.[61] All told, around four thousand plots (*labranzas*) lacked adequate irrigation. According to the hacienda administrator, this situation was not the result of some "mistake in the management" of the haciendas' water resources but rather reflected their overall "lack of water." He explained that he had begged the owner of the San Gaspar hacienda "for the love of God to let some more water pass [downstream], because the haciendas' cane was drying up."[62] Several witnesses confirmed this conclusion. Ignacio Silva, the mayor of Cuernavaca, traveled to the field of La Vega on the Dolores hacienda and reported that "the cane in the labranzas was dry and incapable of producing any fruit." Manuel Luna, mayor of the town of Sochitepec, agreed and wrote on June 4, 1852, that "the field of la Vega was completely dry, that it could not bear any fruit."[63] The mayor affirmed that Ajuría had withdrawn the water from the cane field, which resulted in a loss of 4,000 pesos.

The sales contract for the three haciendas had been signed on October 30, 1851. Ajuría had made a down payment of 50,000 pesos cash and owed another 262,292 pesos to several creditors, whom he promised to reimburse in annual installments. However, the precarious situation of the haciendas left Ajuría unable to repay his debts. The previous owner claimed that this situation abrogated the sales agreement and demanded the return of the haciendas and payment of arrears owed to the creditors. The two parties agreed that the sale should not go through and on November 12, 1852, signed a document before a notary public that formally broke the contract. Faced with the prospect of losing his haciendas, Ajuría took his own life. At that point, the previous owner recovered his properties and entered into a new sales agreement with don Pío Bermejillo, a well-known Spanish financier who had immigrated to Mexico.

Bermejillo, in turn, sold the properties to the Béistegui family. After her husband died, Béistegui's widow, Dolores Arriaga, remarried and yielded administration of the hacienda to her new husband, Marquis Jorge Carmona. Although the three haciendas constituted one of the most beautiful properties in the state, the couple preferred the comfort of their house on Avenue Hoche in Paris to the verdant Morelos tropics.[64] Carmona did succeed in resolving the haciendas' lingering water problems, however. He took advantage of a state concession to draw 1,836 liters per second and used it to turn his haciendas into one of the most important enterprises in the state.⁶⁵ With its water supply ensured, the hacienda finally broke the vicious cycle of bankruptcies and became a solid business that allowed Carmona and Arriaga to live many years in Parisian luxury.

Conclusion

The most detailed diagnosis of the problems facing Mexico in 1910 came from the pen of Andrés Molina Enríquez. He is best known for his critique of the hacienda, which he portrayed as the linchpin of an inefficient agricultural system that not only made inadequate use of the land but also produced a class of great landowners who, in his view, cared more about controlling their workforce and gaining possession of ever greater expanses of property than about implementing modern agricultural techniques or intensifying their use of the land. One aspect of this pattern of social domination consisted in hacienda owners' appropriation of natural resources. Molina considered land and water the foremost among these resources, echoing the ideas of Raoul Bigot and Luis Lejeune discussed above.

However, this chapter has shown that the main problem in Mexico was not a shortage of land or water as such, but rather the absence of an irrigation system capable of letting landowners use the land to its fullest potential. The example of Morelos speaks eloquently to this issue. As a small state well-endowed with water sources, Morelos could not increase the extent of its cultivated land without getting water where it was most needed. Landowners had already developed commercial agriculture during the colonial era and expanded a hacienda-based sugar industry during the Porfiriato to such an extent that the productivity of Morelos sugar haciendas was superseded only by those of Hawaii and Puerto Rico.

Sugar haciendas achieved these results even though they planted cane on a relatively small proportion of their total landholdings in 1910, leaving the majority of their productive land for food crops and fallows. Nearly 90 percent of their property was used for rain-fed agriculture (*temporal*) or was held in reserve or was forested, mountainous, or otherwise unusable for agriculture. Of all these uses, cane was the crop most sensitive to water, which could be delivered only through irrigation in the dry season. The cycle of cultivation extended for between thirteen and sixteen months, while rainy season in Morelos lasted a mere five months, from May to September. Yet cane cultivation depended on increased flows of water during autumn, not only because the fields needed more frequent irrigation at that point but also because more water was required to carry out por mitad and de punta irrigation. Hacienda owners responded by building irrigation networks meant to ensure a reliable supply of water during this critical cycle of sugarcane production.

However, a long historiographic tradition has suggested that peasants in prerevolutionary Morelos had been dispossessed of their lands, which in turn led them to become some of the most tenacious fighters of the Mexican Revolution. Many historians assume that the Zapatista slogan of "land and liberty" expresses the basic elements of the agrarian problem in Morelos. My analysis suggests that access to water was also a significant cause of the agrarian uprising in Morelos, insofar as it represented the primary impediment to the development of cane cultivation there, and hacienda owners responded accordingly. Above all, I propose that access to land did not represent a problem for hacienda productivity, but rather water—specifically, irrigation—did. That explains why Felipe Ruiz de Velasco kicked off his campaign for federal congressman from the Jujutla and Tetecala districts with a speech that highlighted his thirty years of experience as an agronomist, which had taught him that the only way to increase the wealth of Morelos was by constructing new irrigation projects.[66] His observations were confirmed by a cadastral survey conducted in 1903 by Rómulo Escobar that showed Morelos's irrigated land to have a higher value than property anywhere else in the country, including the federal district and the state of Mexico.[67] In this sense, the agrarian thesis for the outbreak of Zapatismo in 1910 represents a misreading of contemporary analysts' judgments about the problem. If we look back to the discussions by Ruiz or Domingo Diez, we find that they considered water as much of an issue as land tenure, yet the historiography of the revolution has emphasized the struggle for land as the predominant issue confronting rural Morelos.

The hydraulic thesis fits the case of Morelos better than the agrarian thesis, and we cannot disassociate the problem of land from the issue of water in Porfirian Mexico. Although rural communities certainly did lose their lands to expanding haciendas in the late nineteenth century, haciendas owned more land than they could use. Why, then, did haciendas appropriate marginal lands from villagers?[68] It seems to make no sense. On the contrary, haciendas needed the food that sharecroppers and *temporaleros* produced on their own lands, not to mention the labor provided by the people of the pueblos. A closer examination of the situation shows that most haciendas did not blindly despoil the rural population; what they truly sought was greater control over water resources that could be directed to sugar fields. This was, in my analysis, the primary challenge facing the Porfirian hacienda owners, all the more so because water rights to rivers had already been completely distributed among several claimants, nearly all of them landowners. Rather than interpreting the disappearance of pueblos such as Acatlipa, Cuachichinola, Sayula, San Pedro, Tequesquitengo, or Ahuehuepan as evidence that haciendas sought to gobble up the communities around them, historians would do better to look at other factors such as the use of the Higuerón, whose waters federal functionaries distributed to landowners in such volumes that they had promised more water than the river itself contained by the time the Porfiriato drew to a close.[69] By this point, landowners, bureaucrats, and villagers all recognized that the haciendas of Morelos required a reliable source of water if they hoped to further expand production. Today, we need to take the resulting resource conflicts fully into account if we hope to understand the roots of social conflict in Morelos.

Notes

1. Louis Lejeune, *Au Mexique* (Paris: Librairie Léopold Cerf, 1892), 160.

2. Horacio Crespo, "La hacienda azucarera del Estado de Morelos: modernización y conflicto," PhD diss., Facultad de Filosofía y Letras, Universidad Nacional Autónoma de México, 1996, p. 188 (emphasis added).

3. Raoul Bigot, *Notes économiques sur le Mexique* (Paris: Boyeua and Chevillet, 1907), 29.

4. George McCutchen McBride, "Los sistemas de propiedad rural en México," *Problemas Agrícolas e Industriales de México* 3, no. 3 (1951): 24. In 1950, another source estimates that the total land under cultivation measured 10.8 million hectares. See Luis Aboites, *El agua de la nación: Una historia política de México, 1888–1946* (Mexico City: CIESAS, 1998), 119.

5. Andrés Molina Enríquez, *Los grandes problemas nacionales* (1909; reprint, Mexico City: Era, 1979), 271.

6. The area with irrigation measured 3,500 hectares in 1869, while in 1909 it approached 10,000. Felipe Ruiz de Velasco, *Historia y evoluciones del cultivo de la caña y de la industria azucarera en México hasta el año 1910* (Mexico City: Editorial Cultura, 1937), 272.

7. Lejeune, *Au Mexique*; Bigot, *Notes économiques*; Domingo Diez, "El cultivo e industria de la caña de azúcar," in *Memorias de la Asociación de Ingenieros y Arquitectos de México* (Mexico City: Imprenta Victoria, 1918); Ruiz de Velasco, *Historia y evoluciones*.

8. Gildardo Magaña, *Emiliano Zapata y el agrarismo en México* (Mexico City: Comisión para la Conmemoración del Centenario del Natalicio del General E. Zapata, 1979); Jesus Sotelo Inclán, *Raíz y razón de Zapata* (Mexico City: Secretaría de Educación Pública, 1981); Eric Wolf, *Peasant Wars of the Twentieth Century* (New York: Harper and Row, 1969); John Womack Jr., *Zapata and the Mexican Revolution* (New York: Knopf, 1968); and Guillermo de la Peña, *A Legacy of Promises: Agriculture, Politics and Ritual in the Morelos Highlands of Mexico* (Austin: University of Texas Press,

1981). For a recent revision to this "*agrarista* thesis" suggesting that the modernization of sugar haciendas ushered in new and harsher labor relations that touched off the revolution, see Paul Hart, *Bitter Harvest: The Social Transformation of Morelos, Mexico, and the Origins of the Zapatista Revolution, 1840–1910* (Albuquerque: University of New Mexico Press, 2005).

9. Domingo Diez, *Bosquejo histórico geográfico de Morelos* (1933; reprint, Cuernavaca: Editorial Tlahuica, 1967), 130.

10. Eduardo de Nájera Informes al Ministerio de Fomento, Archivo General de la Nación (Mexico City), Ramo Nacional Financiera (hereafter AGN-Nafinsa), exp. 21, fols. 251–54; Alejandro Tortolero Villaseñor, *De la coa a la máquina de vapor: Actividad agrícola e innovación tecnológica en las haciendas de la región central de México, 1880–1914* (Mexico City: Siglo XXI, 1995), 339–43.

11. See Tortolero Villaseñor, *De la coa*, 328–53. Also, the Hospital Hacienda irrigated 904 of its 11,859 hectares (7.6 percent), and Chinameca, 520 of 2,018 (25.7 percent); AGN-Nafinsa, exp. 29, fols. 1–38.

12. Manuel Araoz, "Informe sobre las haciendas de caña de azúcar del Estado de Morelos" (Mexico City: Fondo I. Noriega, 1914), held at the Nettie Lee Benson Rare Books and Manuscript Department, University of Texas Libraries, Austin, Texas.

13. Domingo Diez, "Observaciones críticas sobre el regadío del Estado de Morelos," presentation to the Escuela Nacional de Ingenieros, 14 May 1919 (Mexico City: Antigua Imprenta de Murguía, 1919), 25.

14. Diez, "El cultivo," 57.

15. Ibid., 59.

16. Ibid., 62.

17. Diez, "Observaciones críticas," 25.

18. In 1885, Ruiz de Velasco published *Breve relación sobre el drenaje según se practica en la Hacienda de Zacatepec ubicada en el Estado de Morelos* (Mexico City: Imprenta de la Secretaría de Fomento, 1885). In 1910, he released a self-published work, *Las aguas no son denunciables dentro de los límites de la propiedad privada*. Finally, in 1937, he published *Historia y evoluciones*, his most important work.

19. For more on the Escuela Nacional de Agricultura, see Tortolero Villaseñor, *De la coa*; on the Netherlands, see Salvatore Ciriacono, *Acque e agricultura: Venecia, l'Olanda e la bonifica europea in èta moderna* (Milan: F. Angeli Ciriacono, 1994).

20. Ruiz de Velasco, *Historia y evoluciones*, 381.

21. Felipe Ruiz de Velasco, "Bosques y manantiales del Estado de Morelos y apéndice sintético sobre su potencialidad agrícola e industrial," in *Memorias de la Sociedad Alzate (Cuadernos Históricos Morelenses)* (1925; facsimile ed., Cuernavaca: Fuentes Documentales del Estado de Morelos, 2000), 13.

22. "Diario de Felipe Ruiz de Velasco y Leyva," transcribed by Roberto Burnett, 2007, in Felipe Ruiz de Velazco private archive, Cuernavaca, Morelos, 1:49.

23. See Ruiz de Velasco, *Historia y evoluciones*, chap. 5, "Drenaje o avenamiento de terrenos húmedos y depuración de los salinos."

24. Magaña, *Emiliano Zapata*; Sotelo Inclán, *Raíz y razón*; Womack, *Zapata*; Roberto Melville, *Crecimiento y rebelión: El desarrollo económico de las haciendas azucareras en Morelos (1880–1910)* (Mexico City: Nueva Imagen, 1979); de la Peña, *Legacy of Promises*; William Bluestein, "The Class Relations of the Hacienda and the Village in Prerevolutionary Morelos," *Latin American Perspectives* 9, no. 3 (1982): 12–28. One

exception to the nearly universal support for the agrarian thesis that considers the aquatic limits to growth is Arturo Warman, *Y venimos a contradecir: Los campesinos de Morelos y el Estado nacional* (Mexico City: Ediciones de la Casa Chata, 1976).

25. Magaña, *Emiliano Zapata*, 70.
26. Womack, *Zapata*, 45–46.
27. Ibid., 45; see also Bluestein, "Class Relations," 26–27.
28. De la Peña, *Legacy of Promises*, 54.
29. Crespo, "La hacienda azucarera," 189.
30. Diez, "Observaciones críticas," 30.
31. Diez, *Bosquejo histórico geográfico*, ccxxi.
32. Tortolero Villaseñor, *De la coa*, 275.
33. Womack, *Zapata*, 48; Warman, *Y venimos a contradecir*, 58.
34. Ruiz de Velasco, *Historia y evoluciones*, 338.
35. Diez, *Bosquejo histórico geográfico*, 122. Even as late as 1897, only 3 percent of the lands of the hacienda were used for cane cultivation. Domenico Síndico, "Santa Ana Tenango: A Morelos Sugar Hacienda," PhD diss., University of Minnesota, 1980, pp. 94 and 101. For more details on irrigated cane cultivation, see Araoz, "Informe sobre las haciendas."
36. Womack, *Zapata*, 48.
37. The vara is equal to 0.83 meters.
38. Ángel Ruiz de Velasco, *Estudios sobre el cultivo de la caña de azúcar* (Cuernavaca: Imprenta del Gobierno del Estado, 1894), 41–42.
39. Ramón Portillo, *Ideas generales sobre el cultivo de la caña de azúcar en el Estado de Morelos* (Mexico City: Oficina Tipográfica de la Secretaría de Fomento, 1885); Ruiz de Velasco, *Estudios sobre el cultivo*.
40. I would like to thank Robert Burnett for transcribing the diaries and other personal documents in Felipe Ruiz de Velasco's personal archive and for kindly making them available to me.
41. Ruiz de Velasco, "Bosques y manantiales."
42. Ibid., 257.
43. Tortolero Villaseñor, *De la coa*, 345.
44. *Breves apuntes acerca de las diferencias ocurridas respecto de aguas entre las haciendas de Coahuistla y Guadalupe anexa a Santa Inés* (Mexico City: Tipografía de Aguilar e Hijos, 1888), 10–11.
45. Ibid., 18.
46. *Breves apuntes*, 2.
47. Archivo Histórico del Tribunal Superior de Justicia del Distrito Federal (Mexico City), Hac. (17) (14) Coahuistla.
48. *Alegato de buena prueba presentado por los vecinos del Pueblo de San Pedro Apatlaco ante el Juzgado de Morelos en el juicio posesorio que siguen contra la hacienda de Coahuistla, bajo la Dirección del Lic. Anastasio Zerecero* (Mexico City: Imprenta de Andrade y Escalante, 1864), 5.
49. Ibid., 5.
50. Ibid., 10.
51. Ibid., 9.
52. Ibid., 11.
53. Ibid., 12.

54. Ibid., 15.
55. Ibid., 16.
56. The records do not tell us the outcome of the trial, but the *hacendado* most likely would have won, thanks to his enormous political power; indeed, his influence was such that he continued to throw his dependents in the stocks long after this practice had been prohibited by law. Ibid., 8.
57. Tortolero Villaseñor, *De la coa*, 339.
58. *Apuntamientos que presentó a la exma. tercera sala del supremo tribunal de justicia de la nación: El Lic. d. José Fernando Ramírez amplificando los fundamentos de hecho y de derecho verbalmente en sus estrados por la testamentaría de d. Miguel Ajuría con d. Anacleto Polidura sobre la legalidad de la entrega de las haciendas denominadas San Vicente, Chiconcuac y Dolores* (Mexico City: Imprenta de Andrade y Escalante, 1861), 9.
59. Ibid., 46.
60. *Informe en derecho pronunciado en los estrados de la exma. tercera sala del suprema tribunal de justicia de la nación, por el Licenciado Juan B. Alamán por parte de d. Anacleto Polidura* . . . (Mexico City: Imprenta de J. M. Lara, 1860), 114.
61. *Apuntamientos*, 23.
62. Ibid., 31–32.
63. *Informe en derecho*, 52.
64. Clémont Bertie-Mariott, *Un parisien au Mexique* (Paris: E. Dentu, 1886), 149.
65. Tortolero Villaseñor, *De la coa*, 339.
66. "Diario de Felipe Ruiz de Velasco," 10.
67. Rómulo Escobar, "Valor de los terrenos en México," *El Agricultor Mexicano* 16, no. 4 (1903): 70–73.
68. In another study I have shown that hacendados did not liquidate the pueblos of Morelos through the politics of disentailment; indeed, villagers transformed themselves into the beneficiaries of the disentailment. See Alejandro Tortolero Villaseñor, *Notarios y agricultores: Crecimiento y atraso en el campo mexicano, 1780–1920* (Mexico City: Siglo XXI).
69. Only two of these towns can be said actually to have disappeared: Acatlipa and San Pedro. The others either changed their category or become part of the urban zone of Cuernavaca rather than having been destroyed by haciendas. See Crespo, "La hacienda azucarera," 70.

CHAPTER SEVEN

King Henequen
Order, Progress, and Ecological Change in Yucatán, 1850–1950

Sterling Evans

A popular expression in Mexico's Yucatán Peninsula suggests that Yucatán is "the motherland" of fiber. Cordage fiber from the region's native henequen and sisal plants forever changed the history of Yucatán when North American cordage companies discovered that henequen was the best commodity for producing binder twine. The development in the 1870s of the power reaper/binder (or simply, binder), an implement that cut grain stalks and tied them with twine into sheaves to await threshing, allowed farmers to save considerable time and labor in harvesting grain crops. Binders were *the* harvesting implement of choice for North American grain farmers from roughly 1880 to 1950, before combine harvesters (which cut and threshed grain without the need to tie the stalks into sheaves) became affordable. Thus, in the late nineteenth and early twentieth centuries, henequen became king in Yucatán, forever changing the state's economy, society, and landscape. With those changes emerged a remarkable double dependency between the fiber producers in Yucatán and grain producers in the United States and Canada, which I refer to as the henequen-wheat complex. While I have told this more complete story elsewhere, the focus of this chapter is on Yucatán's natural environment, how people understood the landscape, and the ecological changes that Yucatán endured during its henequen boom years.[1] The state's fiber industry came at the cost of deforestation as the woods were cleared to make way for henequen fields and to provide fuel for rasping machines and rail transportation. It also meant the creation of a monocrop plantation agriscape in lieu of the generations-old mixed forests and small corn plots. These developments poisoned many water sources, exhausted the soil, and reduced wildlife habitat.

The popularity of twine binders created an unprecedented demand for henequen and sisal fiber. By 1900 more than 85 percent of all binder twine in North America was manufactured with the Yucatecan fibers.[2] For Yucatán, the demand made henequen king: the number-one cash crop for not merely the peninsula but all of Mexico. But only in Yucatán were geographic conditions optimal for henequen plantations, a surprising fact given that the peninsula is situated on a semiarid limestone karst that makes the land nearly uncultivable. Karst topography is characterized by layers of soluble bedrock, creating a flat but rocky soil surface with occasional sinkholes (cenotes). One study described the area as a "hostile environment" with "inherent limitations," a place lacking in "good soil and surface water" and "covered with a thin layer of soil and a dry, rugged scrub forest" with "no mountains and few hills to deflect the unbearable heat."[3] The peninsula's climatic conditions were key: Yucatán falls in the hot/subhumid category for half the year and the arid/semiarid one with little rainfall for the other half. The rocky, limestone ground prevents the use of plows to break the land, and rain on the peninsula is unpredictable in the dry season—a condition made worse by the soil, which does not hold moisture. When it does rain, the water filters through the thin topsoil and calcareous earth to the water table below. Irrigation is not really an option here, as the soil porosity prohibits the use of ditches to divert water to fields. Rain comes in downpours during the summer, leaching the soil of important nutrients. In most of the plantation zone, industrialized henequen monoculture, as opposed to henequen growing naturally, exhausts the soil over time, with some areas needing to lie fallow for twenty years before being replanted.[4]

The peninsula's original cover was part of the larger dry tropical deciduous forest, or *bosque seco*, that once extended throughout much of Central America and Mexico. Forest ecologists refer to the ecosystem as the "seasonally dry tropical forest"; it has a "smaller stature and lower basal area than tropical rain forests," and "thorny species are often prominent." Seasonally dry tropical forests occur where there is less than 1,600 millimeters (62 inches) of rainfall a year, most of it during the wet season, and during the dry season the vegetation is mostly deciduous, causing a buildup of leaf matter on the forest floor.[5] In Mexico, the bosque seco environment historically has been one of the most heavily used and disturbed by human activity and therefore is one of the most threatened ecosystems in the country, with only 27 percent of the original forest remaining.[6]

This forbidding land and its dense, scrubby vegetation confounded early Spanish colonizers, with one bemoaning that Yucatán was "the country with the least earth that I have ever seen, since all of it is one living rock," and another decrying its "deficiency in mineral endowments." Friar Diego de Landa, a Catholic missionary to the region in the mid-sixteenth century, described Yucatán as "very flat and clear of mountains, so that it is not seen from ships until they come very close." He wrote that "this land is very hot and the sun burns fiercely. . . . [It] is a land of less soil than any I know." But, he explained, "The country is excellent for lime . . . ; it is a marvel how much fertility exists in the soil on or between the stones."[7] Centuries later, this perception persisted. A Yucatecan once mentioned to a US reporter in the 1930s that he had always believed that "when God made Yucatán His original purpose was to use it for hell. He neglected to give it any water or any soil." Those geographic conditions, combined with how it juts out between the Gulf of Mexico and the Caribbean Sea, isolated the peninsula from the rest of New Spain during the colonial era and from the rest of Mexico after independence. The isolation resulted in a distinctly autonomous conditioning of its people or, as one historian described it, "apart and different in both sentiment and jurisdiction."[8] As Yucatecans are fond of saying, "Es país que no se parece a otro" (It is a land like no other).

Yet despite this geography, and in defiance of environmental determinism (which suggests that an area's natural topography will determine its economic potential), the rocky soil and hot, humid, but only seasonally rainy weather with an average temperature of 25°C (77°F) created the ideal ecological conditions for native henequen and sisal. Porous limestone and dolomite foundations hundreds of feet thick allow water to percolate into the calcareous earth and form subterranean channels that drain into Yucatán's famous cenotes. Soil scientists maintain that soils comprising a proper balance of calcareous earth are unusually rich in their natural state, rendering chemical fertilizers unnecessary. Yucatán's soil nutrients originated with the remains of marine organisms, especially mollusks and corals, deposited by prehistoric oceans that covered the region millennia ago.[9] And the north-central part of Yucatán was best of all, with just the right amount of rainfall and the right kind of decomposed limestone, loam, and rocky soil for agaves to thrive; without it, the plants would absorb too much water, making the leaves pulpy and decreasing the quality of the fiber. Thus, the "bad" soil for other crops in Yucatán was perfect for henequen. The region became the *zona henequenera*, 15,000 square kilometers (8,500 square miles) covering roughly one third of the state (map 7.1). In this zone, growers produced 97 percent of the region's twine fiber, with 97 percent of the zone dedicated to henequen production.[10]

King Henequen in Yucatán · 153

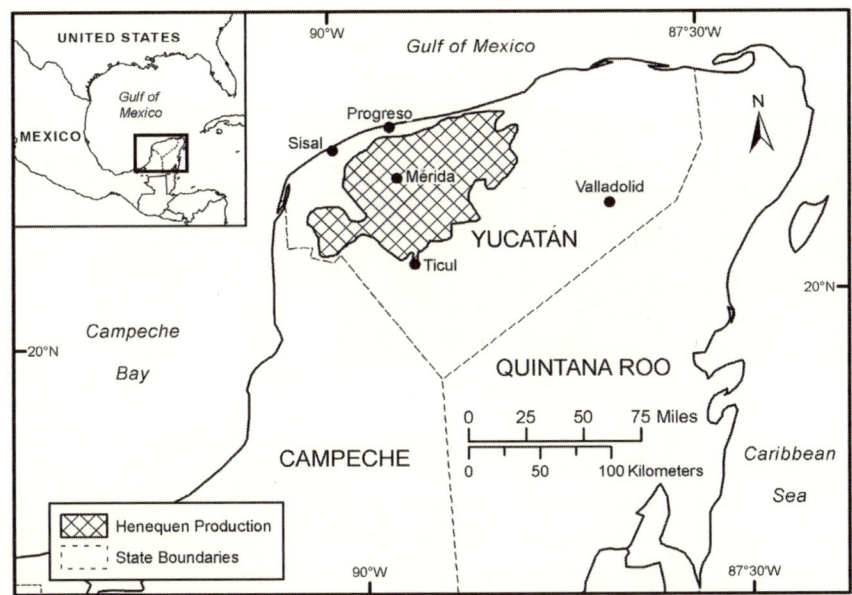

Map 7.1. Map of henequen zone, Yucatán Peninsula.

The Noble Plant

The two fibrous plants that made this area famous are succulents found in the family Agavaceae: henequen (*Agave fourcroydes*) and sisal (*Agave sisalana*), which are related to other Mexican agaves, such as yuccas (*Yucca* spp.) and the century plant, or maguey (*Agave americana*), from which brewers derive the fermented drinks pulque and tequila. Maguey, from Jalisco, is not native to Yucatán and does not thrive there. Henequen and sisal are different species but similar plants. In the fiber industry, however, the generic word *sisal* (named after the small port town of Sisal on Yucatán's west coast) came to refer to the fiber attained from both plants. In Mexico, *el henequén* came to be the generic term used in the industry. Both succulents take seven years to mature and can live for as long as fifteen years, surviving in the peninsula's unique environmental conditions by storing liquid in their long leaves, called *pencas* (figure 7.1). Henequen growers can usually get three cuttings of the spiny, spear-shaped pencas a year before they need to replant the field. Once the henequen is planted, there is little need for weeding, because the plants do not have much competition from other species. Workers harvested the leaves with machetes and then

Figure 7.1. Henequen growing near San Antonio Tehuitz, Yucatán. Photograph by author.

transported them to a rasping shed where a decorticator machine macerated the fibrous pulp from the long leaves. They strung the fiber on wires and left it to dry in the open sun. Once it was ready, they gathered the fiber into large bales, which were hauled to port to be shipped to the cordage manufacturers elsewhere. Every pound of dried fiber yielded up to five hundred feet of twine.[11]

Henequen and sisal were the perfect plantation species for Yucatán, as the Italian journalist Carlo de Fornaro noted in 1915: "Nature through centuries of selection made henequen the ideal plant impervious to inclemencies, drought, to grasshoppers and other destructive pests. . . . The remarkable plant needs practically no cultivating, no irrigation. . . . The Yucatecans call it the noblest plant on the continent."[12] So noble was it that Yucatecan writer Manuel Escoffié called henequen the state's "crown of thorns . . . the crown that brought greatness to Yucatán." Further, the wealth was made possible without the favorable rivers, fertile agricultural lands, or valuable mines "with which the rest of Mexico was gifted." And the terminology is fitting because the word *agave*—the genus named by the Swedish taxonomist Carolus Linnaeus in 1753—is from the Greek word for "noble" or "magnificent." According to botanist Roger Orellana, the name "precisely" fits the agaves "for the durability of their leaves and the great resistance with which they survive adverse conditions and solar radiation."[13]

Because agave plants are native to the region, they occupy an important place in the folklore of Yucatec Mayas—a distinct people from other Mayan groups such as the Lacandón of Chiapas and Cachiquel of Guatemala. They tell the story of a henequen penca that cut the mythological sage Zamná, at which point his servant severed the offending blade and beat it furiously, exposing its fibrous material. Zamná then declared, "Life was born in the company of pain; [and] . . . through the wound was revealed a plant of great usefulness for the people." The ancient Maya discovered a multitude of uses for it. They made ropes, hammocks, sandals, bags, baskets, roofing thatch, hunting bows and traps, fishing nets, needles from its spines, and many other useful household items. They used the plants for decoration, created musical instruments from their long leaves, fermented the juices to make wine, and made cloth ties to bind prisoners of war. Certainly, as historian Renán Irigoyen has argued, the plant and its fibers were tightly interwoven into the economies and the lives of the Maya.[14]

The Spaniards who invaded Yucatán in the sixteenth century observed these indigenous uses of henequen but did not adopt it for much more than naval rigging. And while Friar de Landa noted there "were infinite things that could be made" from henequen, there was no real colonial effort to mass-produce it for commercial or local uses, partly because few Spaniards ever settled there.[15] They found the land unsuitable for European crops and discovered no mineral deposits, although some began cattle operations to export beef and hides to Cuba and cleared small maize plots (milpas) for human and cattle consumption. It was many years later that the colonists' descendants, the Mexicans, borrowed cultivation techniques from Maya people and began growing henequen on small plantations.[16]

By the end of the eighteenth century, the Yucatecans had started to expand production of henequen, primarily for rope, gunnysacks, and hammocks. After Mexico's independence in 1821, large-scale henequen production grew, especially because of an expanding US market for naval cordage. At that point, Mexicans worked to bring more land into production, hiring overseers and local workers, usually Maya Indians working in debt peonage.[17] Recognizing the economic benefits, state officials in 1828 issued a decree requiring every citizen to raise at least ten henequen plants a year in their patios, and they authorized local municipalities to convert community lands to henequen fields (*henequenales*). Two years later, the initiative was so successful that local groups organized a henequen development company, which acquired the Chacsinkín hacienda near the capital, Mérida; sowed nearly 21,000 plants; and thus became the state's first henequen plantation.[18]

The industry grew rapidly in the following decade, with henequen exports surpassing beef and hides, and the trade increased even more from 1830 to 1847. But henequen increases stalled in the mid-nineteenth century when the bloody Caste War raged in Yucatán, started in part by the very increase in henequen production. The Caste War (1847–53) developed in response to the confiscation of Maya lands, impressment into hard labor on sugar plantations, the oppressive control of the Catholic Church, and a racist hierarchical social system that exacerbated the suffering of indigenous people in the region. The ongoing colonialist pattern of land confiscation, ever increasing in this time period because of the expanding production of henequen and Yucatán's entrance into an agrarian capitalist economy, was one of the principal causes of the conflict.[19] The prolonged violence resulted in a demographic collapse in which between 25 and 50 percent of Yucatán's population died because of war, disease, and starvation. The Caste War also effectively ended Yucatán's sugar industry and bankrupted the state treasury, causing local leaders and landowners to seek other avenues to rebuild the state's economy. Those avenues led straight to henequen, which then became "the center of Yucatán's economic life from 1860 to 1940."[20]

There was great hope that henequen would breathe new life into Yucatán's economy. The Mérida newspaper *La Razón del Pueblo* asserted that the plant "will be known as the base of wealth and for public prosperity" and that economic development seemed "to be destined by Providence." Production and exports soon validated this prophecy. In 1873 the state exported 31,000 bales of fiber, but by 1880 that figure had more than tripled, to 113,000 bales.[21] Thus began the henequen boom, which forever changed Yucatán's environment, economy, and society. The golden age of henequen production lasted from 1880 to 1915, a period in which the exponential increase in henequen production coincided with the growing popularity of binders in the United States and Canada. By 1884 exports had grown to 261,000 bales; by 1904, to 606,000 bales; and by 1915—Yucatán's peak production year—to 1.2 million bales. The scale of land conversion matched export production in that more than 70 percent of all cultivable land in Yucatán was devoted to fiber monocropping.[22]

Thus Yucatán entered the realm of other plantation societies across time. The terms *hacienda* and *plantation* both apply to henequen production because each refers to large-scale agriculture, although Mexicans prefer *hacienda*. In English, *plantation* generally refers to large-scale cropping worked by an imported labor force, with a system of authority and force used for the production of commodities to be sold on world markets—a scenario that defined Yucatán.[23] But unlike other plantations in Mexico and Latin America where large agricultural estates were operated by corporations or wealthy, often foreign, landowners, Yucatecan haciendas were family owned but infused with large stocks of foreign capital. And henequen was native to the peninsula, unlike introduced plantation exotics such as sugar, coffee, and bananas. Also unlike those commodities, henequen was not a luxury product; it was the raw material of an industrial product that became a necessity for harvesting grain. Rarely have plantation economies given rise to a significant manufacturing industry the way henequen did for binder twine.

Yucatán's unique geology added to the success of plantations and is an integral part of the state's environmental history. The henequen zone comprises eleven different soil units and sixteen subunits that form a variety of different combinations "truly into a mosaic where the [soil] predominance varies from one area to the next," as geographer Lourdes Villers explained. The majority of the land is made up of Litosol and Cambisol types that "created the fertility that allowed agricultural development."[24] Because of the karst limestone base, henequen and sisal were immune to fungal infections that were the bane of other tropical commodities, especially as plant pathogens could ruin banana and rubber plantations. As Allen Wells has pointed out, the *henequeneros* "were spared Mother Nature's wrath . . . probably . . . because the region was so arid and the thin topsoil retained so little moisture." Nor did henequen ever need fertilizers or pesticides to increase yields.[25]

Henequen's rapid growth also helps explain how it outcompeted other fibers. It was far less expensive than Philippine abaca (manila) or Javanese jute to import into North America, where farmers believed that sisal twine was superior to others. It made the most uniform length twine, was the most durable to withstand the strain of knotters, and was resistant to moisture, rot, and insects.

Exogenous events also led to the rise of Yucatán's henequen boom. In the nineteenth century, international conflict impeded fiber imports to North America. For example, the Crimean War (1854–56) interrupted exports of industrial hemp, the principal fiber used for rope works and bagging, when Russia cut off shipments to western Europe. Later, when binder twine was in great demand, the abaca-rich Philippines became entrenched in a bloody war against the US occupation, causing international fiber shortages and doubling the value of Yucatecan henequen on the world market.

Positivism and Plantations

The incredible growth in the henequen industry coincided almost exactly with the years of the dictatorship of Porfirio Díaz (1876–1911, a period known as the Porfiriato). Díaz waged a campaign to purge Mexico of any obstacles to his dreams of *orden y progreso* for the country. His advisors, known as *científicos* (scientists), were informed by nineteenth-century concepts of positivism—a liberal economic policy characterized by the promotion of agricultural and industrial modernization, a state-directed economy, privatized property, and foreign investment to spur economic development. The doctrine excluded native peoples because many lived and worked on communal lands, leaving many impoverished and marginalized in their rural communities or forced to move to the expanding slums of Mexican cities.

Unlike many other parts of Mexico, however, the development of recognized commons in Yucatán was a relatively recent phenomenon. Campesino and indigenous communities in the northeastern part of the state did not petition for the distribution of *tierras de común repartimiento* (village commons) until 1870. The environment itself had precluded earlier development of communal farming, because agricultural fertility was better in the *monte* (brushlands) than on the land owned by extended families and corporations that became the *henequenales*. It had always been on these uncleared lands that local peoples had had "an extensive knowledge and management" of forest products. Thus, when the government established *ejidos* (communal farms, not to be confused with the postrevolution ejido system throughout Mexico) in Yucatán, officials found that the indigenous groups were occupying *terrenos baldíos* (untitled lands) and that determining where "the ejidos ended and the public lands started" was difficult, as Inés Ortiz argues. This problem had cropped up during liberals' earlier policies in the mid-nineteenth century, and it immediately bedeviled Díaz's modernization schemes. Thus, Ortiz continues, the new "agrarian structure did not guarantee the ecological conditions of the forests that were required for itinerant agriculture and did not consider the dynamic of family groups employed in cultivating *milpas*, hunting, and gathering wood, honey, and natural fruits in the woods."[26]

Instead, the positivist economic thought that informed the Díaz government's policies was to increase production and exportation of Mexican commodities under the model of Ricardian comparative advantage. The smaller family haciendas became consolidated into huge plantations, managed with scientific principles for increased monocrop production. But true to the positivist *mentalité*, the monocrop structure depended on foreign investment—financial resources invested from New York banks and US fiber brokers charging exorbitant interest rates.[27] Nonetheless, a regional trade journal editorialized that henequen could be the agent to make Mexico a modernized country like the United States, that "example of a great agrarian nation . . . whose people are virtuous, patriotic, [and] in love with the land and the honesty of hard work." It asserted that "if Mexico could become [more] agricultural, its future could be assured forever."[28]

Díaz advanced these ideals for Yucatán by rewarding wealthy landowners (known as *henequeneros*) with favors and incentives, including the promotion of a labor regime that would keep henequen profits high. Likewise, in 1896 the federal government issued a decree that allowed individual private property titles on ejidos that enabled the breakup of indigenous lands. The act did not force Indians away from their lands, but it required all properties to be surveyed, an expense the vast majority of campesinos or Indians had no way of paying, thereby forcing them to sell to henequeneros. By 1912, there were virtually no ejidos remaining in Yucatán's henequen zone. Likewise, many landowners invoked an earlier policy that allowed them to acquire vacant lands they had worked for at least four years. On the state level, Yucatecan legislators enacted similar incentives to stimulate the fiber industry.[29] Everything was working to promote the henequeneros and the "economic progress" of Yucatán. In fact, the state went from one of Mexico's poorest to one of its most prosperous in a couple of decades.[30] Henequen was indeed Yucatán's *oro verde* (green gold), as people there called it.

Wealth came quickly for the henequeneros, investors, merchants, bankers, and fiber dealers, all of whom developed into a *nouveau bourgeois* class in Yucatán and made Mérida arguably the most modern city in the country. Yucatán became so wealthy that instead of needing financial support from the federal government it often was able to provide assistance to other parts of Mexico. The boom attracted great numbers of people; Yucatán's population grew from 284,000 in 1877 to 386,000 by 1900, and Mérida's growth rate was even greater.[31]

Demographic growth matched economic growth. A state report in 1916 showed that the cost to raise henequen was $46 per hectare ($115 an acre) and yielded a 53 percent return on capital invested.[32] Some of the aristocratic henequeneros, however, earned returns of up to 600 percent on their ventures, which they then often reinvested into henequen expansion. They operated under the belief in abundance theory, defined as "an almost religious belief that the henequen economy would survive the short-term bust cycles and enjoy lasting success." Thus, they quickly adopted a lifestyle characterized by "hyperconsumption" in a nouveau riche atmosphere surrounding those whom John Kenneth Turner, an American journalist researching the area for an exposé, called the "fifty henequen kings."[33] Indeed, like King Cotton in the American South, King Henequen reigned in Yucatán (figure 7.2). It was, as historians have described it, "order and progress, Yucatecan style." It was in fact Yucatán's Época Dorada—its "Gilded Age."[34]

Figure 7.2. King Henequen. *Source:* Cover of *El Henequén* 2 (31 May 1917).

A small number of henequenero families dominated Yucatán politically and economically, forming an oligarchy that controlled Yucatán during the henequen boom years. One individual, Olegario Molina, came to be the most powerful of all the elite landowners, so powerful that he was known as *el rey de los reyes henequeneros* (the king of the henequen kings) and served as governor from 1902 to 1908.[35] Porfirio Díaz was so impressed with Molina's positivist policies on economic growth and scientific modernization that he appointed Molina to the position of minister of development in the federal government during the last years of his administration. Molina staunchly believed that henequen was Yucatán's economic salvation and that the wealth from fiber exports would spread around the state. His mantra for solving economic problems was "plant more henequen."[36] Able to acquire numerous haciendas at below-market values while governor, he then required other growers to sell their fiber to him so that by 1913 he and his family owned 100,000 hectares (247,000 acres) and controlled 75 percent of the henequen trade.[37]

The success of such a monopoly was dependent on having a steady buyer of fiber. Molina entered into a contractual relationship with International Harvester Company of Chicago and the multinational fiber trader Henry W. Peabody of Boston to clinch the export market. The arrangement allowed Harvester to control 80 percent of all henequen and sisal shipments from Yucatán from 1900 to 1915. The record year was 1910, when International Harvester cornered a staggering 99.8 percent of the market—a classic case study of what economists call a monopsony, or what others have referred to as an "imperial collaborator model" that illustrated International Harvester's "informal empire" and Yucatán's "economic dependence."[38] However, Yucatecans knew they still had the upper hand in the dependency model. As one trade journal in the region put it, "Without henequen, there will be no [grain] harvest."[39]

The Political Ecology of Henequen

All of these agricultural and industrial transformations stemming from the rise of the henequen-wheat complex forever changed the natural environment of Yucatán. As mentioned earlier, the original vegetation of the region was a low deciduous forest and cacti mixed with savannah and thornbush zones on its perimeters. There, henequen and sisal were native to the peninsula and blended with other tropical angiosperms to create a varied landscape on the flat Yucatecan terrain. The area was a "mosaic of vegetation . . . that provided a wealth and diversity of plant life," as one biologist has called it.[40] Changes to this ecological system, mixed with small subsistence milpas, began in the mid-nineteenth century when henequen growers cleared a patchwork of haciendas out of the scrub forest. But with the henequen boom, by 1900 the large-scale henequeneros converted haciendas and cattle ranches into plantations, causing great ecological impact to Yucatán's landscape; forests disappeared and wildlife habitat was seriously compromised.[41]

The shrubby desert landscape radically changed as terrenos baldíos became henequenales. And the change was fast: In 1901, growers converted

87,800 hectares (217,000 acres) to henequen, but during the peak year of 1916, the number more than doubled, to 212,000 hectares (525,000 acres). By the 1940s, the henequen zone constituted nearly 40 percent of the state.[42] Celebrating this transformation as a "victory of man against the desert," Fernando Benítez praised henequen as "the child of the desert" and the spiny leaves as "the green stars of the land." His description, while meant to be an uncritical, laudatory verse, offers poetic insights on the landscape's environmental changes. He wrote that the fields "are as beautiful as the vineyards in Italy or the olive groves in Spain. The plants, all straight and lined up in their rows . . . are ordered symmetrically on the sun-baked plain. [They] have an elegant severity, an economy of lines . . . a synthesis between the formal and the functional."[43] This kind of agriscape geometry prompted the poet Joaquín Lanz Trueba to write in 1917 that the fields were like a "green carpet over the plain, symmetrically aligned in a rigorous order, in prolonged lanes of thousands and thousands of plants that seem to be offering up their hands." Another description suggested that the henequenales appeared as "endless rectilinear rows of bluish-gray spines."[44]

This discourse of order, rigor, symmetry, economy, geometry, and alignment reflects well the Porfirian ideal of order and progress and Mexico's vision for a scientific, economic growth. It describes the much-coveted economic landscape of a modernized Europe and defies Yucatán's natural topography by praising the landscape transformations encouraged by Díaz's modernization policies. Growers would no doubt have applauded Benítez's statement that "every village in northern Yucatán is . . . a victory of man against the desert."[45] And the descriptive panegyric of the monocrop environment was matched only by the language that highlighted the hoped-for wealth from henequen, its capital-generating potential, and its role as the economic savior of the state. An editorial in *El Henequén* lauded "our rich agave—called 'green gold.' . . . That privileged treasure to its arid and . . . rocky ground is . . . akin to a blessing from heaven."[46]

The plant was seen as a golden treasure, like the minerals from other parts of Mexico. But although wealth came to few Yucatecans, the agricultural and economic changes were only a part of the broader landscape transformations in the region. As plantations expanded, so did demand for crops and firewood to support hacienda managers, workers, and their families. With economic growth in mind, federal and state government policies encouraged such transformations and allowed *terrenos baldíos* to be used to expand plantations. The Yucatán legislature enacted a series of policies from 1888 to 1894 that helped spur henequen development and increase *latifundismo* (aristocratic landholding) at the expense of campesinos or indigenous ejidos.[47] Petitions flowed into state authorities requesting permission to clear *monte* for construction, firewood, and milpas. One farmer from near the village of Kanasín wrote directly to Governor Salvador Alvarado in 1916 complaining that nonhenequen agricultural productivity had fallen by 40 percent, necessitating "the search for more *monte* to be exploited for cultivation." He lamented how the increase in population had sent people scurrying to find more wood on vacant lands.[48]

As henequen expanded from the 1880s to the 1920s, the landscape continued changing with the development of a rail network. During the early henequen years, the cut pencas arrived at the rasping mills on the backs of mules and later in carts. But as the industry developed, the henequeneros needed a more efficient transport system. Thus, in the 1880s the state government authorized concessions for movable Decauville narrow-gauge tramlines that used mule-drawn cars to take the harvested henequen from the field to the mills and to haul the dried fiber bales to port. The government also built regular-gauge tracks to connect various parts of the state with the newly built port at Progreso (aptly named in honor of "order and progress"). So useful were these that by 1890 Yucatán had more rail lines than any other state in the republic. Some henequeneros built their own feeder lines, adding another 580 kilometers (360 miles) of rail to the system. By 1923, nearly 6,500 kilometers (4,000 miles) of track connected villages, henequen fields, defibering plants, and the port.[49] The expanded system enabled growers to convert more land into production and further depleted scarce timber resources, as the trains burned wood for fuel.

Meanwhile, technological advances resulted in more mechanized rasping machines (decorticators) for separating the fiber from the leaves of the plants. The need for such devices was great, as the fiber had to be rasped within twenty-four hours of harvest so it would not dry out. Previously, workers had to ret the leaves by hand to render the fibrous pulp—backbreaking work that wasted fiber. The first mechanical decorticators (using rotating knives) were helpful but still cumbersome, slow, and labor-intensive. By the late nineteenth century, however, inventors had designed ones that used steam power, which remained the state of the art for the duration of the henequen boom.[50] Steam required firewood to heat the boilers, which hastened the depletion of monte wood cover, leaving an ever-greater mark on Yucatán's once-forested landscape. In 1917 forested land for firewood was so important for the industry that state authorities classified 240,000 hectares (593,000 acres) as "vacant lands that provide firewood for the rasping of henequen fiber." And as rasping intensified in the 1920s, Governor Felipe Carrillo Puerto decreed that growers could cut wood to fuel their mills on wooded *terrenos baldíos* and mandated that small farmers clearing land were required to provide the wood to henequeneros instead of selling it for use as railroad ties.[51]

Contemporary writers issued warnings about the consequences of such deforestation. In 1878 one concerned Yucatecan advised how "necessary it is to look a bit into the future" to see that "the lack of firewood" would affect people and industry if "the imprudent and disorderly destruction of *montes*" continued. He warned of the dangers of converting too many forests to milpas and of the alarming amount of firewood and charcoal consumed in the henequen zone. Thus, it was beneficial to "avoid all imprudence in cutting the forests" and to "establish a system of replanting trees without waiting for what nature would do spontaneously." To do so would generate more oxygen, "so indispensable for health and life," and would ensure a natural hydrological regime that would guarantee better harvests.[52] Likewise, in 1882 industrial chemist Eugenio Frey conducted a study of Yucatán's forests. In "Utility of Forests," Frey bemoaned the "absurd and ridiculous theory" that forests stood in the way of progress. "On the contrary," he argued, "they are the strongest support and sustainer of it." He noted that he had seen unsuccessful farming in burned-over areas where the soil was left sterile and soon had to be abandoned. From the "history of so many denuded mountains," he gathered data regarding watershed hydrology and how forests served as "absorbents for the heat and emitters of moisture."[53]

Some of this thinking worked to influence policy development. In 1903, Governor Molina established a forests commission and appointed as its director Rodulfo Cantón, who for twenty-five years had been involved with forest conservation. These were perhaps the first steps in "putting a finger in the dike" of the torrent of deforestation, as Víctor Suárez put it, and hence decrease "the immoderate cutting of forests" and "regularize conservation . . . for the benefit of agriculture and public health." It is somewhat surprising that Molina, the largest landowner in Yucatán, would act so prudently on a conservation policy. But if he was convinced that the future of agriculture lay with preserving hydrologic and forest resources, the financial value of conservation would be inescapable. Unfortunately, according to Suárez, the commission's proposals never had a chance to succeed; they were "nullified by the incomprehension and lack of cooperation of the hacendados and farmers."[54]

In 1918, two articles published in *El Henequén* outlined the long-term disadvantages of forest destruction and advocated conservation. The first, "The Cutting of Forests," began by stating, "Every day we are hearing more complaints . . . about the crude war that at times unnecessarily is being waged in our forests." The article lamented the "sad heritage" that forest destruction left to future descendants and the hydrological problems caused by such destruction, and it reminded readers of the important role forests play in providing oxygen and of the ecological problems associated with eroded watersheds that compromised successful agriculture in those areas. Rainfall was better absorbed in the undisturbed monte, the article maintained, a characteristic especially useful in times of drought to sustain vegetation for soil cover and to avoid widespread desertification. It concluded with a call to plant new trees.[55]

The second article, "Yucatecan Trees and Forests," by a proto-environmentalist writing under the pseudonym "Agrófilo" (Agrophile), was published a few months later and had similar warnings. Agrófilo listed his arguments chronologically, starting with the colonial era when Spaniards began "the destruction of the Yucatecan forests" to acquire wood for various purposes. The devastation led to postindependence policies requiring that every citizen in the state's municipalities plant five trees a year, enforced by local officials responsible for counting the new trees. There was also a moratorium on cutting down young trees, to instill a "love for trees" in each community. Thus, it fell to the state government to "prevent the destruction of forests . . . and their annihilation or total ruin." Agrófilo concluded by suggesting that all Yucatecans "oppose the cutting of trees [and] the annihilation of forests" to promote a "better climate and improved lands" and return the landscape to an "ancient and splendid Yucatecan garden."[56]

Others warned of the problems of monocrops. In 1907, a writer in *El Agricultor* stated, "It appears that monoculture production . . . is a very dangerous base for any country." The journal echoed these words in 1923 in the article "Slavery to Monoculture," which told the "undeniable truth" that being beholden to markets in North America stunted Yucatecan economic development; the article warned of the disadvantages of not diversifying agricultural production. It urged the planting of corn, beans, and sugarcane and the grazing of cattle as ways of breaking the monocrop culture.[57]

Such advice was rarely heeded in Yucatán's boom times. Not until 1977 did Víctor Suárez's two volumes on the economic development of Yucatán, with their ecological warnings about henequen production, appear. In his section "Forests and Deforestation," Suárez expressed his concern that the cultivation of corn had been one of the biggest culprits of forest destruction. He showed that the need to feed workers in the henequen sector sparked the rapid development of milpas in the nineteenth century (increasing from 607,212 *mecates* [24,300 hectares] of cornfields in 1883 to 1,269,000 mecates [50,760 hectares] in 1893) and the attendant problems associated with deforestation. The deforested lands were made worse when "torrential rains . . . hit the ground without vegetated cover[,] eroding the land and causing it to lose even more organic cover." He argued that forest clearing and the erosion on the karst-earth that followed had caused the ecological impoverishment of the henequen zone.[58]

Other scholars have been equally critical of the environmental impacts of milpas in the henequen zone. Moisés González Navarro noted that milpa technology had been "primitive": campesinos burnt down forests, planted seeds with a digging stick, and then moved to different land when the corn patch was no longer productive. Friedrich Katz discovered that by 1918 Yucatán corn production was considerably lower than previous years because plantation owners had been converting milpas to henequenales. Not only did this process accelerate ecological transformation and reduce the amount of land operated by rural people, it also meant that Yucatán had to import corn to meet subsistence needs. So great was the need in the 1920s that the federal government intervened to help bring in greater quantities from the United States and other parts of Mexico.[59] Thus began Yucatán's cyclical dependency on the United States: first as a market for its henequen; and second as a source of foodstuffs to feed henequen workers. Worse, for the already impoverished majority of Yucatecans, importing such basics as corn sent prices sky-high.[60] Officials were aware of the problem, as evidenced in an article in *El Sisal Mexicano* that in 1927 encouraged Yucatecans to plant more corn. Rebutting that article, however, the more conservative *El Henequén* argued that there was not enough land available to convert to corn and repeated the line that there was little land in Yucatán good for raising any cereal crops: "If this were in Sonora, OK, but not here."[61]

Conclusions

In the 1920s and 1930s, henequen still reigned as the supreme fiber for binder twine, but the state government raised the price to sell it on the international market in hopes that the higher prices would create greater revenues for education and social services. However, investors in East Africa, Java, and Brazil had learned that sisal grew well in those places and started to compete aggressively on the fiber market. Likewise, much of the henequen country underwent significant redistribution because of agrarian reform legislation carried out during the 1934–40 presidential administration of Lázaro Cárdenas. The president worked with regional officials not only to redistribute henequen lands into 276 new ejidos—accounting for 61 percent of all henequen fields—but also to establish a state credit bank to make loans to the new ejidal farmers. Even during the Great Depression, with many plantations collectivized into ejidos, the zona henequenera remained "*the* agrarian showcase" of Mexico. But the continued predominance of the landed elite, bureaucratic inefficiencies, and divisions among the *ejidatarios* (land reform beneficiaries) curtailed the well-intentioned reforms.[62]

Even had the reforms been more successful, the market for Yucatecan fiber nonetheless started to decline precipitously in the 1940s. World War II did not prove to be much of a market boost for henequen, although the US government did decree a moratorium on the trade of non-Mexican or non-Cuban sisal for twine production to avoid concerns regarding transoceanic shipments.[63] But even that measure came too late. A prophetic postwar article in the newspaper *El Universal* captured the scenario well: "Henequen: Green Gold—The Wealth and Tragedy of the Mayan Land." It described how 40 percent of Yucatán (and 70 percent of its seeded surface) was still devoted to henequen, despite the severe downturn in market demand. The state, the article continued, and 23,800 ejidatarios "were completely dependent on the henequen industry," even as postwar demand for fibers had dwindled.[64] Certainly for Yucatán, the economic tragedy was developing quickly at that point.

Far worse for the henequeneros, and those who depended on that industry, were the changes occurring in North American harvesting. By the 1950s most farmers in the United States and Canada had switched from binders to combines to harvest crops. Combines simultaneously cut and threshed, precluding the need for binding grain into bundles. And despite the fiber industry's efforts to convert to baler twine for hay bales, which provided a temporary spurt in demand, plastic companies had developed polyethylene and polypropylene twines that were cheaper to make synthetically and were just as durable as hard fibers. Certainly, cordage interests continued to promote other sisal products, but nothing came to equal the demand that binder twine afforded during the henequen boom. By then, King Henequen had died a sudden and, for Yucatecans, torturous death.[65]

The decline of henequen, however, has had a beneficial impact on the ecology of north-central Yucatán. In an almost textbook case of ecological restoration, native flora and fauna are returning to the henequen zone. One study demonstrates that abandoned henequen fields are rich in pioneer and later successional species, although original native mature forests are practically nonexistent in the region.⁶⁶ The change is visible and important to local residents. One older local official near the slowly decaying San Antonio Tehuitz plantation told me that in six years after the hacienda's abandonment, local people were seeing "much more of nature" and that many "plants and animals were returning to the area," some of which they had not seen since their childhood days.⁶⁷ Perhaps nature will bat last in Yucatán.

Today the abandoned henequenales are used as sites to diversify the state's agricultural economy. Some projects (such as orange and papaya groves) did not pan out, but studies show that Yucatán's climate and thin soil atop the limestone karst can be used productively for raising aloe vera (with its host of commercial products in the health and cosmetics industries), Italian limes (for consumption and lime oil extract), various other orchard and vegetable crops, and finally, *chaya* (a vegetable gaining in popularity).⁶⁸ All of these products have local, national, and international market potential, but they would never be in demand to the degree that Yucatecan henequen and sisal were during the binder twine years. But perhaps like henequen, they will thrive in a natural setting without the need for much mechanical cultivation, irrigation, fertilizers, or chemical pesticides. Perhaps Yucatán can soon regain some of its past economic sustainability in its unique environmental setting.

Notes

1. For the larger story, see Sterling Evans, *Bound in Twine: The History and Ecology of the Henequen-Wheat Complex for Mexico and the American and Canadian Plains, 1880–1950* (College Station: Texas A&M University Press, 2007).

2. Allen Wells, "From Hacienda to Plantation: Transformation of Santo Domingo Xcuyum," in *Land, Labor, and Capital in Modern Yucatan*, ed. Gilbert Joseph and Jeffrey Brannon, 114–37 (Tuscaloosa: University of Alabama Press, 1991), 115.

3. Allen Wells, *Yucatán's Gilded Age: Haciendas, Henequen, and International Harvester, 1860–1915* (Albuquerque: University of New Mexico Press, 1985), 13–14.

4. Lourdes Villers Ruiz, "Caracterización del medio físico de la zona henequenera y sus potencialidades para el desarrollo agrícola," in *Memorias de la Conferencia Nacional sobre el Henequén y la Zona Henequenera de Yucatán*, ed. Piedad Peniche Rivero and Felipe Santamaría Basulto, 63–76 (Mérida: Estado de Yucatán, 1993), 66; Allen Wells, "Reports of Its Demise Are Not Exaggerated: The Life and Times of Yucatecan Henequen," in *From Silver to Cocaine: Latin American Commodity Chains and the Building of the World Economy, 1500–2000*, ed. Steven Topik, Carlos Marichal, and Zephyr Frank, 300–20 (Durham, N.C.: Duke University Press, 2006), 310–11.

5. R. Toby Pennington, Darién E. Prado, and Colin A. Pendry, "Neotropical Seasonally Dry Forests and Quaternary Vegetation Changes," *Journal of Biogeography* 27

(March 2000): 262. See also Peter G. Murphy and Ariel E. Lugo, "Ecology of Tropical Dry Forest," *Annual Review of Ecology and Systematics* 17 (1986): 67–88.

6. Víctor J. Jaramillo, J. Boone Kaufman, Lyliana Rentería-Rodríguez, Dian L. Cummins, and Lisa J. Ellingson, "Biomass, Carbon, and Nitrogen Pools in Mexican Tropical Dry Forest Landscapes," *Ecosystems* 6 (2003): 610.

7. Quotations in Roland Chardon, *Geographic Aspects of Plantation Agriculture in Yucatan* (Washington, D.C.: National Academy of Science, 1961), v; Friar Diego de Landa, *Yucatán before and after the Conquest, 1566,* trans. William Gates (New York: Dover Publications, 1978), 1, 93.

8. Interview in *New York Herald Tribune*, 27 August 1938, p. 5. Historian quoted in Edward D. Fichen, "Self-Determination or Self-Preservation? The Relations of Independent Yucatán with the Republic of Texas and the United States, 1847–1849," *Journal of the West* 18 (January 1979): 33.

9. *El Sisal Mexicano* 6 (August 1932): 6; *El Sisal Mexicano* 6 (September 1932): 3; Wayne D. Rasmussen, ed., *Readings in the History of American Agriculture* (Urbana: University of Illinois Press, 1960), 75; Eugene M. Wilson, "Physical Geography of the Yucatan Peninsula," in *Yucatan: A World Apart*, ed. Edward H. Moseley and Edward D. Terry, 5–40 (Tuscaloosa: University of Alabama Press, 1980), 7.

10. Wells, "Reports of Its Demise," 311; Chardon, *Geographic Aspects*, 10.

11. Evans, *Bound in Twine*, 35–36.

12. Carlo de Fornaro, "Yucatan and the International Harvester Company," *Forum* 54 (July–December 1915): 337.

13. Manuel Escoffié, *Yucatán en la cruz* (Mérida: n.p., 1957), 33, 27; Roger Orellana, "Agave, Agavaceae y familias afines en la Península de Yucatán," in Peniche and Santamaría, *Memorias de la Conferencia*, 83.

14. Chardon, *Geographic Aspects*, 14–15; Renán Irigoyen, *Los mayas y el henequén* (Mérida: Henequenera de Yucatán, 1950), 8.

15. Landa quoted in Enrique Manero, *La anarquía henequenera de Yucatán* (Mexico City: n.p., 1966), 15.

16. Wells, *Yucatán's Gilded Age*, 18–21.

17. Chardon, *Geographic Aspects*, 15.

18. Víctor Suárez Molina, *La evolución económica de Yucatán a través del siglo XIX: Apuntes históricos* (Mexico City: Ediciones de la Universidad de Yucatán, 1977), 1:32–33.

19. Terry Rugeley, *Yucatán's Maya Peasantry and the Origins of the Caste War* (Austin: University of Texas Press, 1996), xiv.

20. Wells, *Yucatán's Gilded Age*, 25–26; Fred Carstensen and Diane Roazen, "Foreign Markets, Domestic Initiative, and the Emergence of a Monocrop Economy: The Yucatecan Experience," *Hispanic American Historical Review* 72 (November 1992): 555.

21. *La Razón del Pueblo*, 23 October 1869, quoted in Renán Irigoyen, *¿Fue el auge del henequén producto de la Guerra de Castas?* (Mérida: n.p., 1947), 38. Export figures are in Jaime Orosa Díaz, *Felipe Carrillo Puerto (Estudio biográfico)* (Mérida: Fondo Editorial de Yucatán, 1982), 15; and Chardon, *Geographic Aspects*, 29.

22. Statistics assembled from Fornaro, "Yucatan and the International Harvester Company," 338; *Diario Yucateco*, 10 January 1911, p. 5; Manuel Pasos Peniche, *Historia de la industria henequenera desde 1945 hasta nuestros días* (Mérida: n.p., 1974), app. 1; and Jeffrey Brannon and Eric Baklanoff, *Agrarian Reform and Public Enterprise*

in Mexico: The Political Economy of Yucatán's Henequen Industry (Tuscaloosa: University of Alabama Press, 1987), 43.

23. Chardon, *Geographic Aspects*, 4–5, 8; Leo Waibel, "The Tropical Plantation System," *Scientific Monthly* 52 (February 1941): 157. For more on the distinction, see Eric R. Wolf and Sidney Mintz, "Haciendas and Plantations in Middle America," *Social and Economic Studies* 7 (1957): 380–412.

24. Villers Ruiz, "Caracterización del medio físico," 64–65.

25. Wells, "Reports of Its Demise," 304.

26. Inés Ortiz Yam, "Desamortización y ejidos en Yucatán, 1870–1915," paper presented at the 56th annual meeting of the Rocky Mountain Council for Latin American Studies, Santa Fe, New Mexico, 4–7 March 2009.

27. Eric Baklanoff, "The Diversification Quest: A Monocrop Export Economy in Transition," in Moseley and Terry, *Yucatan: A World Apart*, 202–44 (quotation on 209). On "comparative advantage," see David Ricardo, *On the Principles of Political Economy and Taxation* (London: John Murray, 1819).

28. Editorial, *El Henequén* 1 (15 July 1916): 2.

29. Federal and state policy in Chardon, *Geographic Aspects*, 33–34; Suárez Molina, *La evolución económica de Yucatán*, 1:111; and Jaime Orosa Díaz, ed., *Legislación henequenera en Yucatán (1833–1955)* (Mérida: Editores Fomento de Yucatán, 1956), 145.

30. Orosa Díaz, *Legislación henequenera*, 9; Allen Wells, "Henequen and Yucatán: An Analysis in Regional Economic Development, 1876–1915," PhD diss., State University of New York–Stony Brook, 1979, pp. 2, 35.

31. Gilbert Joseph and Allen Wells, "Corporate Control of a Monocrop Economy: International Harvester and Yucatan's Henequen Industry during the Porfiriato," *Latin American Research Review* 17, no. 1 (1982): 71; Fornaro, "Yucatan and the International Harvester Company," 344. Population statistics are in Moisés González Navarro, *Raza y tierra: La Guerra de Castas y el henequén* (Mexico City: Colegio de México, 1970), 172.

32. Secretaría de Agricultura y Comercio (Yucatán) to M. E. Boyle, Austin, Texas, 4 April 1916, Archivo General del Estado de Yucatán (hereafter AGEY), Poder Ejecutivo, caja 522.

33. Investment statistics in Gilbert Joseph, "Revolution from Without: The Mexican Revolution in Yucatán, 1915–1924," PhD diss., Yale University, 1978, 61. Quotations from Wells, "From Hacienda to Plantation," 135; John Kenneth Turner, *Barbarous Mexico, 1911* (1911; reprint, Austin: University of Texas Press, 1969), 8.

34. Wells, "From Hacienda to Plantation," 135–36; Joseph, "Revolution from Without," 61; Wells, *Yucatán's Gilded Age*, 61–88.

35. Fernando Benítez, *Ki: El drama de un pueblo y una planta* (Mexico City: Fondo de Cultura Económica, 1962), 83.

36. Gilbert Joseph, *Revolution from Without: Yucatan, Mexico, the United States, 1880–1924* (New York: Cambridge University Press, 1982), 49; Gonzalo Cámara Zavalo, *Reseña histórica de la industria henequenera de Yucatán* (Mérida: Imprenta Oriente, 1936), 68; Wells, *Yucatán's Gilded Age*, 56.

37. Wells, *Yucatán's Gilded Age*, 74–75; Joseph, *Revolution from Without*, 36.

38. The contract is in IH to Montes, 2 October 1906, file 021-48, International Harvester Company/Navistar Archives, Chicago, Illinois; Wells, "Henequen and Yucatán," 72; Fornaro, "Yucatan and the International Harvester Company," 338; Joseph and Wells, "Corporate Control," 71–72.

39. *El Henequén* 2 (31 July 1917): 1.
40. José Salvador Flores, "Caracterización de la vegetación en la zona henequenera," in Peniche and Santamaría, *Memorias de la Conferencia*, 78–79.
41. Wells, "Reports of Its Demise," 304.
42. Joseph and Wells, "Corporate Control," 70; Pasos Peniche, *Historia de la industria henequenera*, app. I; *El Universal*, 19 April 1946, p. 15.
43. Benítez, *Ki*, 26, 47–49.
44. Joaquín Lanz Trueba, "En un henequenal," reprinted in *El Henequén* 3 (1 January 1918): 17; Joseph and Wells, "Corporate Control," 70.
45. Benítez, *Ki*, 26.
46. *El Henequén* 2 (15 November 1917): 8.
47. González Navarro, *Raza y tierra*, 191–94; announcement about laws allowing conversion of "vacant" lands is from local Socialist Party, Agrarian Commission, 2 December 1916, AGEY, Poder Ejecutivo, caja 537.
48. Faustino Escalante to Salvador Alvarado, 16 November 1916, AGEY, Poder Ejecutivo, caja 537.
49. Louis Crossett, *Sisal: Production, Prices, and Marketing*, Trade Information Bulletin no. 200, Textile Division Supplement to Commerce Reports, US Department of Commerce, Bureau of Foreign and Domestic Commerce, Prepared as Part of Investigation of Essential Raw Materials Authorized by the Sixty-seventh Congress (25 February 1924), 5. Rail development information is from Orosa Díaz, *Legislación henequenera*, 53–188; and Joseph, *Revolution from Without*, 34.
50. On decorticator development, see Benítez, *Ki*, 64–70; and Orosa Díaz, *Legislación henequenera*, 25–53.
51. *El Henequén* 2 (15 October 1917): 4; Joseph, *Revolution from Without*, 247.
52. Quoted (anonymous) in Suárez Molina, *La evolución económica de Yucatán*, 1:210–11.
53. Eugenio Frey, "Utilidad de los bosques" (1882), reprinted in *El Agricultor* 2 (1 April 1908): 57. See also Lane Simonian, *Defending the Land of the Jaguar: A History of Conservation in Mexico* (Austin: University of Texas Press, 1995), chap. 4; and Emily Wakild, chap. 9, this volume.
54. Suárez Molina, *La evolución económica de Yucatán*, 1:125.
55. "La tala de bosques," *El Henequén* 3 (15 February 1918): 5.
56. Agrófilo, "Los árboles y bosques yucatecos," *El Henequén* 3 (15 November 1918): 10–11.
57. *El Agricultor* 1 (1 March 1907): 37; "Esclavitud a la monocultura," *El Agricultor* 10 (1 June 1923): 5.
58. Suárez Molina, *La evolución económica de Yucatán*, 1:125.
59. González Navarro, *Raza y tierra*, 179; Friedrich Katz, "El sistema de plantación y la esclavitud," *Ciencias Políticas y Sociales* 8 (January–March 1962): 118; Ben W. Fallaw, "The Economic Foundations of Bartolismo: García Correa, the Yucatecan Hacendados, and the Henequen Industry, 1930–1933," *Unicornio: Suplemento Cultural de ¡Por Esto!* 9 and 27 October 1997.
60. Wells, "From Hacienda to Plantation," 131–32.
61. *El Sisal Mexicano* (March 1927): 3; *El Henequén* 1 (April 1927): 60.
62. Wells, "Reports of Its Demise," 315. On the Cárdenas years, see Ben Fallaw,

Cárdenas Compromised: The Failure of Reform in Postrevolutionary Yucatán (Durham, N.C.: Duke University Press, 2001).

63. *Revista del Comercio Exterior* (March 1943): 3.

64. "El henequén: Oro verde—la riqueza y tragedia de la tierra maya," *El Universal*, 19 April 1946, p. 5.

65. On combines and synthetic twines, see Evans, *Bound in Twine*, chap. 7. For more on the decline, see Wells, "Reports of Its Demise," 316–18.

66. Aliza Mizrahi, JoséMaría Ramos Prado, and Juan Jiménez-Osornio, "Composition, Structure and Management Potential of Secondary Dry Tropical Vegetation in Two Abandoned Henequen Plantations of Yucatan, Mexico," *Forest Ecology and Management* 96 (September 1997): 273–74.

67. Egidio López López, interview, San Antonio Tehuitz, Yucatán, 6 June 1999.

68. On alternative crops, see the section "Cultivos de diversificación" in Peniche and Santamaría, *Memorias de la conferencia*, 484–580.

CHAPTER EIGHT

Class and Nature in the Oil Industry of Northern Veracruz, 1900–1938

Myrna I. Santiago

Edward L. Doheny, Cruz Briones Rodríguez, and countless American drillers met in northern Veracruz in the early 1900s, but their experience of place differed so much an observer might have never guessed they shared the same geography. In 1900 Doheny, an oil magnate who multiplied his fortune many times over in northern Veracruz, for example, was moved by the "beautiful and awe-inspiring scenery[:] . . . rivers of clear blue-green water; . . . a forest so dense[;] . . . [and] jungle-covered country which extends clear to the harbor of Tampico."[1] In 1913 he affirmed categorically and without a hint of irony that the petroleum companies "have been a blessing to the communities in which they have operated," congratulating himself for paving Tampico's streets with his asphalt and turning the port into "one of the happiest communities of any city in the world."[2] Briones Rodríguez, an oil worker, was more succinct and critical in his recollections. He described northern Veracruz as "the devil's collection of plagues."[3] American drillers working in Mexico left little testimony, but their actions were recorded by photographers and travelers who witnessed a life defined by risk and danger. The differences were rooted in the class relations that ruled the industry; that is, class position profoundly shaped how the men viewed and experienced the natural world around them.

Class is not yet an explicit concern of environmental history. The field is growing robustly among Latin Americanists, but labor does not figure much in the literature.[4] Similarly, environmental history is not prominent in labor and working class history.[5] But looking at environmental history from a labor perspective and looking at labor history from an environmental

173

perspective is fruitful and rewarding. Insofar as historians are practitioners of the craft of disaggregating generalizations in favor of specifics, disentangling the "humans" of environmental histories into constituent groups (for example, classes) sheds light on the very dynamics that explain changes (or continuities) in the interactions between humans and nature. That is, although humans as a species modify, destroy, or protect their environments, not all of them share the same views or experience, wield the same amount of power, and play the same role in the process. Doheny, Briones Rodríguez, and nameless American drillers worked and lived in the same place at the same time, yet they might as well have inhabited separate universes because the social class they belonged to created wildly divergent realities for each one. While Doheny's testimony was plentiful, workers were more circumspect about how they felt or what they thought about nature and their place within it. Their testimony is parsimonious. Yet there is plenty of information about their fortunes, and that material can help the historian reconstruct the men's lives. Doing so with ecology in mind can lead us to see how class relations are embedded in environmental history, as the occupational ladder in the work "environment" becomes the point where workers and the natural world meet and interact. Thus, a focus on the social organization of labor reveals that the labor hierarchy determined, in many ways, everyone's experience in and of nature.

To be specific: those at the top of the socioeconomic structure wielded more power over nature than did those at the bottom. Likewise, the upper classes exerted significant control over the lives and energies of human beings of the lower classes in any given environment.[6] That dual authority positioned the upper echelons of the companies as masters of nature's creatures, subordinate men included. Yet there were gradations of domination down the occupational ladder among workers themselves. Drillers and craftsmen occupied an intermediate position between the big bosses and the less skilled bulk of the workforce. They exercised some level of control over their environment through their skills and supervisory roles, but they were exposed to extreme bodily danger on a daily basis. Laborers—Mexicans all and the lowest rung on the ladder—were the most vulnerable. They were subject to high levels of occupational risk, toxic environments, tropical disease, and the vagaries of weather. Those differences underlined the intense class conflicts that characterized the oil industry for decades, leading Mexican oil workers to develop a strong sense of nationalism alongside a blistering critique of foreign capital. In time, the environmental, ideological, and class turbulence in oil country became an issue of national importance. Thus, Mexican oil workers played a crucial if not fully recognized role in the single most important display of Mexican nationalist fervor in the twentieth century—the nationalization of the oil industry decreed by President Lázaro Cárdenas in March 1938.[7] The nexus of labor and environmental history, then, deepens and enriches our understanding of both the Mexican Revolution and modern Mexican history and adds a new, environmental, dimension to the history of the politics of energy production in the world.

A Sense of Place, Before and After

The Mexican oil industry was born at the opening of the twentieth century in northern Veracruz. The exact location was the Huasteca, so named after the dominant indigenous population, the Huastecos, or Téenek.[8] Until 1900, the Huasteca was a tropical rainforest, the northernmost tropical rainforest in the Americas. Its most prominent feature was the mass of trees that covered the territory, but the landscape was composed of more than just trees. There were streams and waterfalls and the confluence of two of Mexico's most important rivers, the Tamesí and the Pánuco. The seasonal rains, which included hurricanes in the late summer and fall, flooded all waterways, forming and feeding numerous lakes and lagoons between the two ports that flanked the Huasteca, Tampico (in Tamaulipas) and Tuxpan. The precipitation also maintained the marshes and bogs that surrounded the lagoons and lined the rivers and streams, as well as the mangroves that hugged the coast along the Gulf of Mexico. Fish, shellfish, and mussels abounded in that environment. They provided food for numerous species of birds, both local and migrant, and for amphibians and reptiles, from frogs and turtles to caimans. Inland, the forest provided habitat for a diverse population of flora and colorful or fearsome fauna, including *guacamayas*, snakes, *jabalí*, monkeys, jaguars, and insects capable of inflicting much suffering upon humans.[9]

The human inhabitants of the Huasteca were also diverse. Although their numbers were not great, the population was a mixture of indigenous farmers, mestizo *rancheros*, and a few *hacendado* families of colonial Spanish descent trying to transform the rainforest into pasture. That desire had a history dating to the sixteenth century and had been the cause of incessant conflict with native peoples. Through the colonial period and the whole of the nineteenth century, peasant farmers and cattle ranchers had battled over the ecology of the Huasteca. At the end of the century, they had reached a stalemate. Then the oilmen landed.[10]

In 1900, Texas was awash in oil. That fact persuaded two very astute men that the neighbor to the south might also be equally rich in fossil fuels. After all, nature did not recognize fictitious borders drawn by statesmen. Both men came to the Huasteca as experts in the exertion of power over nature and workers. Both had a history of causing dramatic ecological change. The first was Doheny, an American entrepreneur who owed his millions to the transformation of the Los Angeles scrublands and beaches into a forest of wooden oil derricks and pools of petroleum. The second was Weetman Pearson, an Englishman whose profession, by definition, transformed the environment: he was a civil engineer. Pearson had already left a deep ecological imprint in Mexico. He had been the successful director of the drainage works for the Valley of Mexico, the man who claimed to put an end to the floods that had plagued Mexico City since Cortés destroyed Tenochtitlan in the sixteenth century.[11] In four short decades, the oil companies the two men founded and others that followed altered the Huasteca forever.

The oilmen were never conflicted about using their power over nature. They appreciated the landscape before them as unique and beautiful, to be sure. Pearson, for instance, took friends on tours of the "jungle."[12] Doheny, meanwhile, built a special herbarium for Huasteca specimens in Los Angeles to re-create "the jungle" in his arid backyard.[13] For both entrepreneurs, the very existence of so much foliage was proof of the absence of energetic men to make the land "productive." They identified human effort, that is, labor, with a specific, manufactured landscape, one incompatible with a rainforest. It never occurred to them that the Huasteca was already a man-made landscape after thousands of years of human occupation. What the oil tycoons saw instead was a "wilderness," a land going to "waste" because the local population was lacking in ambition, wanting in application and clearly unacquainted with the notion of "progress."[14] Progress was how the early twentieth-century oilmen understood capitalist economic development and what justified ecological change.[15]

As soon as they incorporated their companies in Mexico, the oilmen initiated the great transformation of the Huasteca rainforest. Flooding the forest with thousands of workers from the countryside and abroad, the oil barons erected an impressive industrial apparatus. It included fourteen refineries, two thousand miles of pipeline, dozens of pumping stations, thousands of storage tanks for crude oil, miles of roads, railways, telegraph and telephone lines, thousands of wells, one airport, and three full-service oil ports of different sizes: Puerto Lobos, Tuxpan, and Tampico. As the number of companies chasing black gold multiplied into the hundreds, the forest fell to the workman's machete, its verdure suddenly replaced by blackness. Oil spilled from wells and broken pipelines blanketed foliage, waterways, sand dunes, swamps, and beaches. The "gusher" wells that made Mexico famous exploded and burst into flames—"blow-outs," as they were called, spreading oil, fire, and fear far and wide.[16] Fuel-soaked rivers and streams caught fire, at times reaching oil tankers and detonating them like matches on dynamite.[17] Pollution on that scale had no precedent in the history of the Huasteca. Neither did that level of environmental destruction. The oil industry spared no ecosystem. No mangrove, marsh, coastal sand dune, or estuary escaped its stranglehold. The rainforest was no exception. On the ground, the industrial apparatus of oil, including extraction, refining, transportation, and shipping, translated into an ecological rampage envisioned as progress. In forty years of exercising power over nature, the oilmen destroyed the rainforest. In the process, they also enforced a class regime that determined relations not only among humans but also between human beings and nature.

Class and Nature

The same men who dictated the fate of the forest designed the system of class divisions and relations that would rule in the Mexican oil industry and shape how workers experienced life in the tropical rainforest. The founders of the industry established a rigid occupational hierarchy embedded with the same racial categories that they used to classify the workers in the United States and oil fields everywhere.[18] The top echelons—executives and geologists—were white, Americans or Europeans. The half-dozen Mexican engineers the companies hired were never promoted to executive status.[19] The master craftsmen and the drillers, the indispensable working-class men of the industry, were also exclusively white, as in the United States. The oil barons barred Mexicans from these positions. Mexican craftsmen were hired as assistants, never as masters. The executives reserved the overwhelming majority of Mexican workers for manual labor. Thus the oil companies organized and managed some 2,500 to 4,000 foreigners and possibly upwards of 45,000 Mexicans at the peak of employment in 1921, to tap Mexico's black gold.[20]

The class hierarchy the executives crafted governed not only the organization of the workplace but also the social and environmental spaces outside work. All facilities in the industry were segregated according to skin color, with "whites-only" dining halls, dormitories, hotels, housing complexes, infirmaries, and recreational clubs. The quality of the physical structures, the spatial locations, and the services offered in each improved according to the employment category of the worker. In addition to the sting of social categorization based on artificial constructions, that hierarchical structure meant vastly different encounters with nature for each occupational layer. For executives, the tropical rainforest was a malleable location that they could alter through their control of labor for profit, comfort, and recreation. Working-class men, both foreign and Mexican, by contrast, encountered nature through work.[21] Neither group of workers assumed control over the environment they inhabited like their bosses did. They interacted with their natural environment through the course of the day based on their working conditions. Yet their commonalities ended there. As recognized masters of their craft, foreign workers were steps above in the occupational hierarchy, and that status afforded them protections laborers lacked. Coupled with the premiums the oil barons conferred upon foreign men because of their nationality and race, the privilege this small group of men enjoyed was considerable, extending well beyond substantially higher wages. For the lowest rung on the occupational ladder, Mexicans all, extracting oil from the Huasteca never ceased to be a risky proposition, with nature representing a hostile and dangerous force that battered them in many and often unexpected ways.

Lords of the Fields

The oil executives were confident men accustomed to using and displaying power and control over men and nature alike. They celebrated their prowess thus:

> Discovery and development of the known oil fields in Mexico were the achievement of British and American pioneers, who came into this region at a time when it was a little-known, pest-infested, tropical wilderness. . . . In the face of almost insuperable obstacles, they made remarkably rapid progress. In less than 10 years Mexico had begun to attract worldwide attention as an oil producer. Development of the famous "Golden Lane," one of the world's greatest known oil fields, discovered in 1910, placed Mexico in the first rank among oil-producing countries. The transformation was profound. Tampico, a sleepy little fishing port, became almost overnight a thriving city. . . . In the oil fields, where formerly the tropical jungle supported only a few Indians, 50,000 oil field workers, largely Mexicans, found immediate, continuous employment.[22]

Yet industrial infrastructure did not describe the whole picture. The oilmen changed the face of the tropical rainforest for other reasons as well. To "make living bearable" in the tropics, the executives ordered the remaking of the landscape in the image of the one they had left abroad: company men replaced the rainforest with imitations of the English countryside or Southern California. Pearson's firm El Aguila, for instance, surrounded employee homes with "English gardens" planted with exotic flora such as rosebushes and begonias.[23] The Americans, for their part, assigned Mexican workers to surround executives' homes with white picket fences and Los Angeles landscaping: palm trees uprooted and replanted in neat rows, citrus trees for shade, and imported seed for green front lawns and grassy backyards.[24]

To relax from the arduous tasks of bringing progress to tropical landscapes and peoples, the upper echelons of the oil companies engaged in recreational activities that reaffirmed their mastery over nature. With Mexican guides and porters to lug equipment, they hunted the fauna already under pressure from habitat loss as wells replaced trees: jaguars, pumas, ocelots, wild turkeys, and other aptly named "game." Alligators were harpooned. Prize fishing for shark or tarpon, a marine fish that swam upstream through increasingly oil-polluted river waters to spawn and could weigh over one hundred pounds, was all the rage. Other sporting activities among oil bosses entailed hiring laborers to eliminate the habitats of local flora and fauna and replace it with "slick greens" for golf.[25]

By virtue of their class position, then, the oilmen reshaped nature to fit their desires, whether in the form of a profitable industrial site or as a familiar, comfortable, and recreational space. In production and recreation, at work and at play, oil executives played the role of masters of nature rather "naturally," as befit members of a social class accustomed to command. Elsewhere in the class divide, nature looked and felt considerably different.

Masters of Their Craft

Drillers and craftsmen, the middle rung in the occupational hierarchy of the oil industry, knew nature intimately but not with the intimacy of domination their executive bosses manifested. These men were hands-on, immediate agents of environmental change, digging up the earth's entrails with their coarse tools, channeling fossil fuels through man-made contraptions and inventions, and transforming crude oil into usable liquid energy. They worked in the great outdoors for eight to twelve hours per day (depending on the decade), wrestling "resources" from the earth through considerable physical effort and technical skill, that is, exercising control over their bodies and craft. In the process, they met the forces of nature head-on, and they were not always the winners. Their value to the oilmen lay in their skill, so they merited investment in a harsh natural environment. The experience of this small group of men in the Huasteca rainforest is revealed by looking at health, safety, and leisure patterns.

Skilled workers from outside the Huasteca did not know necessarily that health conditions in the tropical rainforest were difficult. The bosses, however, were well informed and careful about providing protection for the men who would make them rich. The Huasteca hosted a variety of microscopic life that caused humans trouble: dysentery, malaria, yellow fever, smallpox, and bubonic plague were all part of the landscape. To combat them, the executives made sure their craftsmen enjoyed the best sanitation money could buy at the dawn of the twentieth century: potable water, indoor plumbing, fans and ice to tame high temperatures and humidity, mosquito netting, window screens, clean bedding, and showers and baths. By all accounts, such public health measures were quite successful in keeping foreign workers as healthy as possible.[26]

Safety, however, was a separate issue. Oil was an inherently dangerous enterprise, flammable from extraction to consumption. Accidents routinely occurred, including many directly related to the natural chemical composition of Huasteca crude. The petroleum was heavy in hydrogen sulfide, a gas capable of poisoning any creature that inhales it. Drillers were keenly aware of the dangers, as the gas hissing was the first sign of having tapped oil. When the whistling hydrogen sulfide or the gushing oil caught fire, the risks increased exponentially. American worker testimonials recalled deaths of foreign and Mexican workers killed under such circumstances.[27] The photographs of executives and workers taken when the Huasteca well Cerro Azul No. 4 came in epitomize the class differences among foreigners in their relationship and interactions with nature. The higher echelons look very bright in their light-colored and spotless suits and hats, while the workers, "all white American citizens," are totally black, drenched in oil from head to toe, their whiteness limited to their eyeballs.[28] For the drillers, then, nature was unpredictable and dangerous. They did not presume to be masters of the ecology surrounding them. At work, their position was a defensive one, where mastery over their craft was not only what made them a living but also what provided them protection, preventing them from becoming the next victims of the nature of crude oil.

The impulse to shield themselves from danger was most obvious in another way: white working-class men transferring risks to Mexican workers.

The "wages of whiteness" that American social structure conferred on fair-skinned workers in the United States also applied in Mexico.[29] Foreigners became supervisors in the Mexican oil fields, a position in the labor hierarchy that allowed them to avoid some dangerous tasks by assigning such tasks to their Mexican subordinates. Thus, foreign white working-class men in the Mexican oil industry also received an environmental wage, paid by the companies' efforts to keep them disease-free and by the bodies of Mexican men.

Nevertheless, the class differences among the foreigners were deep and significant in environmental terms. Although the craftsmen and the executives shared the social privileges of whiteness in Mexico, the two did not mingle. Craftsmen were entitled to membership in corporate social clubs by virtue of nationality and skin color, but they did not take advantage of the opportunity. There is no record that they played golf, joined prize-fishing expeditions, or engaged in outdoor adventures with their superiors. By all accounts, they spent their scant off-duty hours in the male entertainment centers that sprang up throughout oil country: casinos, bars, and brothels.[30] Given how hazardous oil extraction was, craftsmen seem to have decided that indoor recreational activity was more appealing than chasing tropical rainforest fauna.

Laboring Mexicans

If the occupational status of craftsmen meant some protection from the natural dangers of oil extraction, the men at the bottom of the ladder enjoyed no such considerations. Such an assertion may seem contradictory in light of the fact that Mexican men held nature in their hands, literally and often. They chopped down the trees, dug up the mangroves, cleared away the marshes, and filled in the swamps as needed for the infrastructure of oil extraction and refining. Like the foreign craftsmen, Mexicans knew nature through work. Yet precisely because their labor entailed sheer physical exertion in intimate and extensive bodily contact with the natural world, they were the humans most exposed to the risks and dangers of oil production in a tropical environment. If master craftsmen survived the grueling task of petroleum extraction largely healthy and minimally mutilated, hundreds of Mexicans did not. Work for laborers was a daily struggle against the weather, injuries and disease, toxic chemicals, fire, and workplace hazards of all types, including those passed on to them by working-class men steps above in the hierarchy.

The lowest-ranked workers, for starters, were affected by weather more than anyone else. Company housing for Mexican workers was consistently built in the floodplain, as whatever high ground existed was reserved for lodging foreigners. That location meant that seasonal rains and hurricanes—an inconvenience and annoyance to those higher up in the labor hierarchy and the terrain—were disastrous for Mexicans. Their quarters were always the first to flood and be carried away by raging rivers.[31]

Mexican workers also suffered the most from the biological specimens that crowded the forest they were uprooting. Malaria, the fallout of the rainy season, downed laborers by the thousands, killing countless every year. Epidemic disease (including yellow fever, influenza, and bubonic plague) ravaged Mexicans. In large part, squalid living conditions accounted for those illnesses: company-issued housing for Mexican workers lacked the most basic public health necessities, including running water and toilets that were routine among foreigners. Not until the labor unions and local authorities instituted basic public health services in the 1920s did epidemic diseases abate among Mexican working-class men and their families.[32]

Similarly, the men who did the heavy lifting and menial labor in the construction process risked life and limb on a daily basis. Demolishing a rainforest was terribly dangerous work. Broken bones were common as Mexican workers chopped down trees. Being crushed to death under falling tree trunks or branches was also not unusual.[33] Laying pipeline to transport crude oil from the wells to the refineries or loading docks posed serious dangers as well. Workers pulled muscles, dislocated shoulders, sprained ankles, tore ligaments, or twisted joints with regularity in this task. Others, such as carpenters, blacksmiths, and storage tank builders, reported hand, foot, and head injuries on a regular basis. Many more men complained of sore muscles and heat exhaustion from working up to fourteen hours in high temperatures and humidity.[34]

Chemical agents also affected low-level workers more than anyone else. At work, Mexican men, like their foreign working-class brethren, were exposed to the full toxicity of hydrocarbons in all their permutations: crude oil and distillates such as gasoline, kerosene, solvents, and other refined products. The "irritating gases" they inhaled on a daily basis affected their overall health. Refinery workers, for instance, exhibited symptoms of mild poisoning: nausea, heartburn, headaches, eye irritation, sore throats, tremors, and difficulty breathing.[35] But while many craftsmen and virtually all executives could escape "the stench of petroleum" after the work shift by living on breezier higher ground, Mexican laborers could not. In the camps, the companies erected laborers' quarters next to wells or tanks, exposing workers to steady chemical emissions.[36] In Tampico the story was similar. Workers' houses were next to plants that released toxins into the air and water, polluting both and endangering the health of workers and their families.[37] The location of housing engineering by executives, then, spared them considerable exposure while they condemned Mexicans to life in toxic neighborhoods. Pollution, therefore, was a class issue as well.

Although fire was a risk everyone shared, Mexican workers experienced its dangers most. On the worksite, Mexican laborers were the ones ordered to become fire fighters whenever a well or a tank ignited. Needless to say, the men received no training for that job. Equipment for fire control was all but nonexistent throughout the period of foreign ownership of the industry, so the men confronted the flames with nothing more than shovels, thin metal chest shields, and "wet sac[k]s around their heads and hands."[38] Exposure to fire, moreover, followed Mexican workers home. Their housing quarters, by company design, were built in the shadow of storage tanks, making them susceptible to tank accidents. When one storage tank blew, there went the neighborhood.[39]

Lastly, individual Mexican workers were subjected to the dangers foreign workers passed on to them. The most common example involved work in confined spaces, such as measuring the amount of oil in a storage tank or diving into a recently emptied boiler or still to clean its hydrocarbon residues. In every instance, containers were extremely hot and exuded poisonous compounds that could kill a man within minutes if vigilant safety precautions were not followed. There is no evidence that supervisors offered to Mexican workers the experimental protective respiratory equipment that circulated prior to nationalization.[40] So it was that the men at the bottom of the occupational hierarchy, Mexican oil workers, were the bull's-eye for natural phenomena of every possible kind. Their experience of and in nature was quite distinct from that of foreign white working-class men and utterly remote from that of the oil barons. The reality of class was that the unequal distribution of power among the different groups of men in large part determined the way those men moved through their common environment. All of them lived in the same place, but they all inhabited very different spaces.

Environment and Class Warfare

The conditions Mexican workers faced in the oil industry did not go unchallenged. On the contrary, Mexican workers were notoriously riotous and militant from the 1910s through the 1938 nationalization. Scholars recognize that the petroleum industry was a site of intense class conflict in the first decades of the twentieth century.[41] Labor conditions made Mexican workers quite angry, "naturally." I suggest that this class struggle had environmental dimensions. Likewise, the battles between the Mexican revolutionary state and the oil companies, well documented in the historiography, were also fights over nature. At issue was who would control nature and for what ends. The Mexican oil workers were key players in these conflicts, which ultimately led to nationalization of the industry in 1938, although the credit does not typically accrue to them.[42]

Workers did not speak the language of environmentalism, of course. The times availed them of radical discourses coming out of anarchism and the nationalism of the Mexican Revolution itself, which coincided with the oil boom of 1910–21. The language adopted by the oil workers' movements assailed oilmen as "bloodsucking" capitalists bleeding workers for profit and as shameless imperialists bent on extracting every last drop of Mexico's petroleum wealth. Such ideas made clear the connection between the exploitation of nature and the exploitation of men.[43] As radicals and nationalists, Mexican oil workers denounced the negative effects that oil had on the land.[44] But their militancy was not about the land. It was about economic conditions and discriminatory treatment. Yet they also placed concerns over health and safety "very near the surface," as American workers did contemporaneously.[45] Therein lay the environmental aspects of this particular labor struggle: the men's health was affected by the toxic hydrocarbons they were exposed to at work and at home and the microscopic rainforest life that produced illness in men living in suboptimal conditions. Those were the aspects of nature and the interactions with the natural environment that Mexican oil workers highlighted, the ones that affected them directly in daily life. Indeed, scholars who have catalogued oil workers' demands have found that while wages and ill treatment topped the list, health and safety were next.[46]

Health and safety issues thus became a hidden environmental battleground ensconced in class relations and labor struggles in the oil industry. The 1924–26 strikes that won recognition for the unions and the first collective contracts included extensive health and safety issues that revealed a broad definition of occupational health. Gulf company strikers, for instance, demanded compensation for accidents that resulted in death or injuries such as the loss of fingers, legs, eyes, ears, or teeth, as well as face burns; a hospital with "modern comforts and advantages"; and prostheses for those who lost limbs at work. They also formally requested individual safety equipment: gloves, helmets, chest shields, and boots. Furthermore, they wanted free health care and full pay in case of illness contracted on the job. That included "even," as they labeled it, those illnesses contracted on account of "poor climate."[47] Men from the Pierce refinery made similar demands, including double pay for work in "unhealthy" locations.[48] The men from Cerro Azul were equally adamant. In 1925, their demands included environmental elements: double wages for work inside bodies of water and oil containers, and when pipelines broke.[49] Men at El Aguila's Potrero del Llano camp submitted a similar document in 1926, demanding compensation for accidents, a hospital, safety equipment, and lower temperatures for tank cleaners or double wages for working inside tanks over 55°C (122°F).[50]

The companies fought back hard against the workers on all counts. Forced to sign contracts because of strikes and disruptions in production, they failed to deliver on contractual obligations. Instead, they fired hundreds and closed facilities. In 1932 only three refineries remained open in Tampico. Those who kept their jobs endured wage cuts of up to 50 percent.[51] They did receive some safety equipment and free care at existing clinics, but the companies found ways to dismiss claims for health issues, including docking pay for sick days and summarily firing injured workers.[52] If the men wanted any compensation for health problems, they had to sue before the arbitration boards, the official bodies established by the 1917 Constitution to resolve employer-employee conflicts. Arbitration files, as a result, bulge with such cases in both Veracruz and Mexico City.

The arbitration cases reveal the radically different interpretations that the workers and the companies had developed regarding health and safety. While the companies held workers individually responsible for their bodily integrity, workers subscribed to a more complex and integrated approach. They held the companies responsible for workers' health and safety on the job, the neighborhood, and the rainforest in general—in other words, all the spaces where the men interacted with the natural world. Workers did not disentangle spaces. Oil effluents permeated all three. Furthermore, the companies recruited workers from other ecosystems and brought them into the tropics, in essence leading the men to cross ecological boundaries they might not have trespassed otherwise. The assaults workers encountered at work, at home, and in the tropical rainforest were one package, the direct consequence of oil work. Thus, the men held the companies responsible for their health and safety throughout. As José Ramírez argued before the arbiter, "Well, doesn't the businessman or owner pay from his own pocket the repairs that have to be done to the machines when these wear out in the production process?"[53] Workers deserved the same treatment as machines, at the very least.

All those experiences crystallized in the final confrontation between the companies and the workers, the November 1936 union contract. The unified oil unions, representing some 13,000 to 18,000 men, submitted their most ambitious list of demands ever. Chapter 8 alone, entitled "Of Illnesses and Medical Care in General," for example, had forty-three separate clauses, while chapter 9, "Compensation, Safety and Industrial Hygiene," demanded in clauses 56 and 131 that the companies work to prevent industrial accidents and take "all precautions that science demands."[54]

The proposed contract illustrated the negative ways in which workers encountered nature in its many guises. It included clauses that demanded double salary for dangerous jobs, including work more than ten meters (thirty-three feet) above ground as well as those involving environmental hazards, such as temperatures exceeding 100°F or areas of "excess gas," pipeline repair, tank and still cleaning, oil spill remediation, and work in the tropical rain.[55] Workers demanded salary-and-a-half for any work that required handling toxics, corrosives, acids, sulfur, phosphorus, dynamite, gunpowder, and "similar substances."[56] Furthermore, the men demanded recognition of certain ailments as occupational diseases, including malaria and tuberculosis, on top of comprehensive health care not only for active workers but also for retired men and their families.[57]

The negotiations were extremely hard-fought, yet ultimately they failed. After 120 days of talks, the workers went on strike at the end of May in 1937. They shut down 178 oil installations for 13 days. With gasoline lines stretching for miles, the president sent the contract to state officials for resolution. The conflict reached the Supreme Court while wildcat strikes rocked the industry. In a decision that was heard around the world, the court ruled against the companies in March 1938. The companies rejected the ruling, and Mexican petroleum workers called for a general strike, prompting the president to take drastic action. Cárdenas chose expropriation, closing a chapter in the history of oil in Mexico and opening another one in the history of oil companies abroad.[58]

Conclusion: Hidden Histories of Class and Nature

Humans interact with nature as a species. They utilize, modify, destroy, or protect nature together. But just as class shapes and molds attitudes, tastes, worldviews, and relations among humans, it also shapes and molds how humans experience their natural environments. Taking class into account in this way complicates and reveals hidden histories of nature and class.

Thinking about class sharpens our understanding of the relationship between labor and nature and gives us a more fluid and complex view of nature itself. The notion also suggests that every class evokes a different definition of nature rooted in its social practices and relations.[59] The oil barons, an undeniably important group of the upper class in the twentieth century, showed through their praxis a definition of nature that conforms best to the American contemporary popular view of it: as wilderness, a place to be tamed, enjoyed, or exploited, depending on the purpose or occasion. But the same was not true for working-class men in the oil industry. For them, nature exhibited additional dimensions: microorganisms, chemical compounds, fire, heat, weather. That is not to say that the great oil capitalists of the early twentieth century were not aware of those other facets of the natural world or were not affected or even hindered by them. It is just that their class position removed them from close proximity to such uncomfortable features and sheltered them from the most negative effects. Working-class men suffered nature differently, in their work exposures, in their safety and health, in the pollution in their neighborhoods and the shifts of weather cycles.[60] Disassembling class into its occupational hierarchy, moreover, allows us to see just how such skill-and-power arrangements made nature operational in the daily life of workers. Thus, finding ways to bridge labor and environmental history allows us to highlight aspects of nature that would otherwise be obscured, to participate in the continuous cultural reinvention and reinterpretation of nature.

Linkages across discrete fields can unveil other hidden histories. The Mexican oil industry during the period of foreign ownership is a good example. Labor, environment, class, and nature intersected in explosive ways, both literally and figuratively. Mexican workers confronted the oil companies over wages, hours, and working conditions, as most workers of the world still do. Those struggles and discourses, nevertheless, had nature imbedded in them, pieces that might not be obvious at first glance but are integral to such battles and constitute a critical part of the history of how working-class men and women have lived and interpreted their subjectivity in their environments. The radicalism and nationalism that Mexican oil workers embraced and manifested over decades was in fact tinged with environmental concerns. Indeed, the same is true of President Lázaro Cárdenas's nationalization decree. As several environmental historians have documented, Cárdenas was quite conscious of the importance of nature in the economic life of the nation. He took conservation and what he called "the salvation and protection of nature" seriously.[61] No one would go as far as calling Cárdenas Mexico's first "green" president or his ideology a "green nationalism," but historians have begun to recognize his environmental sensibilities. A joint labor and environmental history approach shows how much the oil workers pushed him to deliberate on the relationship between labor, nature, and nationhood.[62] What has yet to be fully explored is how working-class (and peasant) voices and struggles in general informed Cárdenas's praxis regarding the management of nature. Without a doubt, hidden histories lie in wait there. The invitation to uncover them is open to all interested historians.

Notes

1. Doheny quoted in PanAmerican Petroleum and Transport Company, *Mexican Petroleum* (New York: PanAmerican Petroleum and Transport Company, 1922), 16–17.

2. Testimony of Edward L. Doheny, United States Senate Committee on Foreign Relations, *Investigation of Mexican Affairs*, 66th Cong., 2nd sess. (Washington, D.C.: Government Printing Office, 1920), 2:236–38.

3. Interview with Cruz Briones Rodríguez, conducted by Lief Adleson on 28 November 1976 in Tampico, Tamaulipas, Proyecto de Historia Oral (PHO), Instituto Nacional de Antropología e Historia, 4/52.

4. On Latin America, see Steve Marquardt, "'Green Havoc': Panama Disease, Environmental Change, and Labor Process in the Central American Banana Industry," *American Historical Review* 106, no. 1 (February 2001): 49–80, and "Pesticides, Parakeets, and Unions in the Costa Rican Banana Industry, 1938–1962," *Latin American Research Review* 37, no. 2 (2002): 3–36.

5. Christopher Sellers, "The Dearth of the Clinic: Lead, Air, and Agency in Twentieth-Century America," *Journal of the History of Medicine and Allied Sciences* 58, no. 3 (July 2003): 255–91.

6. Elite control over nature and large numbers of humans is one of the hallmarks of civilization and one of the causes of environmental degradation since antiquity, according to Sing C. Chew, *World Ecological Degradation: Accumulation, Urbanization, and Deforestation, 3000 BC–AD 2000* (Walnut Creek, Calif.: Altamira Press, 2001). See also Angus Wright, *The Death of Ramón González: The Modern Agricultural Dilemma*, 2nd ed. (Austin: University of Texas Press, 2005).

7. Myrna Santiago, *The Ecology of Oil: Environment, Labor, and the Mexican Revolution, 1900–1938* (Cambridge: Cambridge University Press, 2006), 340–41.

8. Bernardino de Sahagún, "¿Quiénes eran los huaxtecos?" in *Huaxtecos y Totonacos: Una antología histórico-cultural*, ed. Lorenzo Ochoa, 133–34 (Mexico City: Consejo Nacional para la Cultura y las Artes, 1984).

9. Santiago, *Ecology of Oil*, 19–27.

10. Ibid., 27–30, 37–57.

11. Ibid., 64–66.

12. Bess Adams Garner, *Mexico: Notes on the Margin* (Boston: Houghton Mifflin, 1937), 118.

13. Margaret Leslie Davis, *Dark Side of Fortune: Triumph and Scandal in the Life of Oil Tycoon Edward L. Doheny* (Berkeley: University of California Press, 1998), 88.

14. Santiago, *Ecology of Oil*, 15, 76, 101, 115.
15. Carolyn Merchant refers to the changes that capitalism wrought as an "ecological revolution." See Carolyn Merchant, *Ecological Revolutions: Nature, Gender, and Science in New England* (Chapel Hill: University of North Carolina Press, 1989).
16. Santiago, *Ecology of Oil*, 102, 108–9, 115, 118–19, 125–29, 133–34.
17. *El Mundo*, 20, 28, and 30 January 1927.
18. On the United States, see Nancy Lynn Quam-Wickham, "Petroleocrats and Proletarians: Work, Class, and Politics in the California Oil Industry, 1917–1925," PhD diss., University of California, Berkeley, 1994. On the organization of labor in oil fields abroad, see Robert Vitalis, *America's Kingdom: Mythmaking on the Saudi Oil Frontier* (Stanford: Stanford University Press, 2007); and Miguel Tinker Salas, *The Enduring Legacy: Oil, Culture, and Society in Venezuela* (Durham, N.C.: Duke University Press, 2009).
19. Testimony of Edward L. Doheny, *Investigation of Mexican Affairs*, 220, 228–29.
20. Jonathan C. Brown, *Oil and Revolution in Mexico* (Austin: University of Texas Press, 1993), 319; Informe, 18 November 1921, Departamento del Trabajo (hereafter DT), Archivo General de la Nación (hereafter AGN), caja 326, exp. 3.
21. See Robert White, *The Organic Machine* (New York: Hill and Wang, 1995).
22. Huasteca Petroleum Company, *Expropriation* (New York: Huasteca Petroleum Company and Standard Oil Company of California, 1938), 1–2.
23. H. S. Wood to DeGolyer, Tampico, 1 November 1917; U. S. Wood to Messrs. Peter Henderson & Co., Tampico, 11 September 1917; and H. S. Wood to DeGolyer, Tampico, 11 September 1917, Papers of Everett Lee DeGolyer (hereafter ED), Southern Methodist University (hereafter SMU), Box 117, Folder 5377.
24. The irony of the Los Angeles landscape the oilmen reproduced in Mexico is that it was itself the product of men like Doheny who transformed it from a desert and semidesert into their own interpretation of a Mediterranean landscape. *Boletín del Petróleo* 25, no. 4 (April 1928), photographic section; *Boletín del Petróleo* 23, no. 4 (April 1927), photographic section.
25. Charles W. Hamilton, *Early Day Oil Tales of Mexico* (Houston: Gulf Publishing Company, 1966), 47, 130; DeGolyer Diary, 27 and 29 February 1916 and 5 March 1916, SMU, ED, Box 105, Folder 5.
26. John Spender, *Weetman Pearson, First Viscount Cowdray, 1856–1927* (London: Cassell and Company, 1930), 106; Interview, Doheny Mexican Collection, Occidental College, Los Angeles, California, Labor File I, #958, #2724.
27. Transcribed interview with W. M. Hudson, Oral History of the Texas Oil Industry Collection, 1952–1958, Dolph Briscoe Center for American History, University of Texas at Austin, Tape 79, p. 35.
28. Mexican Petroleum Company of Delaware, *Cerro Azul No. 4: World's Greatest Oil Well* (New York: DeVinne Press, n.d.).
29. See David Roediger, *Wages of Whiteness: Race and the Making of the American Working Class* (London: Verso, 1991).
30. Hamilton, *Early Day Oil Tales*, 24–25, 34, 40.
31. Santiago, *Ecology of Oil*, 187–88.
32. Ibid., 169–70, 189–93.
33. *Aurelio Herrera vs. El Aguila*, Archivo General del Estado de Veracruz (hereafter AGEV), Junta Central de Conciliación y Arbitraje (hereafter JCCA), caja 65, exp. 65,

1927; Informe del Inspector del Trabajo A. Araujo, AGN, Departamento del Trabajo (hereafter DT), C 489, exp. 9, 1922.

34. Gerente General to Presidente Municipal, 20 August 1919, Archivo Histórico del Ayuntamiento de Tampico (hereafter AHAT), exp. 85-1919; Gerente General to Presidente Municipal, 24 March 1920, AHAT, exp. 78-1920, no. 1693; Leopoldo Alafita Méndez, "Trabajo y condición obrera en los campamentos petroleros de la Huasteca, 1900–1935," *Anuario* 5 (October 1986): 187–96.

35. W. A. Jacobs and C. W. Mitchell, "Métodos industriales usados para la eliminación de los gases tóxicos, altamente sulfurosos, desprendidos del petróleo mexicano," *Boletín del Petróleo* 21, no. 1 (January 1926): 1–2.

36. Verna Carleton Millan, *Mexico Reborn* (Cambridge, Mass.: Riverside Press, 1939), 215.

37. Santiago, *Ecology of Oil*, 195–96.

38. Arthur B. Clifford, "Extinguishing an Oil-Well Fire in Mexico, and the Part Played Therein by Self-Contained Breathing-Apparatus," *Transactions of the Institution of Mining Engineers* 63, no. 3 (1921): 3.

39. Santiago, *Ecology of Oil*, 194.

40. One piece of protective equipment was a gas mask, but the instructions for its use were in English. "Rules and Precautions to be Observed When Using the Proto-Self-Contained Breathing Apparatus," AGN/DT, caja 224, exp. 24, 1919.

41. Alan Knight, *The Mexican Revolution*, vol. 1, *Porfirians, Liberals and Peasants* (Lincoln: University of Nebraska Press, 1990), 406–7.

42. Santiago, *Ecology of Oil*, 330–41.

43. *Sagitario*, 25 October 1924.

44. Santiago, *Ecology of Oil*, 271–73.

45. David Rosner and Gerald Markowitz, eds., *Dying for Work: Workers' Safety and Health in Twentieth Century America* (Bloomington: Indiana University Press, 1986), ix.

46. Armando Rendón Corona, Jorge González Rodarte, and Angel Bravo Flores, *Los conflictos laborales en la industria petrolera*, vol. 1, *1911–1932* (Mexico City: Universidad Autónoma Metropolitana, 1997), 131–33, 171.

47. Pliego petitorio, "Dificultades: Mexican Gulf, 1924," AGEV, JCCA, caja 40.

48. Rendón Corona, González Rodarte, and Bravo Flores, *Los conflictos*, 1:230.

49. Petitions, Huasteca Strike in Cerro Azul, January 1925, AGEV, JCCA, caja 48.

50. *Sindicato de Obreros del Petróleo de Potrero del Llano vs. El Aguila*, 1927, AGEV, JCCA, caja 59, exp. 24.

51. Santiago, *Ecology of Oil*, 310–11.

52. Enclosure No. 1, Visit of President Cárdenas to Tampico and Its Effect on the Local Strike Situation, 11 January 1935, Record Group 59, General Records of the Department of State, Records of the Department of State Relating to Internal Affairs of Mexico, 1930–1939, National Archives, College Park, Maryland, 812.001—Cárdenas, Lázaro/40.

53. *José G. Ramírez vs. East Coast Oil Company*, AGN, Junta Federal de Conciliación y Arbitraje, C 22, exp. 15/928/130, 1928.

54. Collective contract proposal, AGN, Departamento Autónomo del Trabajo, caja 154, exp. 4, 1937.

55. Ibid., clauses 53–55.

56. Ibid., clause 56.

57. Armando Rendón Corona, Jorge González Rodarte, and Angel Bravo Flores, *Los conflictos laborales en la industria petrolera y la expropriación*, vol. 2, 1933–1938 (Mexico City: Universidad Autónoma Metropolitana, 1997), 155–58.

58. Santiago, *Ecology of Oil*, 330–39.

59. On the multiple meanings of nature in the United States, see William Cronon, ed., *Uncommon Ground: Rethinking the Human Place in Nature* (New York: W. W. Norton and Company, 1983).

60. The literature on environmental justice addresses the issue of pollution and neighborhood. See, for example, Laura Pulido, *Environmentalism and Economic Justice: Two Chicano Struggles in the Southwest* (Tucson: University of Arizona Press, 1998); and Robert D. Bullard, *Dumping in Dixie: Race, Class, and Environmental Quality* (Boulder, Colo.: Westview Press, 2002). To examine how the concept is being applied to Latin America, see David V. Carruthers, ed., *Environmental Justice in Latin America: Problems, Promise, and Practice* (Cambridge, Mass.: MIT Press, 2008).

61. Quoted in Simonian, *Defending the Land of the Jaguar: A History of Conservation in Mexico* (Austin: University of Texas Press, 1995), 87; Christopher R. Boyer, "Contested Terrain: Forestry Regimes and Community Responses in Northeastern Michoacán, 1940–2000," in *The Community Forests of Mexico: Managing for Sustainable Landscapes*, ed. David Barton Bray, Leticia Merino-Pérez, and Deborah Barry, 27–48 (Austin: University of Texas Press, 2005); Emily Wakild, "'It Is to Preserve Life, to Work for the Trees': The Steward of Mexican Forests, Miguel Angel de Quevedo, 1862–1948," *Forest History Today* (Spring/Fall 2006): 4–14.

62. Santiago, *Ecology of Oil*, 289–90, 351.

CHAPTER NINE

Parables of Chapultepec
Urban Parks, National Landscapes, and Contradictory Conservation in Modern Mexico

Emily Wakild

Perhaps the oldest nature preserve in the Americas, the verdant oasis of Chapultepec Park, lies within the depths of Mexico City, a metropolis so large it literally chokes on itself. The 850 hectares of arboreal integrity reach outward to the skyscrapers and sinewy highways that grasp its foliage. The park provides one of a few areas where birdsongs drown out taxi horns and squirrels scamper across branches of ancient *ahuehuetes* and fast-growing eucalyptus trees.[1] The park is not wild by any definition, yet the sculpted lakes, meandering paths, and sprawling lawns invite visitors to rest and reflect, shrouded by nature that is scarce elsewhere in the city. And visit they do. Paddleboats, roller coasters, and horse-mounted security guards move through crowds of competitive bicyclists, amorous teenagers, and gleeful toddlers. Both a destination and a source of the city's happenings, the park begins traditional parade routes and bridges centuries of cultural traditions and natural abundance, evident in King Nezahualcóyotl's baths, the prominent Chapultepec Castle, the sitting president's residence, the museum of anthropology (and national history, modern art, and more), and a zoo full of animals ranging from native *axolotl* salamanders to exotic zebras. Nowhere else can visitors sample the country's past and present, nature and culture with such ease.[2]

As it surrounds the park, the city provides a counterpoint brimming with environmental tragedies. From below the ground up past the rooftops, the tangle of humanity and lifeless construction evokes sublime chaos—a topography at once drawing in and repulsing observers. Roads, pipes, and bridges constantly shift and tear, with the sagging earth below causing potholes, leaks, and explosions. In earthquakes, shoddily constructed buildings shake and implode, while the subway and colonial churches prove their resilience. Through a perpetual process of collapse and renewal, the urban landscape strips down weakened edifices and builds up anew, like a forest floor. The city ecosystem is hardly organic, yet it pulses with life and cycles through opportunity and catastrophe intimately connected to its rural surroundings. Hundreds of rural migrants arrive daily, pushed into the city by depleted soils and depressed agricultural prices, augmenting the collective tally of resource use. And the list of paradoxes mounts.[3]

The existence of a naturalized oasis such as Chapultepec inside the archetypal uncontrolled urban environment reveals the contradictions of a larger national relationship with, and fascination for, nature. Chapultepec presents a set of parables for the broader history of nature conservation in modern Mexico. Orderly nature in the beloved park depicts the triumph of reason over the unruly and delirious culture of Mexico City. In the same way, the development of nature protection on a national scale drew upon modern science as it expressed the nation's timeless and organic existence. As a longstanding precedent for valuing natural spaces, considering Chapultepec within the nationwide array of parks illuminates how domestic and foreign notions of nature contributed to making parks hubs of state power. But Chapultepec further reveals that parks are more than political constructions; they reflect and shape cultural trends in a dynamic society. Chapultepec and an extensive network of national protected areas exist and function as hybrid landscapes sustained by the societal changes that alternatively reinvigorate and compromise their ecological integrity. Parks have histories that tell us more about the societies that created them than of the natures they enclose.

This chapter examines the history of Mexican parks to offer some explanations for their existence. Over the course of the twentieth century, the Mexican government created nearly two hundred national parks, nature reserves, and protected natural areas covering almost 10 percent of the federal territory.[4] They ranged from urban parks of a few hectares to vast marine reserves encompassing millions of hectares. Much of this conservation captured a spectacular range of nature valued differently over time by multiple, and sometimes competing, constituencies. The nation's territory spans fifteen latitudinal degrees from desert to tropical forest and includes complex topography reaching from dry islands in the Sea of Cortés to glaciated alpine peaks above 5,600 meters. The corresponding variety of plants and animals is immense: more than 65,000 species have been described and over 200,000 are thought to exist. They range from charismatic jaguars to monarch butterflies, from mysterious boojum trees to stately cypress.[5] When judged solely on the diversity of habitats, few nations can boast as much value, and Mexico belongs in an esteemed group of nations termed "megadiverse" by biologists.[6]

Although impressive, neither the size nor the content of this network of protected areas constitutes its most remarkable feature. Instead, the chronology of actions taken to create protected areas demonstrates that nature protection is not a feat reserved for so-called developed countries. Mexican conservation history shows us that a nation struggling with social revolution, haphazard industrialization, and rural-to-urban migration (among other challenges) created a coherent (yet at times contradictory) and resilient conservation system with an ability to accommodate conflicting values. The attributes of Mexican conservation paralleled the contours of its changing political economy, involved a commitment to modernization, and fostered institutional overlaps in park administration. Through national parks, biosphere reserves, and even the continual rejuvenation of Chapultepec, federal administrators strove to harmonize their commitment to expanding economic development with a countervailing tendency to conserve the natural world. These interlocked imperatives—to produce and conserve—are not unique to Mexico, but the particular features of their relationship there help to round out global debates over nature conservation and its meaning.[7]

Parks as Modernizing Nature

Parks are more than property designations; at their core they expose cultural yearnings for more intimate connections between human societies and natural systems. The notion of a park replicates and reinforces the idea of nature as separate from humanity even though nature inside even a large park rarely lives autonomously. Parks do not merely exist to conserve; they are created by and create space for people. Forests, waterways, wildlife, and vegetation require periodic tending, management, and protection, all of which contribute to a certain aesthetic desired by park aficionados. Even the act of prohibiting people from a park to let nature alone constitutes an act of management. Probing the blurry boundaries between nature and culture in a park reinforces this false dichotomy. The more pressing historical question is, how does a park differ from any other cultural construction that relies on the materiality of nature? What makes a park different from a maquiladora that produces microchips or a farm that grows asparagus or a city that houses millions, all of which also rely on relationships between nature and culture? And what does the difference between a park and a farm (or a factory, or a city) tell us about the culture that classifies these multiple functions differently?

Parks are modern political forms that rely on uneven but reciprocal relationships between nature and humanity. Conservation and development are cut from the same cloth, but they are not the same. Parks promote the wholesale acknowledgment that a reality exists beyond human construction, one that is natural in the expansive sense of the word.[8] Parks differ from factories and farms because the former prevents—or seeks to prevent—the latter on the same territory. Yet globally (and in Mexico), parks did not seek to prevent farms and factories altogether but instead fostered a relationship between citizens (working in farms and factories) and their ability to access parks. In doing so, parks have served as a meager counterweight to the forces of development that ultimately undermined their very integrity. As long as advocates of farms and factories outnumbered park promoters, parks had little chance to be more than emblems of an imagined past relationship to nature rather than challenges to the resource consumption that compromises their existence.

Before I turn to Mexico specifically, the global context of conservation merits consideration. The notion of a nationally constituted conservation area reflects a common modern phenomenon, and many societies have deliberately sought to conserve natural resources to sustain their social and cultural structures. Conventional timelines suggest that the concept of a nation's park originated with the industrialization of the United States; indeed, Congress created Yellowstone, the world's first national park, in 1872.[9] These parks glorified a national landscape alongside the construction of national transportation systems and the rise of consumer culture. The US parks marked distinct, majestic lands distant from urban populations and drew people west. Through tours, publicity, and artistic renderings of the Grand Canyon, Yellowstone's geysers, and other features, US parks became the source of a visual culture that gave nature a modern façade and recognizable referents.[10]

US parks came to have political capital, public resonance, and staunch support with government leaders (such as President Theodore Roosevelt) and successful private businessmen (such as John D. Rockefeller Jr.). Various champions envisioned national parks not as elite reserves but as places where people who labored inside a factory or behind a desk could reconnect with nature in a relationship configured to be fleeting. People with direct, intensive relationships with the land were less interested in setting aside nature for protection and more interested in subjecting nature to development. This is not to say that farmers, ranchers, or hunters saw no benefit in conservation (indeed, they were often nature's most vocal champions), but the park ideal itself held particular resonance with urban folk.

The subsequent park template evolved to include setting aside land for nature to exist without human interference. The 1964 US Wilderness Act clarified this ideal by defining wilderness as "an area where the earth and community of life are untrammeled by man, where man himself is a visitor who does not remain." In practice, such a goal has proven unreachable as park employees struggle with the consequences of invasive species, air pollution, and climate change in addition to the legacies of past use.[11] Nevertheless, the image of pristine nature etched its place onto the collective consciousness.

National parks, according to the author Wallace Stegner and the documentarian Ken Burns, constituted "America's Best Idea," and one that rapidly found force in countries worldwide.[12] Despite uneven levels of development, growing support for conservation compelled nations to reserve areas for nature. A nation that created parks showed its commitment to modernization as tangible proof that the rest of the nation was no longer wild. Political leaders readily created parks in British settler societies: after Yellowstone, Australia created Sydney's national park in 1879, Canada created Banff in 1885, and New Zealand created Tongariro in 1887.[13] Led by the International Union for the Conservation of Nature (IUCN), scores of scientists and policy makers advocated nature protection in an increasingly amenable world system. The IUCN triumphantly (and prematurely) explained in its founding documents in 1949 that the "protection of nature now takes its place amongst the leading concerns of mankind."[14] In the following decades, such recommendations were heeded, and today over 12 percent of all the earth's land—an area greater than the size of the African continent—is dedicated to conservation.[15]

Yet people displaced by parks point out that such accomplishments came at a cost. According to its critics, conservation has been an elite project articulated by and enacted for wealthy foreigners at the expense of local traditions and indigenous peoples.[16] Certainly the daily use of resources—pastures, woodlands, waterways, wildlife, and more—has been restricted by parks, and the people who bear the brunt of these restrictions are those least able to afford them. Detractors point out that the consolidation of conservation funding into the hands of a few environmental organizations has meant reduced national sovereignty, uneven land use decisions, and a system of global conservation that absolves consumers of guilt for the destruction they indirectly cause.[17] More insidious transformations have also occurred, such as the ways in which native peoples "choose" to exchange conservation for development and in doing so generate a symbolic understanding of ecological conservation based on external demands.[18]

However compelling, couching debates as conservation versus rural life sidesteps the elephant in the room. The greatest pressures on both large expanses of contiguous nature and remaining concentrations of indigenous peoples come not from one another but from the industrializing processes of resource use and agribusiness that have relegated parks and native peoples to the periphery. There is no inherent contradiction between farming and conservation or between wildlife and pastureland; healthy compositions of both can be created hand-in-hand.[19] The greater affront is an urban industrial society that seeks production at the expense of the remaining reserves of biological and cultural wealth and removes human sustenance from its natural context by the overuse of chemical inputs and fossil fuels. This is not to say that local conflicts over parks do not occur or are not meaningful to those involved. The root of the crisis lies not in the local relationships between conservation and communities but instead with the larger social forces that encourage the conversion of land from a dynamic array of heterogeneous uses into an irreversible urban environment of concrete and steel. More accurately, conservation happens in tandem with development, and circumscribing global conservation as an elite enterprise or an anti-indigenous force overlooks variations in how conservation has been conceived in places such as Mexico City.

Interpretations of parks as bastions of wilderness or colonialist enclaves largely leave out the actions and activities that took place in Latin American countries. Parks there emerged from a distinct developmental trajectory and never looked like the refuges and sanctuaries idealized by American romantics. Hybrid, proximate, and inhabited parks in Mexico embodied a different ethic that alters traditional timelines of nature conservation. Mexico hosts one of the oldest and most innovative models of conservation in terms of integrating communities and conservation, but such a system earns scant recognition from the international community. Less-rigid restrictions and more-immediate park locations led to the dismissal of many conservation programs. Several Mexican parks were never recognized or were delisted in the 1970s by the dominant global conservation organization, the IUCN, whose classification system placed the majority of Mexican parks in the least revered classification.[20] Although parks have functioned as a marker of modernization, little attention has been given to parks that coexist with seemingly contradictory events and institutions—such as social revolutions or communal property. Considering Mexico's conservation trajectory will help explain these contradictions.

The Mexican Timeline: The Ebb and Flow of Conservation

Support for conservation in modern Mexico has its origins in the late nineteenth century. A group of advisors during the regime of Porfirio Díaz (1876–1911) developed the scientific expertise and progressive philosophy that allowed them to conceive of nature protection as a necessary accouterment for their rapidly modernizing country. In this period, immense foreign capital outlays paid for the construction of railroads and factories, while resident scientists began to notice changes to the climate that accompanied the forest clearing and resource extraction under way. The most vocal of these scientists, Miguel Ángel de Quevedo, spoke of the need to refrain from logging around the Valley of Mexico, which surrounded the capital.[21] Various steps toward conservation proceeded, including the designation in 1898 of a forest protection area surrounding the mining concessions in El Chico, Hidalgo (later a national park); dune construction in Veracruz; and reforestation on eroded hillsides around major cities nationwide.[22] At this time, conservation advocates were largely foreign-trained engineers familiar with the ideas of both national parks in the United States and burgeoning forestry management schemes in Europe. The 1910 revolution distracted these scientists from conservation efforts, although it did not stall them completely.

Slowly and piece by piece, the components for a national conservation system were put into place. President Venustiano Carranza declared the first national park in 1917 on the lands of a former monastery called Desierto de los Leones in the Federal District.[23] The same year, revolutionaries compromised on a new constitution that situated their aims for social reform in a modernizing ethos with radical provisions for education, labor, and natural resource management, including agrarian reform and petroleum expropriation. Article 27 in particular constructed a framework for returning lands to peasants as communal property, established the obligation of the government to provide land, and framed agriculture as a noble, revolutionary act that required modernization. As fighting turned into political stability, conservation advocates put into place additional policy pieces. The Forestry Code of 1926 and the Six Year Plan for 1934–40 each contained a rationale for conservation and called for a system of national parks.

This series of "firsts" included parks and reserves, laws and codes, and a scientific community cultivating public opinion toward conservation.

Although these essential components accumulated over time, it took the right government to embark on a full-scale program to create national parks. As shown in figure 9.1, this conservation leadership emerged during the administration of President Lázaro Cárdenas (1934–40), who guided the parks to life. Forty in total, these parks delivered immediate, tangible results by drawing on Chapultepec as a progenitor for parks elsewhere. The new parks averaged a relatively small size of 38,000 hectares (although many were much smaller) and were mainly located in close proximity to the capital and protected coniferous forests.[24] Cárdenas created a new Department of Forestry, Fish, and Game, which administered the parks according to biological principles of the time. They also had cultural functions linked to tourism, including protecting notable natural features (such as snow-covered peaks) and maintaining scenery (including historic sites). The parks formed one component of a landscape designed to keep people invested in the land and provide for their continual, productive sustenance as laborers and citizens.[25] Indeed, by design, parks complemented alternative land uses such as farming and ranching by sheltering water sources and cultivating forests to stall erosion. Quevedo explained the logic, claiming, "Hacendados and campesinos alike affirm that agriculture in [Toluca, Lerma,]

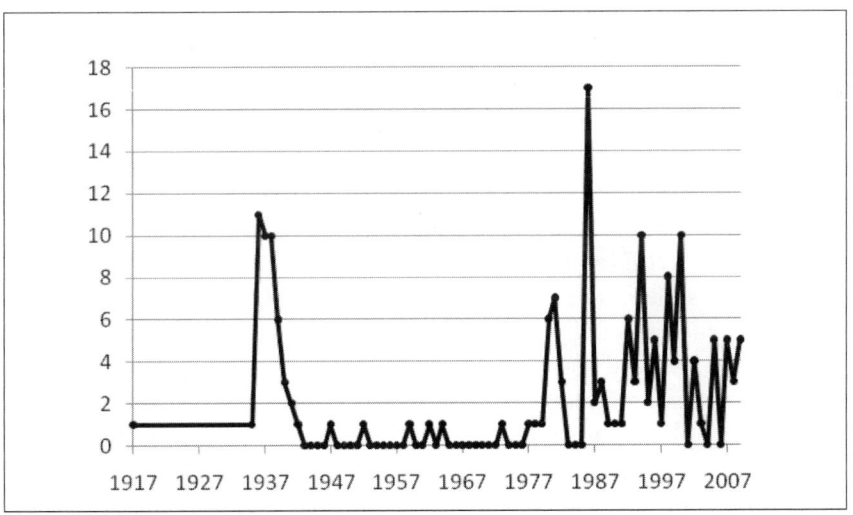

Figure 9.1. Total number of national parks created, 1917–2009. *Source: Diario Oficial,* various dates, 1933–34; Comisión Nacional de Áreas Protegidas (CONANP), http://www.conanp.gob.mx.

and other Valleys of the State of Mexico is not affordable because of the disturbances to the climate and hydrology as a consequence of the immense logging of forests in the surrounding mountains."[26] Quevedo went on to clarify that the "method to save the scarce vegetation that remains on the surrounding peaks of the mountains of this state has been through the creation of national parks." Conservation areas were in tune with Cárdenas's platform for the social development of rural people.

After the whirl of conservation during the Cárdenas era, it was as if the music was silenced and the band went home. Only nine parks were created in the next thirty-six years (figure 9.1). Why did such an abrupt shift in conservation strategies occur? By 1940, economic transformations and political choices had been made, including the mobilization of production to supply the United States with raw materials for World War II. National politics shifted to a more centrist and developmentalist stance. One park provides a fitting example of this trend: Colima National Park, which protected two volcanoes in the state of Colima. In 1942, industrialist Enrique Anisz asked for and received an amendment reducing the park's size to benefit his nearby paper factory. Anisz argued that his company would bring economic benefits to the area, including hydroelectric power, agricultural development, and logging activities. The forestry engineers in the Ministry of Agriculture (then responsible for the parks) supported Anisz's proposal and amended the park decree to include a zone of forestry reserves with a specific designation for the cellulose industry.[27] Several years later, in 1955, local furniture industry representatives requested a similar concession for the use of the forests. Productivist arguments took precedence over conservation and nibbled away at earlier achievements. Despite such reductions, many of the national parks in the most prominent places (including the volcanoes of the central valley) did manage to persist.

Historians have called the era roughly between 1940 and 1980 the "Mexican Miracle" for its dramatic rates of economic expansion and enormous social shifts, including rapid urbanization and population growth. The stable one-party political system and burgeoning state extended patronage to supporters and co-opted detractors. The federal bureaucracy created a clumsy but nonetheless effective set of incentives to increase production and the use of natural resources. The results were staggering. In 1950, Mexico had 25.8 million people, 43 percent living in towns of 2,500 or more inhabitants, making it still a nominally rural country. In 1990, there were more than 100 million Mexicans, 75 percent of whom lived in towns of 5,000 or more. In this same period, literacy rates rose from 56 percent to 87 percent, and average life expectancy increased twenty years.[28] This constituted a miracle for increasing humanity indeed, but the pressures on the natural resources necessary to sustain such growth edged out most efforts at conservation and gave rise to urban conglomerations such as Mexico City. Little to no park creation took place, and previously conserved areas were subjected to logging, grazing, and other unchecked resource extraction in the context of lax enforcement.

But the miracle began to splinter because of low rates of taxation, greater demands for government spending, and the political costs of loans tied to the discovery of oil in the Gulf of Mexico during the late 1970s.[29] Stability came to a crashing end with the global economic crisis in 1982. Politicians began to downsize the state by selling off paragovernmental and state industries as a means to repay debts and trim the public sector. This involved negotiating trade agreements, especially the North American Free Trade Agreement (NAFTA) between the United States, Canada, and Mexico.[30] The concurrent rise of the economic philosophy known as neoliberalism advocated reducing the role of the state in the economy, and in 1992 reforms ended the agrarian reforms of Article 27 put in place by the revolutionary constitution and enacted largely by President Cárdenas.

Conservation surged again in the 1980s and early 1990s. Between 1977 and 2002, the Mexican federal government declared ninety-seven reserves, with an average area of 160,000 hectares, more than four times the size of the earlier parks (figure 9.2). The total amount of land protected increased nearly tenfold.[31] Rather than national parks built around communal lands and historic landscapes ringing the capital, these new areas marked sites of recognizable biological importance. Called biosphere reserves in a gesture toward the global scientific community rather than a token of national identity, the rare desert biomes of the North and exotic tropical forests of the South earned protected-area status in vast extensions. Their worth could be confirmed by biologists, who by then had the scientific framework of biodiversity for valuing nature. The new reserves occupied frontiers with comparatively smaller human populations.[32] However, the new conservation areas did not replace or subvert the older ones. In places such as the Chichinautzin Biological Corridor south of Mexico City, new designations enhanced preexisting parks.[33] Often these overlapping designations lacked coordination because the social values that gave rise to conservation in the 1930s were markedly different from those that permitted conservation in the 1980s.

The century of conservation shows that Mexicans began formally and institutionally protecting nature before notions of biodiversity, ecological integrity, and ecosystem services were fully formed. Nearly a third of Mexico's

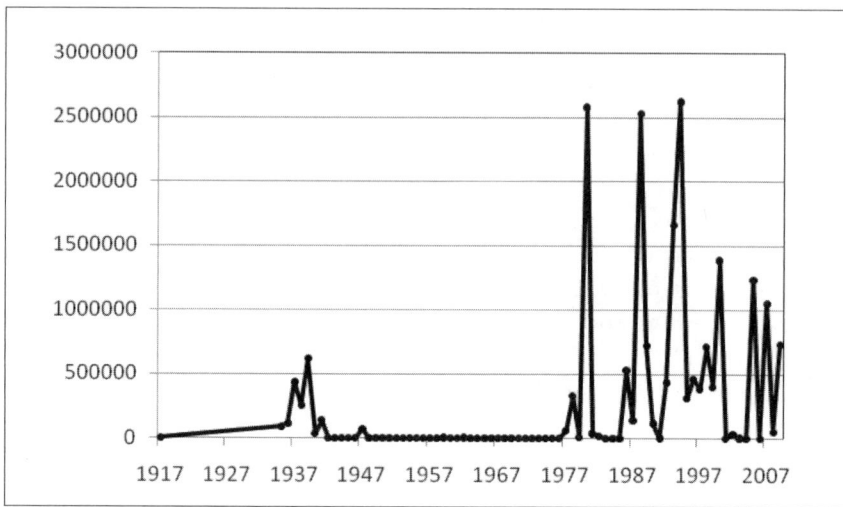

Figure 9.2. Total area given protective status (hectares), 1917–2007. *Source*: *Diario Oficial*, various dates, 1933–34; Comisión Nacional de Áreas Protegidas (CONANP), http://www.conanp.gob.mx.

nature reserves were designated long before an international conservation community existed. The country's first national parks protected integral stands of trees around the capital, a landscape typical and national rather than remote and wild. Nor did Mexicans produce parks to allow wealthy foreigners easier access to wildlife. Wildlife conservation played a minimal role compared to the emphasis on forests and drainage systems, in part because of the relatively depauperate mammalian communities. This is not because Mexican scientists had not studied natural systems and the effects of human-induced changes to them. Scientists held positions of power in the federal bureaucracy and shaped rational approaches to land management dating back to logging assessments in the nineteenth century. Parks were linked to scientific justifications, but their larger purposes remained expressions of solidarity and public good will.

Early on, government planners saw parks as methods of keeping people on the land by honoring local communities and (generally) property rights. This set Mexican conservationists apart from their contemporaries in Southeast Asia and sub-Saharan Africa.[34] Forestry engineer and federal department head Manuel Corona remarked in the early 1940s, "National parks are sites destined to provide recreation to the inhabitants of the nearby communities."[35] The effects of retaining local populations on the tangible conservation of nature are less clear. When leaders of the community of

Tziscao learned in the 1950s that their community would be encompassed by the Lagunas de Montebello National Park in Chiapas, for example, they insisted that their communal property rights coexist with federal institutions. Researchers there have found that this blend of local and national control contributed to forest retention and allowed local residents a stake in management.[36] In other areas, remnants of this system of protection endure in the present and reassert themselves into village life in important ways. For example, the village of Tepoztlán in Morelos refused the 1995 construction of a golf course on their commons by arguing both that the area was a national park and that the lands were communal property and not for sale.[37]

The comparatively early and ubiquitous existence of parks reveals codified, formalized relationships between the government, its citizens, and the resources they share. Enrique Beltrán, one of Mexico's first representatives to global conservation meetings, readily articulated the balance this implied. Beltrán advocated a network of designated areas that would balance local and national needs. He proposed a zoning system with three characteristics: zones for general relaxation, including hotels, restaurants, and parking lots; intermediate zones with open access but with few amenities; and restricted zones reserved for study and investigation. Put forth in the 1960s, Beltrán's template described criteria now employed in biosphere reserves worldwide.[38]

Nature conservation's emergence in Mexico presents an uneven and fragmented story. On the one hand, the system is extensive; on the other, it is limited. Protected areas exist in every state and contain an immense range of bioregions that, as we have seen, protect nearly 10 percent of the national territory. This is an average amount compared with more than 17 percent of land in the United States and less than 9 percent in Canada.[39] As some biologists have pointed out, Mexico could certainly do more to protect nature.[40] While a significant amount of land is protected, it does not meet the need suggested by the amount of diversity nationwide.

However, the question remains, why did Mexicans choose to protect as much nature as they did? Given the nation's revolutionary history, its explosive population growth, and its proximity to the largest economy on the planet, one might expect nature reserves not to exist at all in Mexico. The expansive network of protection and its historical depth does not have a scientific explanation but a cultural one. When the idea of the land as a symbolic and material bequest to the rural poor emerged as a central component of revolutionary ideology, national parks simultaneously marked value on the landscape and confirmed prosperity that lay ahead. Parks proved complements to revolutionary reforms for a time. Decades later, when politicians recognized the flaws in the land reform sector, they offered nature reserves as signs of compromise.

The two crests of conservation—national parks in the 1930s and biosphere reserves in the 1980s—reveal a system of federal conservation that responded to civic rituals and political functions in a broader modernizing agenda. In their time, presidents created protected areas for reasons that fit their political programs. These creations generally emerged as positive, neutral, celebratory acts of little controversy. Each protection surge corresponded to vast social changes: first the creation of a state system and then its dismantling. These trends expose conservation's cultural context and influences, some borrowed from abroad and others longstanding. A more intimate story of the resilience of Chapultepec further illuminates the cultural meanings of Mexican parks.

Chapultepec:
Urbanization and the Search for the Commons

Despite its popularity, Chapultepec is not a national park, a biosphere reserve, or any formal nature protection area. Yet it has been persistently referred to as the nation's park.[41] Aztecs inhabited the forested hill by 1299, Nezahualcóyotl endowed it with a palace in 1428, the Spanish king exempted it from the rewards given to Hernán Cortés in 1530, and Viceroy Luis de Velasco declared it a site of public recreation in 1537.[42] Velasco conferred further recognition in 1558, dedicating Chapultepec as "a public amusement made beautiful by its trees."[43] The city government began to administer the park in 1818, guiding it through episodes of celebration and decline. The park was extensively redesigned in 1906–7 and reopened to the public in 1910. It was enlarged twice in the twentieth century (in projects completed in 1964 and 1984) and was restored and redesigned again between 2003 and 2006.[44] Despite this prolonged administration and historical continuity, the park bears no UNESCO recognition.

If longstanding habitation and civic structures gave this park value, why is Chapultepec worthy of inclusion in a discussion of nature protection? Chapultepec reveals the values that structure the ways Mexicans have thought about the environment in a cherished public locale. The ancient space served as an inspiration for broader conservation actions that expressed collective responsibilities and confirmed a common heritage. Chapultepec also reflected struggles conservation faced as the use, degradation, and restoration of the park followed the city's fluctuations in a pattern similar to how parks in general followed the nation's shifts.

Most modern histories of Chapultepec interpret political events in the nineteenth century as the most significant ones. Standing out among these events are the occupation of the castle by the US Marines in 1847, the reinvigoration of the gardens during the French occupation in 1863–65, the conversion of the castle into a presidential residence by President Sebastián Lerdo de Tejada, and the structural reforms during the Porfiriato that widened promenades, sculpted lakes, and erected statues of national figures.[45] After the 1910 revolution, President Cárdenas's decision to inhabit the more humble Los Pinos (also in the park) gestured toward a more inclusive meaning for the space, as did numerous Arbor Day celebrations in the period.[46] During the presidency of Manuel Ávila Camacho (1940–46), the castle was converted into the museum of national history, changing it from the abode of a single powerful man into a site of pilgrimage for millions of tourists.[47] Changes to the cityscape around the park, visible in figures 9.3 and 9.4, provide stark testimony to the resilience of the park and its forest.

Figure 9.3. Chapultepec Castle and surroundings, ca. 1932. *Source*: Fundación ICA, Acervo Fotográfico.

Figure 9.4. Chapultepec Castle and surroundings, 1968. *Source*: Fundación ICA, Acervo Fotográfico. Note the increase of the horizontal and vertical density of buildings and streets and the reduction in vegetation outside the park.

Chapultepec's lessons for conservation derive from the city's growth in the twentieth century. First, laments over losses of ahuehuete trees show that conservation efforts bridged longings for the past with scientific understandings of nature's value. Editorials in the 1940s mourned falling trees as artifacts, while journalists in the 1980s warned of looming extinction; they all hoped to retain a fragment of nature's past in the forest. Next, public desires for amenities within the park developed alongside this longing for the past. City administrators, urban residents, and even children demanded a growing list of mechanical rides, exotic animals, and redesigned museums. These demands kept the park aligned with public recreation aims but involved compromises between natural and social spaces. And finally, a constant insistence on public accessibility intensified during periods of economic crisis. The outcry over initiatives to charge entrance fees exposed a public sense of entitlement to the nation's natural spaces. I interpret this demand as a search for an urban commons, a site of public recreation and unity despite displacement. As the promise of land reform disappeared, the long-standing urban commons proved resilient and receptive to new cultural meanings. The ways journalists, residents, and administrators defended the past, yearned for the future, and insisted on access while discussing Chapultepec confirm the contours of nature conservation nationwide.

Editorial pages of the city's major newspapers discussed the deterioration of the park's ahuehuete trees, symbols of support and stability. Students of the national forestry school designated the noble trees the "national tree" in 1923, but in the 1940s concerns surfaced over the trees' rapid disappearance.[48] One newspaper article lamented, "What were once marvels of nature are slowly being transformed into firewood," and another mourned, "Like everything in life, Chapultepec is dying."[49] The loss of the forest amplified pressure to conserve the park and raised proposals to rid the park of amblers, vagrants, and indigents. Some complaints turned into indictments of the managers of the park, who paid unreasonably low salaries to the park's gardeners, who then gathered branches to heat their meals at the expense of the park's public integrity.[50] As a growing sense of scarcity threatened public patrimony, the city's most poor and vulnerable found their activities in and around the park restricted to preserve nature.[51] The city's growth amplified concerns over the park at the same time as it heightened the pressures on its use.

Blame for neglect and abuse extended to a silent executioner of the thousand-year-old ahuehuetes: the overdrawing of the city's water supply. Capital residents disagreed on who should shoulder the blame for the lack of water. One writer offered that the *ejidos* south of the park hoarded water and slowly choked off that source of life.[52] Another suggested that the influx of urban migrants caused an abundance of wells to perforate the valley's floor and decreased the water available to the trees.[53] Still another charged that a general neglect by city authorities "that would shame [President Benito] Juárez and [Father Miguel] Hidalgo" had caused the park to be treated like a garbage dump and the stately trees to be abandoned.[54] Ignorant rural people, covetous urban migrants, corrupt or lazy authorities: each accusation of destruction was at once specific and misguided. Competing constituencies treated Chapultepec as their patrimony and, cognizant or not, compromised the life of the stately trees in their independent and individual quests for modernity and comfort. In a sense, then, the ahuehuetes symbolized the tragedy of the commons of the 1940s: as each citizen sought his or her own development, the collective resource of the park wilted away. But the park did not disappear, despite the continued attrition of the once-vibrant trees, which by the 1970s were described in the appropriate scientific parlance to be "in danger of extinction."[55]

Other changes to the park's landscape occurred. In the 1920s and 1930s, the zoo was a major draw for park visitors, and it remained a site of fascination although its contents changed. After a major restructuring and redesign in the mid-1970s, the zoo no longer showcased mostly national fauna but instead reflected the largesse of all the world's biomes.[56] The most direct sign of Mexico's entry into the premier sites of animal care came with the successful breeding of its pandas in 1980. With the most successful program of artificial insemination outside of China, the zoo produced eight pandas in the 1980s after receiving its first pair as a gift from China in 1975. Panda development was an expensive endeavor (one estimate for the care of the first panda to live more than eighteen months was more than 50 million pesos) with little economic but enormous psychological value.[57]

After the zoo rejuvenation, a plan to remodel the entire park and add another section emerged from the government of the Federal District. The plan, developed in consultation with the public, firmly articulated many of the ecological challenges of the park, in particular the need to "protect the flora and fauna that exist within."[58] The Mexican Forestry Society proclaimed in 1974 that the park represented a confluence of natural, cultural, and historic characteristics, all contributing to public well-being.[59] One editorial writer concurred, claiming that parks are "in their grandest sense instruments of popular culture; they form part of the tenacious human fight to rescue nature as an inherent marker of life."[60] Parks solidified the human connection to nature even though the culture that supported them also valued the sources of that nature's demise.

Praise for an urban commons coincided with increased demands for modern services. Some changes were subtle, such as the growth in public lighting or the shift from botanical gardens to sculpture gardens, but others marked a larger social fascination with technology over natural wonder.[61] Rather than expanding the garden areas, administrators extended the variety of activities available to youths, especially mechanical rides. A preliminary renovation of the park, completed in September 1964, added a new area with a large artificial lake with boats and the tallest and longest roller coaster in the world.[62] These amenities increased the demand for more. Further expansions of the scope of the park meant a contradictory pairing of tradition and modernity: thrill rides alongside exotic animals and ancient trees. The contrast was not always stark, because picnic benches, a roller-skating rink, and outdoor chess tables also appeared. The paradox of these activities—which combined a desire to hold on to the past while racing to the future—captured the malleable popularity of the park. For example, in 1981, more than two million people celebrated Easter week in the park, bringing picnics to enjoy and relatives in tow. The most popular sites were the anthropology museum, Chapultepec Castle, and the amusement rides, including the roller coaster, spook house, and bumper cars.[63] The variety of activities that took place in the park led planners and administrators to use zoning regulations (much like those of biosphere reserves) to circumscribe what activities belonged in which places. Each zone harbored proof of the sophistication of the city (and really, the nation) and was designed to showcase worthy elements. Modern ornamentation combined with longstanding relics of nature produced an enduring site of public recreation.

After nearly seventeen years of use, the original rides needed refurbishment, and enterprising park authorities designed a recreational activity to keep the park grounded in its legacy: fishing in the lakes. Starting in January 1981, the lakes were opened to fishing on Mondays, Wednesdays, and Fridays from six to ten in the morning. Anglers had to purchase a license and then could cast freely during the scheduled times. Journalist Jorge Vera Ávila claimed the program proved an "excellent idea by authorities that gives *capitalinos* the ability to know the true excitement of fishing."[64] This coordinated effort to give children the ability to reclaim traditions of the past (however contrived) while enjoying the fruits of the modern world captured the tensions that shaped the park's landscape, just as similar tensions shaped the development of conservation programs nationwide.

Controversies over privatization of the park peaked in the early 1980s, as neoliberalism reached into other aspects of the economy and into ideas for nature reserves. An early 1983 proposal by the zoo's administrator, Marielena Hoyos, advocated charging a one-peso entrance fee to the zoo as a preliminary step toward having it "function like a commercial enterprise."[65] In keeping with measures of the time, Hoyos commented that commercializing the park would make it self-sufficient and more orderly. The proposal came after several animals, including a leopard cub, went missing from the zoo (most likely stolen by employees and sold). Yet the neoliberal solution to the park's budget woes met with immediate public outcry.[66] City council delegate Fernando Ulíbarri Pérez expressed his commitment to provide sufficient funding without cutting off public access. The administrative and fiscal debate expanded over the summer and included the convocation of several public forums on the park. The forums gave voice to the idea that the problems facing Chapultepec—piles of garbage, mistreated animals, and trampled vegetation—were manifestations of the larger unsolved social problems of illiteracy, displacement, and unemployment. The debate underscored the difficulties inherent in administering a park with an ambiguous status as neither a national park nor a state or strictly federal entity. Outcries even escalated into a plea for President Miguel de la Madrid to intervene and save this patrimonial refuge for the popular classes. One editorial opined that if he let the reforms pass, his legacy would never recover.[67] The final compromise was telling: access to the park, the lakes, and the zoo remained free of charge but the modern rides all raised their fees.[68] In this way, access to the most "natural" of the amenities was preserved, while those taking advantage of the energy, engineering, and machinations of modernity remained costly and inaccessible to millions.

The programs for rescue and rehabilitation of the forest followed the contours of political and social change in the city. The public attention—by visitors, administrators, columnists, and children—to the condition of the park demonstrated the persistent quest for keeping Chapultepec a public space that sheltered the past, reached for the future, and strove to remain accessible in a city unable to meet demands for public recreation. The search for an urban commons nurtured unity among popular groups as the metropolis gradually encroached on the countryside. The park reclaimed its place as a site of celebration—for fishing or thrill rides, strolling in the shade, or reviving links to past ages—and the public defended it against privatization, against restricted access, and at times even against each other. This commons represented people cut off from wild nature and searching for a compromise between preserving the roots of the past and making the amenities of the future available. Nonetheless, the park offers a tragic warning. The zoo expanded its collections, but the ahuehuetes have all but disappeared. Development of the city and pressures of resource use from air pollution to water subsidence have effectively killed the best example of nature within the park, despite the efforts to link the trees to national culture.

Conclusion:
Capital Organicity and Geographies of Resilience

Let us return to the questions raised at the outset, using Chapultepec Park to elaborate on national trends in conservation. Several contradictory tensions—including laments for the past, desires for modernity, and yearnings for a publicly accessible common natural area—reveal the ways parks were designed to appeal to a broad array of citizens and adapt to evolving values. Protected areas functioned as expressions of responsibility and demonstrations of nature's resilience when enmeshed in efforts at social reform that changed orientation over time. Just as Chapultepec complicates the urbanity of Mexico City, so too do national conservation areas blur distinctions between natural and artificial. But the distinctions still have meaning. The range of conservation areas demonstrates that zoos are not equal to parks, just as fiberglass palms and plastic flamingos are not nature. These fabricated items do not and cannot exist without human interference. The same is not true for a jaguar, a gray whale, or a monarch butterfly. Synthetic birds and phony trees provide disappointing simulacra of their wild kin. Similarly, Chapultepec's place in the capital city where vendors sell iced beverages is not the same as Iztaccíhuatl-Popocatépetl National Park, where ice remains from the last glacial maximum. Artificial replicas exist as vestigial images of the wild, reaffirming to the society that produces them the intertwined relationship between humanity and nature.

Mexican nature conservation distinguishes parks from zoos through a matrix of different designations and separate zones within those designations. These distinctions are based on human laws—not the laws of nature—and they reveal that parks are exceptional because nature takes priority within a park and is there permitted to self-replicate, if not function entirely apart. The idea of a park suggests that people have authority but not control. In parks, flora and fauna are left to flourish of their own accord, while in zoos they are completely provided for. And parks—most assuredly national parks—imply a sense of stewardship to a public that relies on the wild in a spiritual and cultural context. Although Mexican parks are not completely wild, the magnificence of these swaths of nature is the confirmation that they cannot be replicated by humanity.[69] Chapultepec is wild only in a comparative sense: it is fenced, gated, and policed and in this way serves as a small gesture of nature with a greater resemblance to humanity's routines. This park is an escape that corresponds to the nature of the nation although it protects only the latter.

The existence of parks as expressions of public responsibility and natural resilience likewise characterizes Mexican conservation. The creation of a system of protected areas suggests that a common ecology exists and provides the public with the right to visit these areas. The park system extends the reassurance that such places survive in the nation's name. This expression of public good speaks to the popularity of declaring protected areas even during times of political strife. That many of the early national parks were declared around places in need of rehabilitation or restoration demonstrates the social value given to recovery rather than preservation (although that emphasis has shifted). Public campaigns to save, protect, and reforest earn recognition, but the muted emphasis on flexibility also merits acknowledgment.

However lingering their significance, parks are also products of dynamic shifts, especially in the trend toward urbanization. The amassing of immigrants into Mexico City created an ever-larger constituency of users for Chapultepec—by some estimates, more than 13 million a year—while the countryside was gradually neglected. The growth of urban populations has augmented the already large constituency and high popularity of Chapultepec, increasing the audience for urban parks while increasing pressure upon them. Some scholars have suggested that conservation creates refugees, but that is only true in the sense that conservation forms one of the tendrils of a larger process of modernization that draws laborers off the land and into cities. Conservation has created refugees where it has evicted residents, but this was rarely the case in Mexico. Instead, the real enemy of the environment and the poor was not those who sought to conserve nature but rather the extractive industries that exploit the labor of the poor and the resources they might otherwise maintain.

Mexican conservation history suggests that Mexicans have tried to extend a measure of justice from their social reforms onto nature. Despite the rising numbers of citizens who found themselves within the traps of the capital city's burgeoning urbanity, a collection of protected areas continued to function and expanded as emblems of nature set aside in physical spaces. In what seems a contradictory mandate, some environmentalists argue that the only solution to the unsustainable practices of modernity is to foster greater urbanization. Some conservation biologists have begun to advocate the hastening of urbanization as a way to practice "preemptive conservation," because more people living in cities could result in reduced deforestation and land clearing in nonurban areas.[70] When more people live in cities, these scholars claim, more land can return to a state of wildness, and the areas that remain undeveloped can persist. If this is the template for a twenty-first-century conservation plan, Mexico might be leading the conservation curve once again with its vast urbanization and recognition of resilience and restoration. While a vibrant park inside a teeming metropolis embodies contradictions, it might signify the most far-sighted conception of conservation yet.

But a park like Chapultepec, as grand as it is, cannot adequately meet the needs of 20 million people seeking solace in nature on Sundays. Similarly, a park system that protects 10 percent of the national territory hardly conserves enough area to provide ecosystem services and adaptability to a highly diverse industrialized nation. As long as urban areas are studied separately from the realities in the countryside, an obscured picture of the landscape emerges.

Notes

1. Ahuehuete is also known as sabino, Montezuma cypress, and *Taxodium mucronatum*.

2. Rubén M. Campos, *Chapultepec: Its Legend and Its History* (Mexico City: Talleres Gráficos del Gobierno Nacional, 1922), 28–29; John Beardsley, *Mario Schjetnan: Landscape, Architecture, and Urbanism* (Washington, D.C.: Spacemaker Press, 2007), 25–39; Andrea Moerer, "Space and Commemorative Practice in el Bosque de Chapultepec," paper presented at American Historical Association Conference, Atlanta, 6 January 2007.

3. Joel Simon, *Endangered Mexico: An Environment on the Edge* (San Francisco: Sierra Club Books, 1997), 60–90; Rubén Gallo, "Introduction," in *Mexico City Reader*, trans. Lorna Scott Fox and Rubén Gallo, 3–29 (Madison: University of Wisconsin Press, 2004).

4. 9.88 percent of the territory is designated as federally protected land. Comisión Nacional de Áreas Naturales Protegidas, http://www.conanp.gob.mx. I use the term *parks* to refer to protected natural areas, nature reserves, national parks, and so forth. This term reflects widespread popular understandings of conservation rather than separate technical designations. I do not imply that only one definition for conserving nature exists.

5. Anthony Challenger, *Utilización y conservación de los ecosistemas terrestres de México: Pasado, presente, y futuro* (Mexico City: CONABIO and UNAM, 1998), 36.

6. Ibid., 34.

7. For an introduction to these debates, see William Cronon, "The Trouble with Wilderness; or, Getting Back to the Wrong Nature," in *Uncommon Ground: Rethinking the Human Place in Nature*, ed. William Cronon, 69–90 (New York: W. W. Norton, 1996); Mark David Spence, *Dispossessing Wilderness: Indian Removal and the Making of the National Parks* (New York: Oxford University Press, 2000); Roderick Neumann, *Imposing Wilderness: Struggles over Livelihood and Nature Preservation in Africa* (Berkeley: University of California Press, 2002); Dan Brockington, Rosaleen Duffy, and Jim Igoe, *Nature Unbound: Conservation, Capitalism, and the Future of Protected Areas* (London: Earthscan, 2008); and Mark Dowie, *Conservation Refugees: The Hundred-Year Conflict between Global Conservation and Native Peoples* (Cambridge, Mass.: MIT Press, 2009).

8. Michael E. Soulé and Gary Lease, eds., *Reinventing Nature: Responses to Postmodern Deconstruction* (Washington, D.C.: Island Press, 1995).

9. The literature on US parks is immense. For starters, see Roderick Nash, *Wilderness and the American Mind* (New Haven, Conn.: Yale University Press, 1967). For more recent analyses, see Richard Grusin, *Culture, Technology, and America's National Parks* (Cambridge: Cambridge University Press, 2004); and Adrian Phillips, "Turning Ideas on Their Head: A New Paradigm for Protected Areas," *George Wright Forum* 20, no. 2 (2003): 8–32. For a more provocative timeline, see Paul Sutter, "National Parks Beyond the Nation: Global Perspectives on the History of 'America's Best Idea,'" keynote address presented at the National Parks Beyond the Nation Conference, Colorado State University, Fort Collins, 14 September 2011; and Ian Tyrrell, "America's National Parks: The Transnational Creation of National Space in the Progressive Era," *Journal of American Studies* 46, no. 1 (February 2012): forthcoming. Sutter points out

at least five alternative points of national park origin in the United States, while Tyrrell argues that the United States borrowed liberally from European ideas before 1916.

10. Grusin, *Culture*, 8.

11. Donald Worster, "Nature and the Disorder of History," in *Reinventing Nature: Responses to Postmodern Deconstruction*, ed. Michael E. Soulé and Gary Lease, 65–87 (Washington, D.C.: Island Press, 1995).

12. Wallace Stegner, "The Best Idea We Ever Had," *Wilderness* (Spring 1983): 4–5; Ken Burns, *America's Best Idea*, Public Broadcasting Service, 2009.

13. Melissa Harper and Richard White, "The 'Nationalisms' of the First National Parks: Was the Australian Model Different?" in *Civilizing Nature: Towards a Global History of National Parks*, ed. Bernhard Gissibl, Sabine Hoehler, and Patrick Kupper (New York: Berghahn Books, forthcoming).

14. Secretariat of the International Union for the Protection of Nature, *Preparatory Documents to the International Technical Conference on the Protection of Nature*, August 1949, 4.

15. Dowie, *Conservation Refugees*, xx.

16. Spence, *Dispossessing Wilderness*; Neumann, *Imposing Wilderness*.

17. Brockington, Duffy, and Igoe, *Nature Unbound*; Dowie, *Conservation Refugees*.

18. Paige West, *Conservation Is Our Government Now: The Politics of Ecology in Papua New Guinea* (Durham, N.C.: Duke University Press, 2006), xii–xiii; Arun Agrawal, *Environmentality: Technologies of Government and the Making of Subjects* (Durham, N.C.: Duke University Press, 2005), 6–7.

19. Patrick C. West and Steven R. Brechin, eds., *Resident Peoples and National Parks: Social Dilemmas and Strategies in International Conservation* (Tucson: University of Arizona Press, 1991); and Stephan Amend and Thora Amend, eds., *Espacios sin habitantes?* (Caracas, Venezuela: Nueva Sociedad; Gland, Switzerland: UICN, 1992); Ivette Perfecto, John Vandermeer, and Angus Wright, *Nature's Matrix: Linking Agriculture, Conservation, and Food Sovereignty* (London: Earthscan, 2009).

20. International Commission on National Parks, *United Nations List of National Parks and Equivalent Reserves* (Geneva: United Nations, 1971).

21. On Quevedo see Lane Simonian, *Defending the Land of the Jaguar: A History of Conservation in Mexico* (Austin: University of Texas Press, 1999), 67–84; Christopher R. Boyer, "Revolución y paternalismo ecológico: Miguel Ángel de Quevedo y la política forestal, 1926–1940," *Historia Mexicana* 57, no. 1 (July–September 2007): 91–138.

22. Secretaria de Desarollo Urbano y Ecologica, Delegación Hidalgo, *Plan de manejo Parque Nacional "El Chico"* (Mexico City: SEDUE, 1988), 11–12.

23. Simonian, *Defending the Land*, 65, 79. President Sebastián Lerdo de Tejada (1872–76) restricted logging around Desierto de Leones from commercial exploitation in 1876. See Enrique Beltrán, "Génesis y evolución del concepto del parque nacional," unpublished manuscript, n.d., Biblioteca de la Universidad Autónoma de Chapingo, 14; Emily Wakild, *Revolutionary Parks: Conservation, Social Justice, and Mexico's National Parks, 1910–1940* (Tucson: University of Arizona Press, 2011).

24. This average size (37,766.32 ha) comes from the total area of parks divided by the number created between 1935 and 1940, according to data compiled from examination of *Diario Oficial* reports from 1933 through 1945 and http://www.conanp.gob.mx, comprising national parks, natural monuments, areas of natural resource protection, areas of flora and fauna protection, biosphere reserves, and sanctuaries. Expanding

this period from 1917 with the first national park to 1942 marginally changes these figures to forty-five parks and an average of 37,563 ha.

25. Christopher R. Boyer and Emily Wakild, "Social Landscaping in the Forests of Mexico: An Environmental Interpretation of Cardenismo, 1934–1940," *Hispanic American Historical Review* 92, no. 1 (February 2012): 73–106.

26. Quevedo to Secretario de Hacienda y Crédito Público, 3 February 1939, Archivo General de la Nación: Secretaría de Agricultura y Recursos Hidráulicos (hereafter AGN-SARH), caja 1434, exp. 1/2007.

27. See three editorials by Fernando González Gortázar, "Nuestros parques nacionales: Esa ficción, esa urgencia," *Unomásuno*, 4, 18, and 22 October 1983. For the concession proposal, see Enrique Anisz to Lázaro Cárdenas, 15 July 1940, Archivo General de la Nación: Ramo Presidencial, Lázaro Cárdenas del Rio (hereafter AGN-LCR), caja 660, exp. 523.8/21; and Jesús Gonzalez Lugo to Secretaria de Agricultura y Ganadería, 22 June 1955, AGN-SARH 1465, exp. 21/3735.

28. Laura Randall, "Introduction," in *Reforming Mexico's Agrarian Reform*, ed. Laura Randall, 1–14 (Armonk, N.Y.: M. E. Sharpe, 1996), 5; Neil Harvey, *The Chiapas Rebellion: The Struggle for Land and Democracy* (Durham, N.C.: Duke University Press, 1998), 187–90.

29. Stephen Haber, Herbert S. Klein, Noel Maurer, and Kevin J. Middlebrook, *Mexico since 1980* (Cambridge: Cambridge University Press, 2008), 58–65.

30. Ibid., 74; and Carlos Salinas de Gortari, *México: The Policy and Politics of Modernization*, trans. Peter Hearn and Patricia Rosas (Barcelona: Plaza y Janés Editores, 2002), 37–73.

31. The total protected area went from 1,778,848 ha in 1976 to 15,515,769 ha in 2002 and increased to 20,378,172 ha by 2009.

32. Nora Haenn, *Fields of Power, Forests of Discontent: Culture, Conservation, and the State in Mexico* (Tucson: University of Arizona Press, 2005); Lydia A. Breunig, "Conservation in Context: Establishing Natural Protected Areas during Mexico's Neoliberal Reformation," PhD diss., University of Arizona, 2006.

33. *Voices of Chichinautzin*, film, United Nations University, 2007.

34. Dowie, *Conservation Refugees*, 101–5; Neumann, *Imposing Wilderness*.

35. Manuel Corona, "Programa de trabajos que desarrollara la oficina de bosques nacionales y particulares y parques nacionales e internacionales," n.d., AGN-SARH, caja 1425, exp. 1/1609, leg. 2.

36. Kris A. Johnson and Kristen C. Nelson, "Common Property and Conservation: The Potential for Effective Communal Forest Management within a National Park in Mexico," *Human Ecology* 32, no. 6 (December 2004): 703–33.

37. María Rosas, *Tepoztlán: Crónica de desacatos y resistencia* (Mexico City: Biblioteca Era, 1997), 19, 64–65.

38. Enrique Beltrán, "Use and Conservation: Two Conflicting Principles," in *First World Conference on National Parks*, ed. Alexander B. Adams, 35–43 (Washington, D.C.: United States Department of the Interior, 1962), 38–39.

39. Figures for total terrestrial area protected as of December 2006. World Database of Protected Areas, Annual Release December 2007, International Union for Conservation of Nature, http://sea.unep-wcmc.org/wdbpa/.

40. John Terborgh, *Requiem for Nature* (Washington, D.C.: Island Press, 1999), 185.

41. Miguel Ángel de Quevedo, "Las fiestas del árbol," *México Forestal* 15, nos. 3–4

(1937): 9–22; Salvador Novo, *México* (1968; facsimile reprint, Mexico City: Editorial Porrúa, 1999), 305.

42. Novo, *México*, 305–8; Raziel Garcia Arroyo, "Chapultepec en la historia," *El Nacional*, 8 September 1974; Jesús Romero Flores, *Chapultepec en la historia de México* (Mexico City: Secretaria de Educación Pública, 1947), 22–23.

43. Campos, *Chapultepec*, 10–11.

44. Romero Flores, *Chapultepec*, 73; Beardsley, *Mario Schjetnan*, 25–29.

45. In addition to the sources in note 42, see Claudia Agostoni, *Monuments of Progress: Modernization and Public Health in Mexico City, 1876–1910* (Calgary, Alta.: University Press of Calgary, 2003); Emily Wakild, "Naturalizing Modernity: Urban Parks, Public Gardens, and Drainage Projects in Porfirian Mexico City," *Estudios Mexicanos/Mexican Studies* 23, no. 1 (Winter 2007): 101–23.

46. Francisco Mercado to Lázaro Cárdenas, 27 March 1940, AGN-LCR, caja 103, exp. 135.2/179.

47. Cárdenas called for this change, which Avila Camacho implemented. Instituto Nacional de Antropología e Historia, *Museo Nacional de Historia Chapultepec* (Mexico City: Secretaría de Educación Pública, 1955), 4–5.

48. "El ahuehuete o sabino: El árbol nacional," *México Forestal* 1, nos. 9–10 (1923): 1–4.

49. "La agonía de Chapultepec," *El Universal*, 4 January 1944; "Legendaria Bosque de Chapultepec se acaba," *La Prensa*, 30 December 1943.

50. María Elena Soli de Pallares, "En Chapultepec se quema la leña de ahuehuetes," *El Universal*, 4 August 1943.

51. "Perdimos el bosque, conservamos el parque," *El Nacional*, 28 December 1976.

52. José Pérez Moreno, "Corre serio peligro el pintoresco Chapultepec," *El Universal*, 20 March 1944.

53. "Legendario bosque de Chapultepec se acaba," *La Prensa*, 30 December 1943.

54. Miguel G. H., "Descuido y abandono en Chapultepec," *El Universal*, 2 March 1944.

55. Samuel Maynez Puente, "Chapultepec: Riesgo de extinción," *Excelsior*, 4 October 1974.

56. "Remodelación del viejo zoológico de Chapultepec," *Excelsior*, 9 September 1978; "El nuevo zoológico de Chapultepec ha hecho nuevos adquisiciones," *El Día*, 17 April 1973.

57. The first successful birth was a dire disappointment when the baby was crushed and killed by the mother after two days. Rafael Moya García, "Réquiem por la osita panda," *El Universal*, 20 August 1980. See also Matilde Pérez, "Autorizó el DDF un presupuesto de 300 millones de pesos para salvar al bosque," *El Dia*, 11 November 1983.

58. Lucio Méndez Cárdenas, "A más de dos meses de la consulta popular, nada se ha hecho en el Bosque de Chapultepec," *El Dia*, 31 October 1983.

59. "Chapultepec es un conjunto ecológico dice la Sociedad Forestal," *Excelsior*, 31 October 1974.

60. Jorge Coca P., "Chapultepec tiene 286 hectares más para esparcimiento del Pueblo," *El Universal*, 14 September 1974.

61. Rocío Castellanos, "Erigirán un jardín escultórico," *Novedades*, 12 February 1980.

62. Carlos Ravelo, "Otro gigantesco pulmón para el pueblo del DF," *Excelsior*, 12 August 1964.

63. Manuel Gallardo, "Miles de capitalinos acudieron ayer al legendario 'Bosque de Chapultepec,'" *El Nacional*, 19 April 1981.

64. Jorge Vera Ávila, "Una buena actividad para la semana mayor: La pesca en el lago del bosque de Chapultepec," *Excelsior*, 13 April 1981.

65. José Alberto López Sustaita, "Quiere la administradora que el Bosque de Chapultepec funcione como empresa comercial," *El Dia*, 9 June 1983.

66. Guillermo Valencia, "Investigan la 'misteriosa desaparición' de animales del bosque de Chapultepec," *El Universal*, 20 May 1983.

67. José Cabrera Porra, "De la Madrid y Chapultepec," *Excelsior*, 22 August 1983.

68. Teresa Cárdenas, "Definitivamente no cobrarán la entrada a Chapultepec," *El Heraldo de México*, 11 November 1983.

69. I intentionally refrain from using the term *wilderness* in this context because of its lack of resonance with Mexican conservation. See Arturo Gómez-Pompa and Andrea Kaus, "From Pre-Hispanic to Future Conservation Alternatives: Lessons from Mexico," *Proceedings of the National Academy of Sciences of the United States of America* 96, no. 11 (25 May 1999): 5982–86.

70. S. Joseph Wright and Helene C. Muller-Landau, "The Future of Tropical Forest Species," *Biotropica* 38, no. 3 (2006): 287–301.

CHAPTER TEN

The Illusion of National Power
Water Infrastructure in Mexican Cities, 1930–1990

Luis Aboites Aguilar

According to Eric Hobsbawm, the twentieth century will be remembered as a time of revolutionary change that saw the end of a period of seven or eight thousand years since the discovery of agriculture, during which time most of humanity lived on the land.[1] One impact of this transformation was a massive increase in the human population and, above all, in the number and size of cities in the developing world. For example, Mexico City's population had reached one million inhabitants by 1930 and jumped to eight million by 1990, while Bogotá grew by a factor of fifteen between 1938 and 1993 and Santiago de Chile expanded eightfold between 1920 and 1990.[2] As the twentieth century progressed, the urban ruling classes generally chose to support this social transformation. In their opinion, and that of an increasing number of urban dwellers generally, the city represented a powerful symbol of modernity and civilization. These attitudes further validated the impressive spatial redistribution of human society.

This chapter presents a historical overview of a topic that has received little attention from historians or other scholars: the development of water and sewage infrastructure in Mexican cities. It attempts to identify the decisive stages and actors in the history of urbanization by focusing on the use and conceptualization of water. Indeed, one of the primary objects of twentieth-century models of urban development was to provide households with access to running water and sewerage whenever possible. Only a tiny minority had these services in 1930, whereas sixty years later the majority of the population of Mexico City (and the rest of the world) did so.

In Mexico, increasingly widespread access to reliable water services changed social practices and domestic habits. Some of the most obvious environmental impacts associated with this urban transformation include the establishment of new bodies of water near cities and the modification and/or disappearance of agricultural production in the surrounding areas. Most of the impetus for extending piped water and sewers to urban areas came from a national government located in the nation's primary urban center. Its role in this process is less a reflection of the state's putative political "strength" than of the overall limits on provincial cities' abilities to bring about change. Ultimately, this model of national development had a short life span that lasted a mere forty years and derived from the illusory belief that a capital city could bring its civilizing light (not to mention the waters) to a nation in the process of rapid development.

The Urban Confluence:
New Waters, Ideas, Services, and Demands

The rapid concentration of population into small urban spaces was made possible by the transformation of the economy, which became increasingly dependent on services and a bureaucracy that expanded to the beat of an increasingly powerful national state. Cities were also nourished by the overall increase in commerce, the establishment of banks and other financial services, white-collar jobs in businesses, and ever more extensive transportation networks, initially based on railways and later on highway and air travel. Also contributing to urban growth was the process of industrialization, which sometimes proceeded at a breakneck clip.

Twentieth-century urban expansion would not have been possible, or at least would have occurred in a very different way, if not for the technological and economic developments that permitted energy to be generated on a massive scale and transmitted over great distances. While electrical generators of the 1880s were little more than appendages of agricultural, mining, and industrial plants, by the early twentieth century they had grown into enormous corporations specialized in the production of power on a scale unimaginable only a few decades previously. It could be said that cities grew on electricity. In 1905, for example, the Necaxa hydroelectric plant came online 150 kilometers northeast of Mexico City. At the expense of obliterating a beautiful waterfall and the landscape of a handful of villages, the construction of Necaxa gave birth to various bodies of water needed for the hydroelectric operations. The sudden availability of electrical power enabled civil engineer Manuel Marroquín to devise a system for capturing the water from springs in Xochimilco and conducting it 26 kilometers to the Condesa pump house in a public works project unveiled in 1913, during the regime of the counterrevolutionary dictator Victoriano Huerta.[3]

Not only did electricity inspire changes in factories, homes, and offices, it also transformed transportation and urban services generally. Above all, it allowed for an ever-expanding water supply to grow in tandem with urban areas themselves. We are speaking here of a particular *sort* of water, as Jean Pierre Goubert has capably demonstrated. It was chemically pure water, free of bacteria. Work by renowned scientists such as Louis Pasteur and Robert Koch had revealed that water was an efficient vector for pathogens.[4] Simply having a pleasant appearance and smell was no longer enough: people had come to expect bacteriological analyses to guarantee purity and potability as well. Yet water quality could not be guaranteed without eliminating open-air channels and required the application of specific filtration or purification procedures. All this had to be done on an increasingly grand scale. The result was the adoption of a system of water pipes organized as a network that ran beneath the streets and made possible a remarkable innovation: improved access to sources of water and in some cases the introduction of running water into households themselves. If water went in, it eventually had to go out. This opened the way to the so-called system of "wet drainage" (*drenaje húmedo*), that is, the removal of feces and other waste using moving water. Water and sewage had become united.

In some places the government promoted these innovations primarily because existing supply and drainage systems had reached their limits, leading to water shortages and severe public health problems. These old systems usually functioned by collecting water from springs and conducting it through open-air canals or—in places such as Morelia, Querétaro, Chihuahua, and elsewhere—via aqueducts. Within cities, the water reached a few houses and shops; the rest was supplied from public sources by digging shallow wells or paying for the services of water carriers. With the exception of people who had toilets inside their houses, the majority of the population simply threw fecal matter into the open conduits that ran alongside the streets, where municipal workers later collected it.[5] That typhus remained a severe threat in Mexico City as recently as the 1900s is hardly surprising. The outbreaks of disease belied the modernizing and even civilizing images that populated the speeches that authorities invariably gave whenever a new water or drainage project was inaugurated.[6] Comparisons with European and North American cities appeared almost compulsively on these occasions, indicating the level of attention that Mexican experts paid to scientific research into the role of water as a vector for diseases such as typhoid and cholera.[7] Moreover, some well-informed residents of provincial cities carefully monitored developments in the national capital and paid particular attention to advances in federal regulations and legal codes, which often spurred them to seek ways of incorporating such modern improvements in their own hometowns.[8] The wealthy classes were not the only ones interested in these improvements; soon popular groups likewise endorsed these ambitions.

Technological innovations also appeared as entrepreneurs sought out attractive business opportunities. Businessmen approached local and state governments independently, in groups, or in corporations to propose plans for infrastructure projects ranging from the modest and simple to the ambitious and complicated. The German merchant Santiago Graff, for example, recognized that the introduction of running water to Toluca in the 1870s represented a new opportunity to guarantee a regular supply for his brewery and suggested that both municipal government and his business could benefit by shoring up the water infrastructure in the state capital.[9] Likewise, the city of Monterrey, Nuevo León, signed a contract with a Canadian company in 1909 to make improvements to its water supply, but this project soon devolved into a bureaucratic nightmare. A similar fate befell other cities, such as San Luis Potosí, where a private corporation built the giant San José Dam in 1903.[10] In other places, such as Mexico City and Chihuahua City, municipal authorities preferred to keep water infrastructure in the government's hands. The national capital rejected one company's offer to buy the municipal water works in 1885. Administrators' arguments against privatization were not particularly original: they pointed out that a public service could not be made compatible with the profit motive that drove private corporations.[11]

Some urbanites initially objected to becoming "civilized" by using the technology that urban water infrastructures represented. They resented the high cost of laying water pipes and associated service fees and worried that the new services would add to the tax burden on their homes.[12] In part because of this (virtually unstudied) popular opposition and in part because neither the business class nor municipal authorities were interested in pressing the project forward, these public works projects of the late nineteenth and early twentieth centuries had a limited reach. Their modest scope made it impossible to hide their elite nature.[13] One recent study carefully shows that the Marroquín water project and other social services bypassed the working-class Tepito neighborhood in Mexico City, for example.[14] Yet only a few years later (and in yet another episode historians have yet to study),

Tepito residents began to demand running water, having come to the conclusion that the lack of service represented an emblem of backwardness. Water became yet another marker of social inequality, and its absence became a new source of popular discontent. The riot of December 1922 caused by water shortages in Mexico City must be seen in this context. The emergence of public demand for potable water—for *modern* water, in this sense—had taken its first steps. It was no longer a wish, or a dream, exclusively held by the upper classes.

The "Nationalization" of Drinking Water

Studies showing that running water influenced Mexicans' living conditions probably did not come as a surprise to anyone. A 1916 report concluded that chemical, biological, and bacteriological impurities in drinking water constituted one of the "primary causes of mortality, and in particular of high levels of morbidity, in Mexico City."[15] A few years later, one expert declared that the entire nation's water provision system was substandard. Only Nuevo Laredo, Tamaulipas, and certain parts of Mexico City had infrastructures capable of delivering what could reasonably be called potable water.[16] One 1938 legislative bill even admitted that Mexico had one of the world's highest levels of infant mortality, with 136 deaths per thousand births in 1933. One hundred thousand fatalities per year were caused by water-borne diseases attributable to the deficiency or poor quality of water. Out of a population of 18 million people, only 2.6 million, or 14 percent, had access to drinking water. The situation was most dramatic in the countryside, where only 200,000 (1.6 percent) individuals out of a total population of 12 million had a reliable source of potable water. In contrast, 28 percent of the population of the United States consumed filtered or chlorinated water, 19.7 percent had partially filtered water, and 12.9 percent had access to unfiltered water.[17] Compelling private corporations to invest in water infrastructure and expand service required a titanic effort, as the government of Nuevo León learned when it brought a lawsuit against the local privately owned water company in 1922. A final verdict did not come until 1945. The city of San Luis Potosí also ran into difficulties when it began a long and troublesome process of reversing the privatization of its waterworks in 1938.[18]

The February 1933 articles of incorporation of the Banco Nacional Hipotecario Urbano y de Obras Públicas (National Bank for Mortgages and Public Works, now known as Banobras) explicitly recognized the futility of relying on private corporations to provide public water. The Banco Hipotecario had been proposed by Secretary of Finance Alberto Pani and the economist Gonzalo Robles, among others, and served as a complement to the National Bank founded in 1925 and the National Bank of Agricultural Credit established the following year.[19] It was one of the first examples of the quintessentially twentieth-century Mexican phenomenon of nationalizing water services—and urban services generally—in the nation's cities. In a sign of the times, the bank's founders suggested that the limited resources of municipal governments and private enterprises required that the federal government take responsibility for making improvements to urban infrastructure—a political maneuver that simultaneously strengthened the central government's authority at the local level not only in this realm but also in many other aspects of public life, such as education, land reform, irrigation, health, work, and taxation.[20] In terms of water supply, it had become clear that the earlier model based on a coalition of local officials and private interests had simply not succeeded. Unsurprisingly, one of the bank's first moves was to authorize credit for the reconstruction of the Xochimilco–Mexico City aqueduct. This move signaled that the federal government would become involved in a realm previously occupied by municipal authorities and, in some cases, of state governments: the provision of drinking water. This national model endured for decades and ultimately involved expenditures of multiple millions of pesos.

The world financial crisis of 1929 reinforced the idea that the central government needed to get involved in such projects. Countries such as Mexico that regarded the lack of population as a serious national problem were forced to give up their long-held aspiration of attracting foreign migrants to meet their labor needs, leaving them with no other choice than to rely on natural population growth. This dramatic change helped give currency to "populationist" ideas (articulated especially persuasively by Gilberto Loyo) that held sway in Mexico and throughout the world for a period of three or four decades.[21]

This is the context in which we must understand the attention that President Lázaro Cárdenas (1934–40) paid to the expansion of water and sewage systems. One of his administration's first measures was to increase funding for the Banco Hipotecario and charge it with expanding its lending to small towns. In mid-1935, when the Cárdenas government's radicalism had yet to take wing, one newspaper went as far as to suggest that the Cárdenas years would become known as the "*sexenio* [six-year presidential term] of drinking water."[22] Credit seemed like an ideal solution to address the double problem that federal authorities confronted. On the one hand, water infrastructure was a municipal responsibility that the central government could not simply assume overnight. On the other, the government lacked the economic resources to take on such an obligation. With direct investment out of the question, credit appeared to be the best way for Mexico City to support the municipalities. This strategy was formalized in a 1935 law that authorized the secretary of finance to direct more resources for water and sewer works to the Banco Hipotecario under the rationale that the federal government could not remain indifferent to water and sanitation services.[23]

Federal authorities could take responsibility for the modernization of infrastructures because a technological solution seemed feasible. So-called large-scale hydrology promoted by transnational capital after the late nineteenth century incorporated irrigation projects, hydroelectric plants, and eventually water supply systems for cities. These projects used not only surface water but also underground sources. The key to their success was the increasingly widespread availability of electric and internal-combustion pumps. These new devices enabled extraction of a large volume of subsurface water, which the best minds of the time considered infinite and inexhaustible. There is evidence of the intensive use of subterranean water sources in Sonora's Yaqui Valley and Comarca Lagunera between Coahuila and Durango as early as the 1920s. In Mexico City, the use of aquifers entered a new era beginning in 1928, when autonomous municipalities were abolished in the Federal District and replaced by a single administrative structure. At that point, water shortages in the capital made clear that the Xochimilco aqueduct reconstruction project inaugurated a mere fifteen years before had already become inadequate.[24]

Pumped water was expensive on a per-unit basis and often contained a great deal of sand and salts. Nevertheless, it appeared attractive because the expense of drilling and extraction cost far less than major projects to control the flow of surface water. The idea of building a public works project like the San José Dam in San Luis Potosí simply to expand the public water system seemed an act of folly if deep wells represented a viable alternative. They were much simpler in technological terms, hence less expensive; they also had the virtue of taking little time to establish. From politicians' point of view, the choice was simple. As a result, the expansion of water service in Mexico—and the rest of the world as well—revolved around the discovery and use of subterranean water in the modern sense. Within a few decades, however, the immense fragility of this modern water became clear, a trait that had major implications for the environment, as we shall see.

In the simplest terms, these new networks consisted of one or more deep wells ("deep" was a figure of speech in those years, because they commonly descended less than one hundred meters), a reservoir for storage and water management, and some sort of filtration or purification system. From that central point, the water flowed (ideally by gravity, which was the most cost effective) to the distribution system of lead and, later, iron underground pipes and from there to dwellings. City managers concurrently built drainage systems to remove wastewater. One advantage of such a network was the ability to close some branches temporarily for repair and maintenance while the rest continued to function normally. They were not complicated projects, but they nonetheless required the involvement of outside experts, particularly to help map out the location of aquifers and proper sites for wells. As a result, federal funds provided the credit necessary not only to construct provincial water projects but oftentimes to pay for engineering expertise as well.

In 1935 and 1936, the bank financed public works in Acámbaro, Cuernavaca, Irapuato, Taxco, Tepic, Tulancingo, Zamora, Dolores Hidalgo, Mazatlán, Morelia, Nuevo Laredo, Oaxaca, Pachuca, Querétaro, Guadalajara, Orizaba, and Huamantla.[25] Later, after the difficult years of dwindling institutional reserves between 1938 and 1941, loans were directed to large cities and especially the capitals of states. The idea was to benefit the largest number of people at the least cost. The principal urban areas were therefore the first to benefit from this new and expanding public service.

State governments needed to approve the credit lines granted to municipal governments; on some occasions, those entities, as well as local governments or even residents, paid part of the costs for pipes, labor, or other expenses. One expert noted that federal authorities preferred to invest funds in the 1930s and part of the 1940s without asking for any financial assistance from the people who would benefit from new water services, which resulted in a veritable disaster.[26] The bank soon established *juntas de mejoras materiales* (public works committees) that gave local stakeholders a say in how waterworks would be designed and who would win the contracts to build them. In essence, these working groups displaced the municipal governments themselves.[27] Examination of the negotiations over particulars of how federal projects would proceed within municipalities provides a window onto the intrigues and arrangements between the two levels of government. It also reveals how long-standing arrangements about access to, and distribution of, water were originally forged. In Aguascalientes, for example, there was a prolonged conflict between the council, neighborhood groups (growers and property owners), and federal authorities over ownership of the water itself (that is, whether it belonged to the municipality or to the nation), over the demands of orchard owners versus urbanites, and over the municipality's unyielding refusal to lower fees for services (insofar as the municipality pointed out that the Banco Hipotecario demanded repayment of its loans). Obviously, the overall shortage of water set the stage for this clash, which began in 1931 and lasted for another fifteen years.[28]

Another example occurred in the city of Querétaro, where the state's new governor—the aptly named Agapito Pozo (*pozo* means "water well" in Spanish)—refused in 1943 to ratify a loan from the Banco Hipotecario that his predecessor had negotiated. Both Pozo and the new mayor alleged that the loan, while necessary, imposed onerous terms of repayment and an unacceptable level of administrative oversight.[29] The mutual recriminations that set Mexico City against the states and municipalities constitute a key element of the new relationship that grew up around modern, urban infrastructure. This issue merits closer examination.

The Banco Hipotecario's limited resources rivaled those of the states and municipalities and gave rise to the mobilization of local groups demanding access to public water services, particularly in smaller urban areas. Many communities adopted the venerable strategy of seeking moral and economic backing from local notables but in this case for the modern goal of gaining access to potable water, or at least running water of some sort.[30] At an earlier historical moment, people had used this tactic to obtain schools, roads, or bridges; in subsequent years, they employed it to demand electricity as well. In some places, such as Camargo, Chihuahua, prominent citizens embraced the cause of running water and organized fund-raisers such as galas, dances, and raffles. Perhaps they made donations from their own pockets to win the governor and municipal authorities over to their high-minded cause.

The advent of municipal services also brought change to timeworn relationships between city dwellers and water. In my grandparents' home in Chihuahua city, for example, the arrival of running water in the early 1940s made the well at the stables redundant, and it was sealed off; the same thing happened with the trough. Like other urbanites, my family also stopped buying from the water carriers who sold water from barrels lashed to mule-drawn carts; the water, drawn from the irrigation ditches on the outskirts of town, originated in the Río Conchos.

How did these new water services expand, and how had their advantages become known? What did people observe, who promoted the project, and in what terms? Can we speak about an urban development model that local groups quickly accepted, based on a new class of public services?

The product of efforts by government institutions and social groups (the latter of which we understand far less) laid the groundwork for the expansion of water and sewer systems throughout the nation, especially in medium-sized and large cities. As with the process of electrification, people who obtained running water often complained about high fees and poor service. In 1935, some residents of San Luis Potosí formed a consumers' league to pressure the privately owned water company to do something about water shortages. Two years later, the nascent workers' central known as the Confederación de Trabajadores de Mexico (Mexican Federation of Workers, or CTM) seconded the complaint and created the Emergency Pro-Drinking Water Committee of San Luis Potosí. In November 1937, against the background of the conflict between the state's longtime political boss (*cacique*) Saturnino Cedillo and President Lázaro Cárdenas, the CTM—along with railroad workers, electricians, and factory workers—declared a strike to protest inadequate water service. Cedillo supported the owners of the waterworks, at which point the president intervened by sending top officials to investigate and dispensing 300,000 pesos to improve access to drinking water. Tempers cooled, but water scarcities persisted. Soon afterward, the state government bought the water company using funds from a loan extended by a private bank.[31]

The Expansion of Urban Water Services

The expanding availability of running water changed the relationship between neighbors in the cities and transformed public opinion over time. Water also seemed to promise an overnight solution to longstanding problems of illness and sanitation, meaning that public health once again became a priority. The city of Querétaro, for example, launched a campaign in the 1940s against intestinal diseases caused by tainted water. A century earlier, the relationship between water and maladies such as typhoid was a mystery, but the emerging awareness of public health issues helped produce a social movement for greater access to drinking water. At the same time, population growth created new pressures on urban infrastructure. In 1945 Monterrey had 117 streets without access to water pipes and 65 that lacked drainage. Typhoid mortality had risen 7 percent during the 1930s, and morbidity had shot up from 18 per 100,000 residents in 1938 to 67 in 1944. Elsewhere, a report found that 53 percent of the students at one Mexico City elementary school that lacked adequate water *or* drainage suffered from some form of intestinal parasites, predominantly helminthes.[32] Statistics gathered in the city of Querétaro in 1950 found that 65 percent of households had access to piped water: a mere 27 percent of all households used it exclusively, while 38 percent obtained water from public hydrants. Just a few years earlier, in 1944, studies showed that on most days, the people who lived in the higher elevations of the city had access to running water for only one hour in the morning and a few minutes in the evening. The report added, "The pressure of the water was once very good but has declined to the extent that the cisterns on rooftops cannot get filled."[33] A general estimate in 1947 suggested that only 23 percent of the nation's population had access to piped drinking water service.[34] As people began to recognize that infrastructural improvements were feasible from both technical and financial standpoints, they began to demand them. To a certain extent, they had adopted the idea (formerly the province of politicians, doctors, engineers, and the educated classes) that access to these services constituted a basic element of civilization itself.

The establishment of public housing programs in the late 1940s not only required more water but also promoted the expansion of new services. The Miguel Alemán multifamily housing project inaugurated in Mexico City in September 1949 is a case in point. Almost overnight, federal investment created an unprecedented 1,080 apartments for five thousand people (nearly all of them government employees).[35] This project, which incidentally led to the creation of the Ingenieros Civiles Asociados (the United Civil Engineers, or ICA, which soon became the nation's largest private construction company), can be seen as a model of new ways of inhabiting the city, which entailed among other things the creation of a complex delivery system that supplied each apartment with running water. Every unit had its own bathroom equipped with toilet, shower, sink, and perhaps a washing machine—consumer durables that had begun to be manufactured domestically in the context of import substitution industrialization. The people we now call plumbers must also have begun to take their current form as experts who repaired leaks and broken pipes in exchange for very high fees. In the western part of the city, wealthy suburbs for old-line elites and the nouveau riche grew by leaps and bounds. Their ostentatious homes, often with large gardens and a pool, modernized, as it were, social inequality in terms of access to water. These new homes also generated a spike in the per-capita consumption of water as well as a growing disparity reflected in official statistics: figures for the metropolitan area of Mexico City indicate a distressing gap between the neighborhoods to the north and east (Valle de Chalco, Ecatepec) that consumed fewer than fifty liters per capita, compared to the five hundred to six hundred liters per capita used in the posh western neighborhoods such as Lomas de Chapultepec.[36] But there was yet another inequality that requires scholars' careful consideration: industrial water systems owned by private concerns in cities such as Monterrey and Puebla began to expand without public financing or, significantly, government regulation.[37] Businesses drew from the aquifer almost at will, regardless of recent weather patterns or the effects that a sudden reduction of water supplies might have on the inhabitants of these cities.

Despite their technical simplicity, the new water supply and sewerage systems required a considerable outlay of funds. These expenditures seem to have taken local and municipal governments by surprise, among other reasons because they revealed these governments' financial weakness. Like other urban services, the provision of drinking water had clear-cut fiscal implications that also involved the nation's federal structures. The state of Querétaro provides a good illustration of this situation. In 1951, it initiated public works projects, budgeted to cost 7 million pesos, meant to address water shortages and enhance drainage capacity in the capital city. Yet the entire state budget for 1951 amounted to only 2.7 million pesos, and the municipal government's precarious financial situation kept it from making more than a token contribution.[38] How, then, could the project get under way?

The treasuries of state and local governments became increasingly impoverished after the 1910 revolution, in part because the central government sought to capture the most attractive sources of tax revenue (which also happened to be the ones that did the most to undermine the development of the domestic market). Moreover, land reform had an unintended consequence insofar as the redistribution of land and water led to a drop in property tax collection, which had previously constituted the main source of municipal revenues. Experts pointed out that the sharp reduction in municipalities' income—hence their capacity to borrow—diminished their overall capability to undertake public works projects.[39] For their part, state governments often attempted to increase revenues by raising taxes, but these initiatives ran headlong into strong opposition from employers, landlords, and other sectors. The Querétaro state government gave up on such an attempt in 1945, as did Veracruz in 1965. Sometimes the opposition to tax hikes developed into political movements capable of removing governors from office, as in the case of Oaxaca in 1946 and 1952.[40]

This context lays the foundation for understanding why federal authorities elected to prioritize the provision of water and sewerage to cities within its overall public works agenda. This policy began to take shape with the creation of the Secretariat of Hydraulic Resources (Secretaría de Recursos

Hidráulicos, or SRH) in late 1946. The new institution was supposed to overcome the perceived shortcomings in water and sewer services attributable to fiscal mismanagement or official inattention. A 1947 report noted, for example, that the federal government had invested over 900 million pesos in irrigation projects between 1926 and 1946, whereas the amount earmarked for water and sewerage in that period barely reached 16 million pesos.[41] The emphasis on agriculture and rural development could not have been more evident. And while the Cárdenas administration pushed for greater federal involvement in the credit markets through the Banco Hipotecario, Miguel Alemán (president from 1946 to 1952) favored direct investment through the SRH. Thus, from both flanks (or three flanks, if we take into account the projects that the Secretariat of Health and Welfare undertook in villages with fewer than 2,500 inhabitants), the federal government lay behind the expansion of piped water and sewerage in the late 1940s. A report released by the SRH in 1951 gave greater definition to these new policies and underscored the conviction that only federal power could resolve problems of water shortages and sanitation. The state's heightened role in public services lasted until the 1970s, when it began to deteriorate.[42] The ascendancy of national power does not seem accidental. Beginning with the World War II economic boom of the 1940s, the Mexican population began to move to the cities. For the first time in the twentieth century, overall growth was largely urban, and Mexico City—the country's largest— grew more than any other.[43] Mexico City also gained power, not only because its position as the national capital allowed it to provide credit and technical expertise to other cities but also because it became an emblem of the nation's capacity to engage in the construction of monumental works projects that increased residents' access to water.

Between 1942 and August 1951, the Department of the Federal District and the federal government itself invested a whopping 226 million pesos in the headworks and piping that eventually transported six cubic meters of water per second from springs in the upper Lerma watershed to the Valley of Mexico. The investment was vastly more than state and municipal treasuries could ever hope to dispense; we saw, for example, that Querétaro received the equivalent of 7 million pesos to deal with its water problems in 1951. Mexico City's Lerma works was meant to be paradigmatic in more than one sense. First, it reflected the primacy of the federal government and the capital city itself. Second, it made use of the latest hydraulic technologies. To top it all off, Diego Rivera painted murals inside the distribution chamber depicting an engineer giving drinking water to Indians, in a clear expression of the allure of large-scale state engineering projects.

The inaugural festivities emphasized time and again that the revolutionary government had conquered nature by finding the means to carry water from the basin of Lerma (on the Pacific) to the Valley of Mexico and from there to the Pánuco (on the Gulf of Mexico), a displacement of immense proportions, as we would say today. The work remedied "thirty-five years of shortages during which time the city only sporadically had access to adequate water supplies."[44] But that assertion missed its mark. The supply never met the increasing demand provoked by rapid population growth and changing consumer habits. Even so, Mexico and indeed the entire Mexican Revolution appeared to have fulfilled its promise to the nation's most important city. Little account was taken of the impact that this monumental project would have on the lakes of the upper Lerma basin, some of which had already disappeared by the late 1960s.[45]

Another type of mechanism was developed for the nation's other cities, when the SRH became the leading sponsor of municipal water and sewer projects thanks to massive increases in federal expenditures. The new funding stemmed from something other than federal largesse: officials in Mexico City chafed at the thought of granting local authorities too much autonomy, because of their well-worn reputation for entering into shady business arrangements. As a result, the number of water systems built using federal resources multiplied from 117 in 1947 to 720 in 1965.[46] As we have seen, the Banco Hipotecario around this time also created the *juntas de mejoras materiales* (public works committees; later complemented by the essentially similar *juntas para agua potable y alcantarillado*, or water and sewer committees), comprising federal officials and representatives of local business interests that controlled how the federal money was actually spent. Little is known about these panels' day-to-day activities, although a 1966 review of the system noted that while the committees were self-sufficient in terms of budget authority and operating expenses, they lacked enough income to fund new investments. Furthermore, the SRH calculated that it would take a very long time (at least fifty years) to recover investments in infrastructure if committees' tax receipts remained at the current levels. In other words, municipal water systems were at risk of becoming obsolete and failing to keep up with the rapid growth of urban populations.[47] The water boards' inability to collect revenues was a constant source of concern. Some analysts did not hesitate to blame the "communities" for failing to raise revenues by charging fees for water services. As one expert put it, the lack of revenue resulted from "the commonplace misperception in many communities that authorities must provide water services for free or virtually so." In the end, "political and cultural factors" made people unwilling to pay for municipal services.[48]

In 1965, the federal government announced the launch of a national drinking water project. The influence of the Cuban revolution—and the American reaction expressed in the form of the Alliance for Progress—also played a role in the timing.[49] Mexico announced ambitious goals for infrastructure improvement, such as increasing the proportion of urban locations with household connections to water from 47 percent to 70 percent, which meant expanding the total urban population with access to running water from 8 million to 15.3 million people. Investing more resources than ever before in infrastructure would be necessary. The government budgeted 2 billion pesos (about US $160 million at that time) for the five-year program, which accounted for around three-quarters of the national infrastructure budget; other funds would also come from local contributions (through loans granted by the Banco Nacional de Obras y Servicios Públicos, or Banobras) and state governments. By 1968, more than 840 million pesos had been invested to benefit 3.8 million people in urban locations, particularly those twenty-six localities that had more than 100,000 inhabitants (which included ten state capitals: Aguascalientes, Saltillo, Chihuahua, Durango, Toluca, Morelia, Oaxaca, Culiacán, Hermosillo, and Mérida). Federal authorities also emphasized that the credit provided to manufacturers of asbestos-cement and plastic pipes, which totaled 155 million pesos, demonstrated the fruitful linkages between expenditures on public services and industrial development.[50]

Water and sewer service coverage had made considerable gains by 1970 (table 10.1): running water and adequate drainage had become widely available in the urban centers of the country's largest cities. The nationalizing push between 1950 and 1970 seems to have been the decisive factor in expanding these services, a feat that is all the more impressive insofar as the nation's fastest demographic growth occurred during the 1960s.[51] Clearly, the overall improvement in living conditions and public health services played a pivotal role in the dramatic decline of infant mortality and the rising overall rate of population growth.[52]

In truth, however, the expansion of water and sewer service was inadequate. In 1970, 62 percent of the population still lacked regular access to running water, and 59 percent had no sewage services, in part because investment had flowed into only the largest and most important cities. Another estimate from 1970 suggests that a mere 11.4 percent of urban households had reliable access to modern piped water, drainage, and electrical services.[53] In the countryside, just 2 percent of households enjoyed these three services. Rural-urban inequality—as well as disparities *within* cities— is a constant theme in the study of these public services. For example, a

Table 10.1 Percentage of residences with interior running water and sewer services, 1950–90

	1950	1960	1970	1980	1990
Running water	17.1	32.3	49.4	70.6	77.1
Sewer service	n/a	28.9	41.5	51.0	62.0
Electricity	n/a	n/a	58.9	74.8	87.5
Toilet	n/a	20.9	31.8	n/a	75.3

Source: INEGI, "Censos de población y vivienda," 1950 through 2000, with the exception of the figures for 1980 (which must be regarded with some skepticism), which derive from INEGI, *Estadísticas históricas de México*, 2 vols. (Mexico City: INEGI/Instituto Nacional de Antropología e Historia, 1990), vol. 1, table 3.

1956 law that promoted institutional cooperation in the provision of potable water to municipalities specifically sought to direct more resources to smaller communities that had no other access to credit.[54] Apparently, it did not go far enough to reverse the bias in favor of larger urban areas.

Fortunately, some experts raised their voices to criticize the overall condition of public services in Mexico. In 1966, the director of the College of Engineering of the Universidad Nacional Autónoma de Mexico (UNAM), Antonio Dovalí Jaime, decried the "immense discrepancies" in the availability of water and sewer services and pointed out that waterborne disease and the time-consuming pursuit for fresh water imposed immense human and economic costs.[55] Water's capacity to create civilization, a concept that the federal government had championed from its seat in Mexico City, had been very unevenly distributed in both social and regional terms. In a society marked by deep inequalities, however, it comes as no surprise that differential access to water and sewer services not only expressed and reinforced the country's overall inequality but also added a new dimension to it.

The Decay and Collapse of the National Model, 1970–1990

In much the same way as the original model of funding public works through private and municipal funding began to show signs of exhaustion during the 1930s, the policy of nationalizing responsibility for water infrastructure reached its own point of exhaustion four decades later. This structural crisis took place in the context of ongoing fiscal difficulties, but alongside this age-old problem grew a new set of concerns about environmental degradation. During the 1960s, for example, an enormous array of deep wells came online that pumped water from the Toluca-Ixtlahuaca aquifer, which it then sent coursing through an aqueduct to the nation's capital at a rate of eleven cubic meters per second. The project not only had severe ecological impacts (including landslides, the depletion of springs and wells, and a loss of soil moisture) but also sparked serious political tensions with the area's rural population.[56] The huge pipe to the capital of the country had become a new element of the political landscape.

This "national model" of public works depended on continual federal spending and an overconfidence in the ability of technological innovation to bring water and sewer service to the masses. However, both public expenditure and technological optimism began to crumble in the 1960s as engineers began to deal with the consequences of their previous endeavors. The monumental undertaking of 1967–75 in the Valley of Mexico known as the Deep Drainage Project, for example, showcased the "national model's" favored recipe for development, namely, combining federal spending with cutting-edge technological innovation to produce an engineering megaproject (built in this case by the ICA, the thriving private construction company mentioned above). Unlike the Lerma waterworks project inaugurated in 1951, however, the Deep Drainage Project did not propose to enhance society's (supposed) control of nature, but rather was designed to alleviate the unintended consequences of previous projects intended to control nature—specifically, the Valley of Mexico aquifer. This was a vastly different proposition. The problem of flooding had become increasingly palpable as the city sank farther and farther into the lakebed, indicating that twentieth-century Mexico's across-the-board strategy of underwriting urban and industrial development by pumping water from the nation's aquifers had reached its limit, as it had elsewhere in the world.[57] In the capital itself, the extensive extraction of groundwater through the use of deep wells (which first appeared in 1928) imposed a heavy cost in very little time: the pumping actually sped up the capital city's subsidence. In 1951—the same year that the Lerma project was completed—the federal government created a new agency charged (of course) with studying the subsidence issue and addressing the problem of water supply. In a further irony, Mexico City had endured serious floods between 1950 and 1952.[58]

Although federal expenditures reached historically high levels in the 1970s owing to a combination of inflationary measures, an ever-increasing debt, and the oil boom, numerous signs already pointed to the exhaustion of the "national model." One of them—perhaps significant only in a symbolic sense—was the disappearance of the SRH in late 1976. Its successor institution, the Ministry of Agriculture, no longer had authority over water supply and sewerage. These domains were transferred to a new federal agency whose name reflected the emergence of new priorities and concerns: the Ministry of Human Settlements and Public Works (Secretaría de Asentamientos Humanos y Obras Públicas, or SAHOP).

More important is that the federal government had lost its appetite for managing urban water and sewer services. The transition came in 1975, when it delegated authority over these services to a *fideicomiso* (board of trustees) empanelled by the Banco Hipotecario (by then called Banobras). Five years later the federal government, through the SAHOP, moved to turn water systems over to the states. The central government pleaded a lack of resources and suggested that inefficiencies in the management of the water infrastructure could be addressed by enhanced federalism, that is, by expanding the state governments' administrative ambit.[59] This trend was reinforced in 1983 by changes to Article 115 of the Constitution, which assigned municipalities authority over water and sewer services. According to one expert on the subject, "the implicit goal of these changes [was] to create [a] self-sufficient water and sewerage sector. It [proposed] to minimize federal subsidies or direct expenditures and replace them by making credit available to municipal administrative units."[60] Thus, the federal government unburdened itself of a responsibility that it had acquired in the 1930s. The problem was that the new sources of credit did not cover the gap left by the withdrawal of direct federal investment. Moreover, the national government's abandonment of water infrastructure occurred at a time when state and municipal finances were already stretched thin.

The overexploitation of aquifers and resulting conflicts between neighboring water wells also signaled an impending crisis. Very little has been written about these aspects of urban life, which have led to social tension and even violence. After twenty or perhaps twenty-five years of pumping, aquifers throughout the country verged on depletion. As a result, the proportion of the urban population that enjoyed access to piped water declined for the first time in some places. According to one illustrative study on the city of Puebla, the proportion of urbanites who had water service decreased from 82 percent in 1970 to 65 percent in 1980 and further still to 45 percent in 1987. Municipal officials reported cases of deep wells becoming exhausted a mere five years after they had first come on line; in 1984, for example, twelve of thirty-seven recently drilled wells had run dry.[61]

The diminishing reach of urban services was not confined to Puebla. As a result of "decentralization"—that is, the dismantling of the national model—expenditures on water and sewer infrastructure fell dramatically throughout the 1980s. In 1987, investment had fallen to less than a fifth of its level in 1980. As a result, drinking water coverage in medium-sized cities (those between one hundred thousand and one million inhabitants) fell from 86 percent to 83 percent and sewerage from 75 percent to 68 percent between 1980 and 1987.[62]

An unprecedented wave of urban popular movements broke out at the same time as the federal government stepped away from water services. Major popular mobilizations rocked the cities of Monterrey, Chihuahua, and Durango, as well as Mexico City and its suburbs, as city dwellers orchestrated urban land invasions and demanded better housing and services. Mexico was not alone in this regard. Several studies show that major cities throughout Latin America were shaken by the mobilization of the immense numbers of people who lived in deplorable conditions, many of whom were homeless or lived in rickety housing with no access to urban services.[63] These conditions were the result of a development model that, while integrating the country and many social groups, also segregated them in terms of working conditions, income, and overall standards of living. For housing and access to basic services, the gap was tremendous.[64] In Monterrey, for example, 30 percent of the population lacked a reliable water supply in 1970, and the rest suffered from acute shortages that led to rationing (*tandeos*) and frequent interruptions. The Popular Defense Committee in Chihuahua and Durango and the Land and Freedom Front in Monterrey were perhaps the most aggressive organizations in those years.

Vivienne Bennett has shown that popular mobilizations in São Paulo and Lima forced authorities to increase investment in water and sewer services in the 1980s.[65] A similar response may have occurred in Mexico; all that we can say for now is that surface water extraction reappeared as an option to meet what had become an explosive demand for water in several of the nation's major urban centers. But the new sources of surface water that officials contemplated in the 1980s differed in fundamental ways from those held in older reservoirs by turn-of-the-century projects such the San José Dam in San Luis Potosí (opened in 1902) and the Chihuahua Dam (opened in 1908). The difference was the cost: by that point, the most attractive sources of water lay far away from the cities. The new generation of waterworks planned and built after 1970—such as Mexico City's Cutzamala system, the Cerro Prieto dam, the 135-kilometer Linares-Monterrey aqueduct, and the 180-kilometer aqueduct that transported four cubic meters per second of Colorado River water to Tijuana and Tecate—must be considered in this light.⁶⁶ The federal government paid most of the multimillion-peso price of these projects. Had the government withdrawn resources from some regions to concentrate on the ones with the most explosive social and political conditions? Was the Mexican state plugging holes and putting out fires in the style of the 1930s and 1940s? Whatever the case, the shift to surface water clearly did not mean that authorities would not continue to use (and overuse) the nation's aquifers, which had become the primary source of water for most purposes and remain so today. No wonder reports began to appear with disturbing frequency about falling water tables and the consequent rise in the cost of pumping (the Bajío), saline intrusion (the Hermosillo coastline), and even the appearance of toxic substances such as arsenic in the aquifer (the Laguna). Clearly, the withdrawal of federal funds and the devolution of public services to states and municipalities occurred in a context of severe environmental degradation. In addition to overexploited aquifers, pollution in the form of industrial and urban wastes began to degrade many watercourses in these years, most prominently the Lerma and Blanco Rivers.

Consider for a moment the case of Monterrey. Shortages of drinking water had reached alarming dimensions by the late 1970s. Given the state government's financial inability to solve the problem, local authorities approached the federal government for help, and it agreed to take action. "The Nation Has Come to Monterrey's Rescue," reads a subheading in one book that reproduces the official line.⁶⁷ Although the central government did agree to provide massive funding for Monterrey's water infrastructure, its motivations were anything but straightforward. One scholar who investigated the affair maintained that federal officials hoped their investment would help mend fences with the businesses in the Monterrey Group in the wake of the 1973 murder of prominent entrepreneur Eugenio Garza Sada and the struggle over control of the municipal water system that arose some years later. The state government, with the support of federal authorities, had managed to regain control of the water commission by removing commissioners aligned with business interests, whom authorities accused of favoring wealthy neighborhoods and industrial parks at the expense of the city's impoverished majorities. The outrage and popular mobilization provoked by severe water shortages forced a rapprochement between government and business leaders, paving the way for large-scale federal investment. This is the only way to explain the central state's sudden willingness to fund the construction of two large dams intended to meet Monterrey's demand for new sources of surface water: the Cerro Prieto, completed in 1984; and El Cuchillo, completed a decade later.⁶⁸

The Monterrey case is instructive, and we may hope that other, equivalent studies will appear soon.[69] For one thing, it leads one to wonder about the possible link between the dismantlement of the national model of infrastructure development and renewed investment in surface water infrastructure as a result of aquifer depletion. Could the federal government have already taken the precarious state of aquifers into account when it decided to give responsibility for public services back to the states and ultimately to municipal governments?[70]

Final Considerations

The illusion of national power was predicated on the notion that the federal government had adequate fiscal and legal powers to bring piped water and sewerage services to the entire nation. Although it never met these goals, the government did make significant progress. Two institutional innovations, both of an unmistakably economic variety, lay behind these advances: in the first place, the provision of federal credit via the Banco Nacional Hipotecario Urbano y de Obras Públicas (Banobras); and in the second, direct public investment through the Secretaría de Recursos Hidráulicos and to a lesser extent the Secretaría de Salubridad y Asistencia. Apart from its fiscal resources, the federal government also controlled most of the nation's technological expertise related to water and sewer systems. All of this took place against a background of the unremitting centralization that delivered nearly every source of tax revenue to the federal government, thereby contributing to the impoverishment of the state and municipal treasuries. That the central federal government's involvement in water infrastructure projects began to be dismantled in the mid-1970s, shortly before the outbreak of the 1982 debt crisis, is perhaps no coincidence.

Moreover, federal involvement in public works created significant environmental changes. Increased water consumption in cities cannot compare with the explosive increase in agricultural uses (the total acreage under irrigation leapt from one million to three million hectares between 1930 and 1980). Nevertheless, the expansion of urban water service contributed in at least three ways to environmental degradation: it aggravated the overexploitation of aquifers; increased the pollution of surface water; and subordinated rural needs to the requirements of urban growth. These trends first appeared in the 1930s, when deep wells seemed to offer a solution to the nation's water problems. Deep wells grew increasingly widespread over the next half century, until their environmental costs could no longer be ignored. Political organization, taxation, urban expansion, technological change, and environmental degradation seem to go hand in hand (see appendix, table 10.2).

To contend with the diminishing federal role in public services, municipalities established specialized agencies of their own to manage water and sewerage infrastructure. By 1995, an estimated 805 such organizations had sprung up throughout the country.[71] The water system returned to its birthplace: the city council. In concert with the global pro-privatization airs of the 1980s, new voices began to call for the participation of private capital in public service sectors. These voices had little new to say. The arguments that local governments had deployed when contracting the services of private firms between 1880 and 1930—that is, bankrupt municipal treasuries, the inefficiency of public administration, and technological backwardness— were dusted off to justify the privatization of public services in the 1990s. A new generation of private water companies has established a variety of different institutional arrangements with the cities of Aguascalientes, Cancún, Puebla, and Mexico City itself.[72] Most of these companies are primarily interested in raising service fees and put little effort into extending their services to people who do not already have them. We may reasonably assume that at some point these companies will run up against public opposition and be dissolved. Their forebears met such a fate between 1922 and 1945 in San Luis Potosí and Monterrey—as well as in Cochabamba, Bolivia, in early 2000.

The resurgence of water as a municipal service and the appearance of a new generation of private water companies suggest a kind of return to 1930. What appeared modern at that time appeared old or outdated a few years later. Decades later, around 1990, it once again strikes us as modern. To be sure, there are several new issues this time around, including the problem of depleted aquifers and a potential for urban popular movements to reach dimensions unthinkable in 1930. (Consider, for example, the opposition of small towns to projects that would pipe their water to large cities.) A great deal of social learning has also taken place—indeed, it constitutes a veritable cultural change—about the advantages of housing equipped with running water and adequate sewerage. Apparently, there can be no return to a time when such services were considered luxuries. Current trends make it appear unthinkable, inevitable, that these public services will one day become universal. However, the reverses that characterized the so-called lost decade of the 1980s cast doubt on whether such a scenario will ever come to pass.

Appendix

Table 10.2. Population of principal Mexican cities, 1900–90

	1900	1910	1921[a]	1930	1940	1950	1960	1970	1980	1990
Tijuana					16,486	59,952	153,303	283,951	435,454	747,381
Mexicali					18,775	65,749	174,540	263,498	341,559	601,938
Cd. Juárez			19,457	39,669	48,881	122,566	262,119	407,370	544,496	798,499
Chihuahua	30,405	39,706	37,078	45,595	56,805	87,000	192,624	282,155	411,922	534,699
Hermosillo				19,959	18,601	43,519	95,978	176,596	297,175	448,966
La Paz							24,253	46,011	91,453	137,641
Culiacán			16,034	18,202	22,025	48,936	85,024	167,956	304,826	601,123
Torreón	62,266	34,271	50,902	66,001	101,354	188,203	345,929	438,461	689,195	878,289
Monterrey	62,266	78,528	88,479	134,202	190,128	354,114	695,504	1,242,558	1,988,012	2,573,527
Tampico	16,313	16,528	44,822	89,847	110,550	135,419	302,863	298,337	469,286	560,890
Aguascalientes	35,052	45,198	48,041	62,244	82,234	93,358	126,617	181,277	293,152	547,366
León	63,263	57,722	53,639	69,403	74,155	122,726	300,903	470,370	722,384	951,521
San Luis Potosí	61,019	68,022	57,353	74,003	77,161	131,715	206,261	297,012	471,047	658,712
Querétaro	33,152	33,062	30,073	32,585	33,629	49,440	67,674	112,993	215,976	555,491
Morelia	37,278	40,042	31,148	39,916	44,304	64,979	100,828	161,040	297,544	492,901
Guadalajara	101,208	119,468	143,376	179,556	240,721	401,283	867,035	1,480,472	2,264,602	2,987,194
Pachuca	37,487	39,009	40,802	43,023	53,354	58,658	64,571	83,892	110,351	201,450
Puebla	93,521	96,121	95,535	114,793	138,491	226,646	376,250	629,344	1,136,875	1,686,044
Toluca	25,940	31,023	34,265	41,234	43,429	53,481	89,396	149,750	597,350	827,163
Cd. México	344,721	471,066	661,708	1,048,970	1,559,782	2,872,334	4,993,871	8,623,157	12,994,450	15,226,800
Cuernavaca						43,309	77,484	134,117	347,189	483,951
Acapulco						28,512	49,149	174,378	301,902	593,212
Oaxaca	35,049	38,011	27,792	33,423	29,306	46,632	75,196	99,535	167,607	301,738
Tuxtla G.					15,883	28,243	41,224	66,851	131,096	289,626
Veracruz	29,164	48,633	54,225	67,801	71,720	101,246	159,912	253,182	367,339	473,156
Villahermosa			15,819	15,395	25,114	35,418	52,262	99,565	158,216	386,776
Mérida	46,630	62,447	79,225	95,015	96,852	142,858	207,702	263,316	454,712	658,452

Source: Gustavo Garza, La urbanización de México en el siglo XX (Mexico City: El Colegio de México, 2005), tables A1, A2, and A3.

Note: For a discussion of the distinction between population figures for metropolitan zones and cities proper, see Garza, La urbanización de México, 164.

[a] Because of revolutionary instability, the 1920 census was delayed for a year.

Notes

1. Eric Hobsbawm, *The Age of Extremes: A History of the World, 1914–1991* (New York: Vintage, 1994), 9.

2. Francisco Alba, "Crecimiento demográfico y transformación económica, 1930–1970," in *El poblamiento de México: Una visión histórico-demográfica*, vol. 4, *México en el siglo XX: Hacía el nuevo milenio*, ed. Ana Arenzana, 74–95 (Mexico City: Consejo Nacional de Población, 1993), 83; http://www.dane.gov.co/index.php; http://www.ine.cl/canales/chile_estadistico/home.php.

3. Leticia Ruiz Rivera, "Cuando llegaron los gringos: La construcción del sistema hidroeléctrico de Necaxa y sus efectos socioambientales (1900–1930)," licenciatura thesis, Escuela Nacional de Antropología e Historia, 2000; Manuel Marroquín y Rivera, *Memoria descriptiva de las obras de provisión de aguas potables para la Ciudad de México* (Mexico City: Muller Hnos., 1914).

4. Jean Pierre Goubert, *The Conquest of Water: The Advent of Health in the Industrial Age* (Princeton: Princeton University Press, 1989).

5. On the characteristics of these older systems, see (among others) Diana Birrichaga Gardida, "Las empresas de agua potable en México (1887–1930)," in *Historia de los usos del agua en México: Oligarquías, empresas y ayuntamientos (1840–1940)*, ed. Blanca Estela Suárez Cortez, 181–225 (Mexico City: CIESAS, 1998); Rocío Castañeda González, "Esfuerzos públicos y privados para el abasto de agua a Toluca (1862–1910)," in ibid., 107–79; Blanca Estela Suárez Cortez, "Poder oligárquico y usos del agua: Querétaro en el siglo XIX (1838–1880)," in ibid., 15–103; Christina M. Jiménez, "Popular Organizing for Public Services: Residents Modernize Morelia, México, 1880–1920," *Journal of Urban History* 30, no. 4 (May 2004): 495–518; and Stephen Webre, "Water and Society in a Spanish American City: Santiago de Guatemala, 1555–1773," *Hispanic American Historical Review* 70, no. 1 (1990): 57–84. In contrast to the situation in cities, rural localities are considered in Diana Birrichaga Gardida, "Modernización del sistema hidráulico rural en el Estado de México (1935–1940)," in *La modernización del sistema de agua potable en México, 1810–1950*, ed. Diana Birrichaga Gardida, 193–217 (Zinacantepec: El Colegio Mexiquense, 2007).

6. Moisés González Navarro, *Historia moderna de México: El Porfiriato*, vol. 4, *La vida social* (Mexico City: Hermes, 1970), 129–30.

7. Birrichaga Gardida, "Las empresas," 189–92. This author mentions that a presentation Koch gave about cholera in India was translated and published in Mexico a mere one year later.

8. See Hortensia Camacho Altamirano, *Empresarios e ingenieros en la ciudad de San Luis Potosí: La construcción de la presa de San José, 1869–1903* (San Luis Potosí: Ponciano Arriaga Editores, 2001); and Hortensia Camacho Altamirano, "Los discursos del agua potable en la ciudad moderna: Transformaciones urbanas, sociales y culturales en la ciudad de San Luis Potosí, 1879–1920," master's thesis, Universidad Autónoma Metropolitana, 2006.

9. Castañeda González, "Esfuerzos públicos," 167.

10. Enrique Torres López and Mario A. Santoscoy, *La historia del agua en Monterrey, desde 1577 hasta 1985* (Monterrey: Castillo, 1985); Camacho Altamirano, *Empresarios e ingenieros*.

11. Ariel Rodríguez Kuri, "Gobierno local y empresas de servicios: La experiencia

de la ciudad de México en el Porfiriato," in *Ferrocarriles y obras públicas*, ed. Sandra Kuntz and Priscilla Connally, 165–90 (Mexico City: Instituto Mora, 1999).

12. Castañeda González, "Esfuerzos públicos," 125–26.

13. For a different interpretation, see Ariel Rodríguez Kuri, "Desabasto de agua y violencia política: El motín del 30 de noviembre de 1922 en la ciudad de México; Economía moral y cultura política," in *Formas de descontento y movimientos sociales, siglos XIX y XX*, ed. José Ronzón and Carmen Valdez, 167–210 (Mexico City: Universidad Autónoma Metropolitana, 2005).

14. Ernesto Aréchiga Córdoba, *Tepito: Del antiguo barrio de indios al arrabal, 1868–1929; Historia de una urbanización inacabada* (Mexico City: Unidad Obrera y Socialista, 2003), 195–207.

15. Alberto J. Pani, *La higiene en México* (Mexico City: Imprenta de J. Ballescá, 1916), 61.

16. José Ángel Ceniceros, *El problema de la insalubridad* (Mexico City: Botas, 1935), 30.

17. "Proyecto de ley de federalización del servicio público de Provisión de Aguas Potables," 21 October 1938, Archivo Histórico de la Secretaría de Salud, Fondo Salud Pública, S.SJ, caja 50, exp. 27; Martin V. Melosi, *The Sanitary City: Urban Infrastructure in America from Colonial Times to the Present* (Baltimore, Md.: Johns Hopkins University Press, 2000), 223.

18. An overview of this experience of what can be considered the first generation of private water companies can be found in Birrichaga Gardida, "Las empresas."

19. Banco Nacional de Obras y Servicios Públicos (Banobras), *Cuarenta y cinco aniversario de Banobras, 1933–1978* (Mexico City: Banobras, 1978), 13.

20. See María Rosa Gudiño Cejudo, "Campañas de salud y educación higiénica en México, 1925–1960: Del papel a la pantalla grande," doctoral thesis, El Colegio de México, 2009, 7–63, 182–228.

21. Alba, "Crecimiento demográfico," 90–93.

22. Diana Birrichaga Gardida, "El abasto de agua en León y San Luis Potosí (1935–1947)," in *Dos estudios sobre usos del agua en México (siglos XIX y XX)*, ed. Blanca Estela Suárez Cortez and Diana Birrichaga Gardida, 93–194 (Mexico City: CIESAS and Instituto Mexicano de Tecnología del Agua, 1997), esp. 105.

23. *Diario Oficial*, 11 March 1935.

24. José Lorenzo Cossío y Soto, *Guía retrospectiva de la ciudad de México* (Mexico City: Talleres Gráficos Laguna, 1941), 174.

25. In 1935, the Banco Hipotecario saw to it that pipes for drinking water were included in a package of machinery and other items that Germany agreed to barter in exchange for 13,000 metric tons of rice. See Birrichaga Gardida, "El abasto," 106.

26. José Luis Bribiesca C., *El agua potable en la República Mexicana* (Mexico City: Secretaría de Recursos Hidráulicos, 1958), 77–78.

27. Birrichaga Gardida, "Modernización," 207–11.

28. Luis Aboites Aguilar and Valeria Estrada Tena, eds., *Del agua municipal al agua nacional: Materiales para una historia de los municipios en México, 1901–1945* (Mexico City: Archivo Histórico del Agua, 2004), 183–200.

29. Eduardo Miranda Correa, *Las pugnas por el abastecimiento del agua potable y el drenaje en la ciudad de Querétaro, 1940–1970* (Querétaro: Presidencia Municipal, 1996), 17.

30. On the neighborhood cooperatives formed in the state of Mexico to demand running water in the 1930s, see Birrichaga Gardida, "Modernización," 210.

31. Birrichaga Gardida, "El abasto," 129.

32. Letter from the governor of Nuevo León to the Secretario de Salubridad y Asistencia, 4 June 1945, cited in Nicolás Duarte Ortega, "La estatización del agua en Monterrey," in *Monterrey, siete estudios contemporáneos*, ed. Mario Cerutti, 181–207 (Monterrey: Universidad Autónoma de Nuevo León, 1988), 189.

33. Cited in Miranda Correa, *Las pugnas*, 14, 29.

34. José Hernández Terán, *México y su política hidráulica* (Mexico City: Secretaría de Recursos Hidráulicos, 1967), 31.

35. Graciela de Garay, ed., *Rumores y retratos de un lugar de la modernidad: Historia oral del Multifamiliar Miguel Alemán, 1949–1999* (Mexico City: Instituto Mora, 2002).

36. María García Lascuráin, "Calidad de vida y consumo de agua en la periferia metropolitana: Del tambo a la llave de agua," in *Agua, salud y derechos humanos*, ed. Iván Restrepo, 23–162 (Mexico City: Comisión Nacional de Derechos Humanos, 1995), 132.

37. Vivienne Bennett, *The Politics of Water: Urban Protest, Gender, and Power in Monterrey, Mexico* (Pittsburgh: University of Pittsburgh Press, 1995), 45.

38. Miranda Correa, *Las pugnas*, 28, 31.

39. Armando G. Servín, *Las finanzas públicas locales durante los últimos cincuenta años* (Mexico City: Secretaría de Hacienda y Crédito Público, 1956), 49–50.

40. Luis Aboites Aguilar, *Excepciones y privilegios: Modernización tributaria y centralización política, 1922–1972* (Mexico City: El Colegio de México, 2003), 317–18.

41. Secretaría de Recursos Hidráulicos, *Informe de labores de la Secretaría de Recursos Hidráulicos: Del primero de diciembre de 1946 al 31 de agosto de 1947*, 2 vols. (Mexico City: Talleres Gráficos de la Nación, 1947). Note, however, that this figure does not include credit that the Banco Hipotecario granted for public works projects in this period.

42. José Trinidad Lanz Cárdenas, *Legislación de aguas en México (estudio histórico-legislativo de 1521 a 1981)*, 4 vols. (Villahermosa: Gobierno del Estado de Tabasco, 1982), 2:793–94.

43. Gustavo Garza, *La urbanización de México en el siglo XX* (Mexico City: El Colegio de México, 2005), 43.

44. *Obras de provisión de agua potable para la ciudad de México: Sistema Lerma* (Mexico City: Departamento del Distrito Federal, 1951). Various photographs of the murals discussed here can be found in this publication. See also Bribiesca, *El agua potable*, 93–95.

45. Beatriz Albores, *Tules y sirenas: El impacto ecológico y cultural de la industrialización en el Alto Lerma* (Toluca: El Colegio Mexiquense, 1995).

46. Secretaría de Recursos Hidráulicos, *Informe de labores de la Secretaría de Recursos Hidráulicos, 1965–1970* (Mexico City: Talleres Gráficos de la Nación, n.d.), 239 (hereafter cited as *SRH, 1965–1970*). See also Ignacio Bernal, *Apuntes para la historia de la infraestructura en México* (Mexico City: Banobras, 1998), 86. In general, these local nonmunicipal organizations represent an enigma. On the Xalapa *junta*, see Richard R. Fagen and William S. Tuohy, *Politics and Privilege in a Mexican City* (Stanford: Stanford University Press, 1972), 75–80.

47. *SRH, 1965–1970*, 239.

48. Luis Robledo Cabello, "Los programas de inversiones en agua potable," *Ingeniería Hidráulica en México* 23, no. 2 (1969): 230–35, quotations on 231.

49. Secretaría de Recursos Hidráulicos, *Memoria del simposio sobre agua potable y alcantarillado* (Mexico City: SRH, 1966), 158–59.

50. Robledo Cabello, "Los programas," 235.

51. Alba, "Crecimiento demográfico," table 1, p. 77. In the 1970s, the average annual rate of growth reached 3.28 percent, compared to a mere 1.72 percent in the 1930s.

52. Ibid., table 2, p. 78.

53. Bernal, *Apuntes*, 222.

54. Bribiesca, *El agua potable*, 82; the text of the law can be found in Lanz Cárdenas, *Legislación*, 2:305–7.

55. Secretaría de Recursos Hidráulicos, *Memoria*, 64.

56. Claudia Cirelli, "La transferencia de agua: El impacto en las comunidades origen del recurso; El caso de San Felipe y Santiago, Estado de México," master's thesis, Universidad Iberoamericana, 1997.

57. *Memoria de las obras del sistema de drenaje profundo del Distrito Federal*, 4 vols. (Mexico City: Talleres Gráficos de la Nación, 1975), vol. 1; Secretaría de Recursos Hidráulicos, *Simposio: Sobreexplotación de aguas subterráneas* (Mexico City: SRH, 1975).

58. José P. Arreguín Mañón, *Aportes a la historia de la geohidrología en México, 1890–1985* (Mexico City: CIESAS and Asociación Geohidrológica Mexicana, 1998), 49.

59. Lanz Cárdenas, *Legislación*, 2:1211–22.

60. Priscilla Connally, "Programa Nacional de Desarrollo Urbano y Vivienda, 1984: ¿Desconcentración planificada o descentralización de carencias?" in *Una década de planeación urbano-regional en México, 1978–1988*, ed. Gustavo Garza, 103–20 (Mexico City: El Colegio de México, 1989), 116.

61. Jaime Castillo Palma, "Gestión del agua y poder local en Puebla," *Ciudades* 43 (July–September 1999): 25–31, esp. 26.

62. Connally, "Programa," 118.

63. Bennett, *Politics of Water*, 67–76.

64. A wonderful study that describes living conditions and access to water in Ecatepec and the Chalco Valley at the beginning of the 1990s is García Lascuráin, "Calidad de vida."

65. Bennett, *Politics of Water*, 70–71.

66. Secretaría de Agricultura y Recursos Hidráulicos, *Agua y sociedad: Una historia de las obras hidráulicas en México* (Mexico City: Secretaría de Agricultura y Recursos Hidráulicos, 1988), 156, 161.

67. Torres López and Santoscoy, *La historia del agua*, 126.

68. Bennett, *Politics of Water*, 128–63.

69. On Puebla, see Arsenio Ernesto González Reynoso, *Cambios en la gestión del agua y del saneamiento en la ciudad de Puebla, 1988–1994* (Mexico City: Instituto Mora, 2000).

70. I would like to thank Diana Birrichaga for drawing on her wellspring of knowledge to point this out to me.

71. Bernal, *Apuntes*, 228.

72. Rubén Barocio R. and Jorge C. Saavedra S., "La participación privada en los servicios de agua y saneamiento en México," in *El agua en México vista desde la Academia*, ed. Blanca Jiménez and Luis Marín, 289–316 (Mexico City: Academia Mexicana de Ciencias, 2004), 306–7.

CHAPTER ELEVEN

Episodes of Environmental History in the Gulf of California
Fisheries, Commerce, and Aquaculture of Nacre and Pearls

Mario Monteforte and Micheline Cariño

Pearl oysters (*Pinctada mazatlanica* and *Pteria sterna*) have played a fundamental role in the environmental history of the Gulf of California region for nearly five hundred years.[1] Pearls' value derived from geographic and temporal coincidences in which ancient myths and legends that linked pearls to great wealth became enmeshed with changing human interactions with natural ecosystems. Christopher Columbus's accidental arrival in the Americas was the first step in opening the Gulf of California to European eyes, and with it the natural pearl oyster beds that soon became famous for the quality of their nacre and pearls. This discovery laid the foundations for what we call the "Pearl Myth," which drove the exploration of the Gulf and colonial settlement of the Baja California peninsula. The Pearl Myth was tinged with abstract, fairytale-like symbolism that derived from ancient European chivalry novels but was the guiding economic principle and political basis for the colonization of Baja California in the eighteenth century. The market for the Gulf's nacre (sometimes misidentified as "mother of pearl") and pearls emerged in the 1730s and led to the foundation of La Paz, which has been recognized as a major pearling center ever since.

The Pearl Myth was grounded in commerce. For centuries, nobles and magnates collected jewelry and luxury items generously inlaid with nacre and pearls. The world's greatest personal fortunes and national treasures nearly always included such objects, many of which are now prominently displayed in museums, churches, and personal collections. Artisans and factories acquired great quantities of shells used to inlay fine objects with nacre decorations. Their handiwork can be seen throughout the world, in ceilings, doors, columns, and altars of temples and churches; in thrones and furniture adorning palaces and castles; in coffers and boxes of all shapes and sizes; in the formerly enormous button-making industry; in nacre decorations on firearms and other hand weapons; in pins, brooches, combs, hairbrushes, cufflinks, necktie sets, and other adornments and accessories for gentlemen; and in the huge quantity of jewelry with nacre and pearls that adorned royalty, celebrities, and countless women through the ages.

The economic flows generated by pearl oysters prompted settlement in many isolated regions, but that did not keep traders from devastating wild populations in a race to meet global demand. The productivity of famous pearling regions gradually declined to near extinction in places such as the Red Sea, the Gulf of Manaar, the Seas of Timor and Arafura, northwestern Australia, South and Southeast Asia, the Philippines, South Pacific islands and atolls, the coasts and islands of the Antilles, and the Gulf of Panama.

The Gulf of California was the last region to experience exhaustion, even though it was a major world supplier. The introduction of modernized diving suits in 1874 intensified the activity but did not extinguish the natural beds as in other regions, because farming activities in the Bay of La Paz in the early 1900s helped to support the natural repopulation of wild stocks. The Compañía Criadora de Concha y Perla de Baja California (Baja California Pearl and Shell Breeding Company, or CCCP) was the world's first successful experiment in massive cultivation of pearl oysters (*P. mazatlanica* in this case) and allowed its founder, Gastón J. Vives, to build a uniquely successful business.[2]

The CCCP used a method of extensive aquaculture to cultivate pearl oysters.[3] The main product was the excellent nacre of *P. mazatlanica* destined for the European market; pearls were a natural and welcome by-product of the process, and they appeared in greater numbers and quality on the natural oyster beds. Vives marketed hundreds of beautiful pearls in New York, London, Paris, and Venice before local rivalries and the violence of the Mexican Revolution ultimately led to the looting and destruction of the company's installations in 1914. Even so, the CCCP had left a mark on regional socioeconomic structure, and its closure severely restricted employment and other economic activities associated with pearl oyster farming. Moreover, the end of continuous larvae supply and recruitment from reproduction of the millions of farmed pearl oysters had significant repercussions for the regional ecosystem—a blow made all the more severe when the new revolutionary government lifted previous restrictions on the fisheries. Less than fifteen years later, the wild stocks had become severely exhausted.

The environmental history of pearling in the Gulf of California follows a familiar global pattern. At different historical moments, with different trajectories, and each in a particular social context, the pearl fishery has experienced severe impoverishment. Nearly everywhere, the pearl industry garnered incredible profits for a few wealthy individuals at the expense of the health and welfare of thousands of divers. Later, after a few pioneering efforts at extensive aquaculture,[4] pearl oyster farms succeeded in culturing pearls themselves (rather than the entire oyster, as Vives had) in Japan, Australia, India, and the Tuamotu atolls as these countries enforced bans and other protections for natural populations. Cultivated nacre and pearls have supplied the global market ever since the late 1950s, except for northwestern Australia, where pearl culture relies on the extraction of wild adults of the species *P. maxima* (the giant member of the Pteriidae, which can grow up to thirty centimeters in diameter). However, the Australian government and pearl industry manage this resource under strict rules and impose draconian sanctions on offenders.

In 1952, Kokichi Mikimoto patented the special instruments that are used in the surgical induction to produce free round pearls, and he quickly became rich by cultivating the Japanese *P. martensii* on his farms. His intelligence and gift for self-promotion, along with enviable business acumen, allowed him to reach agreements with Australia and French Polynesia. Before long, Japanese technicians were in charge of pearl production throughout the world. The shift from natural to cultured pearls provoked long and ferocious debates among naturalists, jewelers, consumers, and even the public.[5] Whatever the relative merits of arguments on both sides, overexploitation of natural beds had produced such severe shortages by the early 1900s that entrepreneurs and local governments had already begun to urge scientists to develop aquaculture technology. By the early 1960s, many producers had succeeded in developing their own techniques independent of Japanese control, and most pearl farms replaced divers (who worked in dangerous conditions for little pay) with skilled workers whose specialized knowledge commanded substantially better compensation.

Today, the billion-dollar market for nacre and pearls has acquired a new shape. There are up to 1,500 farms of large pearl oysters (*P. maxima* and *P. margaritifera*) that produce forty tons of luxury pearls (aka "South Seas pearls") annually and 3,100 farms of *P. martensi* concentrated in Japan producing about twenty-four tons of the smaller pearl known as Akoya.[6] The remarkable development of science and modern technology, as well as easy access to specialized expertise, ensures the feasibility of virtually every pearling venture. Management strategy is broadly similar on all farms, although each one has its peculiar characteristics owing to geography, environmental conditions, the bioecology of target species, idiosyncratic operational issues and management styles, and the local availability of equipment and materials.

As an example of how a natural resource was transformed from food and simple ornamentation for native peoples into a global commodity initially harvested in the wild and later farmed artificially, the pearl industry is one of the most important episodes in the environmental history of the Gulf of California. The natural vocation of *Pinctada mazatlanica* and *Pteria sterna* (that is, their capacity to produce nacre and pearls, and their positive response to aquaculture) influenced socioeconomic structures, cultural practices, and political and group dynamics in the region. In the ongoing search for sustainable development alternatives for renewable resources, the environmental history of nacre and pearls serves as an important object lesson.

The Pearl Myth

The Pearl Myth can be defined as the anticipation of overnight wealth through the discovery of rich natural beds of pearl oysters.[7] It appeared in Europe even before Columbus found pearls on the coasts of the Antilles and developed in parallel with the epic tales (*cantares de gesta*) in vogue in France and Spain in the early seventh century, in which heroes met kings, princesses, magicians, and dragons en route to the discovery of fabulous treasures comprising gems and precious metals in imaginary lands. *Juglares* and *trovadores* (medieval minstrels) performed these tales from town to town. The *Cantar de Roldán de Roncesvalles*, one of the most ancient and best known, alludes to a group of warriors named the Califerne from Afrike. During the early 1500s in Spain, Garci Rodríguez de Montalvo became famous for his five-book saga about Amadís de Gaula and his son, Esplandián. The fourth book, *Las Sergas de Esplandián*, depicted an island called "California" that lay to the east of the Indies, where Esplandián met Queen Calafia and her Amazons. Rodríguez wrote vivid descriptions of the landscapes and those warrior women, mounted on wild beasts and adorned with gold, gems, nacre, and pearls.

Rodríguez's California bore a striking resemblance to the island of Cihuatán, a place of mythic importance for the Nahuatl Indians of central Mexico. It lay off the west coast, where the sun god (Tezcatlipoca) arrived at dusk and women who died giving birth enjoyed an eternal life in peace and beauty. Incidentally, the largest American members of the family Pteriidae, *Pinctada mazatlanica* and *Pteria sterna*, live in that direction: the rocky reefs of the Mexican Pacific coast, from Oaxaca to the Gulf of California.[8] During the explorations that culminated with the conquest of Mexico in 1521, Hernán Cortés noticed the beautiful pearls and nacre that the noble Indians used in their adornments. At his no doubt anxious questioning about the origins of such marvels, native leaders responded by vaguely pointing "over there," toward Cihuatán, which Cortés probably associated with Calafia and her Amazons.

In 1533, Fortún Ximénez, the pilot of *La Concepción*, mutinied against his captain and ran the ship aground on the southern coast of the Baja California peninsula.[9] When the surviving mutineers returned to Mexico City, they reported to Cortés about the sight of Californio Indians (Pericúes) wearing long hair braided with beautiful pearls and large, startlingly bright nacre shells adorning their bodies. Cortés immediately proceeded to explore. What the Spaniards found was surprisingly similar to the imagined land of Esplandián.[10]

Along with the Pearl Myth, other legends such as those of El Dorado and the Seven Cities of Cíbola played an important role in promoting the effort to conquer, explore, and colonize the New World. In many cases, the exploitation of natural resources would grow to industrial levels. Generally, native labor was the main source of production, and often the labor conditions bordered on slavery. Scientists and technicians created apparatuses and machinery to improve extraction. Many living organisms suffered the consequences of overexploitation, as did those human communities that interacted with or depended upon them.

By the twentieth century, oysters had been fished for more than three hundred years by numerous fleets, or armadas,[11] in the Gulf of California. Pressure on the fishery increased in the nineteenth century when a British company known as the Mangara introduced the diving suit in 1874. By 1934–35, any investment in pearl oyster fisheries was no longer profitable. Finally, in 1940 the Mexican government dictated a permanent ban, declaring *Pinctada mazatlanica* and *Pteria sterna* as "endangered." The status changed in 1994 to "under special protection," and the ban has remained in place ever since.[12] At present, both species are in the process of being declared "strategic species strictly limited for farming and pearl culture by

Mexicans, with obligation of repopulation activities and research, and to maintain quality standards," a statute quite similar to the one that the Group d'Intérêt Économique (GIE) established in French Polynesia for pearl oysters (*Pinctada margaritifera cummingi*) in the 1960s.

The Pearl Myth has had other expressions in the history of science and technology. Pearl formation had intrigued the human imagination from Pliny onward, and in Europe, all kinds of explanations prevailed until the seventeenth century: that they were tears of the moon or droplets of mist falling into the oyster's valves or that they resulted from the intrusion of foreign matter or worms or from some mysterious disease. However, the Chinese understood the mechanism of pearl formation as early as the fifth century. They produced the famous Buddha pearls by implanting figurines of Buddha made of stone, marble, porcelain, ivory, or lead on the interior face of the valves of the nacre-bearing freshwater mussels *Cristaria plicata*, *Hyriopsis schlegeli*, and *H. cummingii*. News about these pearls appears to have motivated the first scientific studies in Europe of the phenomenon and, eventually, research to discover how to induce pearls artificially. In 1761, the Swedish scientist Carolus Linnaeus succeeded in producing pearls from the European freshwater mussel, *Margaritifera margaritifera*, and by 1904, William Saville-Kent in Australia had defined for the first time the procedures for round pearl induction in *Pinctada maxima*. Mikimoto had already produced Mabé pearls in Japan with *P. martensii* during the late 1890s, but his technique was not really aquaculture, for he picked animals from wild stocks and held them in cages after implanting the half nucleus. His round pearls started to appear on the market in the late 1920s.

Mikimoto's contribution to pearl culture technology has stirred controversy. In 2008, the Australian scientist Denis George demonstrated in a posthumous publication that in the late 1890s Saville-Kent was already managing in Australia a small pearl farm using *P. maxima* when he received Mise and Nishikawa, two functionaries involved in bilateral fishery agreements between Australia and Japan in that time. Back in Japan, they adapted the techniques to the Japanese pearl oyster, *P. martensii*, also mastering the surgery method of round pearl sometimes known as "Mise-Nishikawa," which Mikimoto later patented as his own. Professor George's statements have yet to penetrate the scholarship on pearl culture. For instance, Joseph Taylor and Elizabeth Strack (in an article published around the same time as George's work) do not mention the role of Saville-Kent in the development of Japanese pearl culture, nor do they question the traditional idea of Japanese technological innovation.[13]

Regardless of who developed modern aquaculture, the invention and transmission of more efficient technologies allowed pearl farms to spread. Numerous non-Japanese technicians acquired expertise in the surgery procedure, and non-Japanese manufacturers learned to fabricate the instruments. Jewelers and merchants recognized that natural beds were gradually collapsing and concluded that the future of their industry depended on artificial reproduction. This turn of events introduced new aspects into the Pearl Myth, suggesting that pearl merchants could overcome scientific challenges and take a prized position in the multibillion-dollar international market. At the beginning of the twenty-first century, however, the Pearl Myth has a different meaning in light of increasing global production and a concomitant decline in the quality and price of pearls. Along with the thousands of pearl farms existing in Japan and the Indo-Pacific, China has developed the massive cultivation of *P. chemnitzi* in laboratories. These low-cost Chinese pearls have undercut Japanese and Indian production in the "small pearl" market and have begun to compromise the luxury market of larger pearl oysters as well.[14] New scientific and technological advances occur at regular intervals, and a huge amount of information is now accessible in books, manuals, publications, congresses, and courses, as well as through media such as the Internet. Hatchery laboratories can now produce pearl oysters by the millions, and technologies have improved and become widely available for farming practically all the commercial species of *Pinctada* and *Pteria*.

Pearl Oysters and Colonialism

Baja California is the world's second longest peninsula, measuring approximately 1,200 kilometers in length. It has an average width of 140 kilometers, bounded to the east by the Gulf of California and to the west by the Pacific Ocean. Apart from its aridity, a defining feature of the peninsula is its relative isolation from the continent. Its environmental characteristics have had considerable influence on the peninsula's diverse native societies, most notably the so-called Californio Indian groups (Guaycuras, Pericúes, Cochimíes) that were the first residents of the region and numbered between forty thousand and fifty thousand before the arrival of the Spaniards.[15] In their arduous process of adaptation, indigenous people developed multiple strategies to exploit the limited natural resources and deal with the constraints of this environment. Through their deep knowledge of the environmental characteristics, they attained a subsistence level of existence and achieved social reproduction. They moved periodically from inland caves and oases to the coastal areas and perfected strategies to use their resources sustainably, such as gathering particular foodstuffs only in specific seasons and maintaining the balance of local ecosystems. Marine fauna constituted the largest component of native peoples' diet, because capturing certain species along the coasts, such as mollusks, fish, turtles, and marine mammals, was easy. Mollusks were profusely collected from shallow bottoms of bays and coastal lagoons. The natives also went underwater to collect prized species that attach to the rocks, such as pearl oysters, which had both nutritional and ornamental value. Malacological analysis of the *concheros* (ancient shell deposits coinciding with Californio settlements or feeding spots) has shown that people feed upon the larger (older) adults, thus ensuring natural reproduction and conservation of species.[16] The Californios cherished the beauty of nacre and pearls and used them as personal adornment or as offerings in tombs.

The appetite for pearls motivated explorations and efforts to conquer the peninsula. Hernán Cortés reached the coast in May 1535 and dubbed the bay known today as La Paz the "Bahía de la Santa Cruz."[17] The colony founded by Cortés lasted just a few months. He confirmed that the wealth of pearl oyster beds contrasted with the poverty of the land and advised that exploiting them would be extremely difficult, because it would require depending on the local Indians' diving skills and coping with their constant rebellions.

During the following seventeen decades, the Spanish explorers vainly tried to establish a colony and to exploit the rich pearl "placers" of the Gulf of California.[18] The Viceroyalty of New Spain favored the demarcation and exploration of the Californian coast without burdening the finances of the Royal Treasury by granting licenses for pearl fishing in the Gulf on the condition that license holders explore the peninsular coasts and establish a port-of-refuge for the Manila Galleon. They also needed to pay the *quinto real*, as the 20 percent tax on the extraction of precious commodities such as silver was known. Some licensed pearl merchants—such as Sebastian Vizcaíno (who made the first trip in 1596 and the second one in 1602), Tomás de Cardona (1611), and Pedro Porter y Casanate (1640)—became important figures.[19]

Combining the efforts of exploration and colonization with the workings of pearl fisheries did little to increase European knowledge about the California peninsula, but it consolidated the fame of the Gulf as one of the most important pearling regions in the world. At the end of the seventeenth century, King Charles II ordered another expedition commanded by Admiral

Isidoro Atondo y Antillón, who reported that the exploitation of wild stocks had exhausted the resource in a matter of decades. The expedition's account of arid lands inhabited by "untouched" Indians sparked the interest of Jesuit friars, who hoped to build a utopian City of God on earth. During the Jesuit missionary era (1697–1740), the friars prohibited collection of pearl oysters, allowing wild stocks to rebound. The Jesuits also protected the Indians against Spanish soldiers who wanted to impress them into diving for pearls. In this way, the friars hoped to eliminate overexploitation in the isolated peninsular territories and avoid the abuse of the Indians.[20]

Despite the prohibition, Manuel de Ocio, a soldier in the service of the San Ignacio Mission, challenged the missionaries' authority as soon as he received news that the sea had deposited hundreds of pearl oysters on the beach. After picking them up, he resigned his commission and devoted himself to the fishery on the central Gulf coast, a task he undertook so intensively that the wild stocks were depleted within eight years. Ocio then shifted his attention to the south, where he founded the first civilian colonial establishment of the Californias, the Real de Santa Ana, in 1748. He invested his profits from pearls into silver and gold mines in the Sierra de San Antonio. This pattern established the first regional economic structure. Ocio worked the terrestrial silver deposits in the summer and the marine ones in the winter, all while becoming a rancher and merchant. By that time, the native population had collapsed, forcing Ocio to bring workers from other regions of the country, thus finally accomplishing the goal of colonization.[21]

In the late 1760s, King Carlos III sent Marquis José de Gálvez as a royal supervisor to the north of New Spain to drive the Jesuits from the peninsula and enforce Bourbon rule. Gálvez concluded that the only way to fund these reforms was through exploitation and commercialization of the most valuable natural resource in the California peninsula: pearl oysters. He planned the creation of a company that would export large quantities of nacre and pearls to the Orient. Unfortunately, Ocio and other merchants had pillaged the natural beds.

In April 1811, a few months after the outbreak of Mexico's independence movement, the Courts of Cádiz ordered the development of the Californias under the assumption that the prosperity of the region would derive from marine resources. The charter declared that "the fishing of pearls in all the domains of the Indies, for all the subjects of the Monarchy," was entirely free of taxes, as were "the contracts made between pearl merchants [*armadores*] and divers."[22] These statements became law when Emperor Agustín de Iturbide later eliminated the *quinto real* after independence.

The development and colonization of Baja California soon became a burden to the newly independent Mexican nation, thanks to its penury and constant demand for resources. The central government appointed a commission—the Junta de Fomento de Las Californias—that produced seven proposals on governance and economic reconstruction. The sixth recommended opening trade with Asia in pearls, fine fish, and leather products.[23] Although this proposal came half a century after Marquis de Gálvez announced his plans, both initiatives envisioned primary roles for pearl oysters. Both were influenced by the Pearl Myth, which construed pearls as the region's most valuable resource.

The productivity of wild stocks had cycled between abundance and exhaustion ever since the sixteenth century. After a few years of intensive fisheries, the profitability decreased, resulting in periods of rest that helped the rebound of fishing grounds. This is why the Pearl Myth consistently appears in documents from the colonial period until the mid-twentieth century—when the resource finally reached exhaustion—regardless of the actual productivity of pearl placer beds. It is therefore hardly surprising that when the general commander of Baja California, Manuel Victoria, proposed in 1830 to remedy the deplorable situation of the territory, he suggested (among other measures) dedicating ships to increase coastal trade, gather pearls, and hunt whales, otters, and seals.[24] Pearl oyster populations returned to a phase of abundance, and the myth became reality. It helped prompt the resurgence of the regional economy and society, as well as the definitive foundation of La Paz.

Nacre and the Development of La Paz

In the first decades of the nineteenth century, new commercial circuits opened up as nacre acquired commercial value. Until then, the shells of pearl oysters had been considered a waste and were abandoned on the beach. However, in 1830, the impact of nacre as a new and valuable natural resource revitalized the exploitation of pearl oysters, and with it the socioeconomic structure of many pearling regions. The Gulf of California was no exception.[25]

In 1830, the French merchant Cipryen Combier visited La Paz and nearby coasts, where he spotted a great quantity of shells abandoned on the beaches. He shrewdly used them as ballast for his ship and sold them later. Shells soon became the principal object of pearl oyster fisheries. The recovery of commerce as about a hundred *armadas* requested fishery licenses between 1838 and 1868 sparked a new wave of migration to Baja California.[26] Conflicts often broke out between *armadores* and divers, who lived in perpetual debt and often skipped out without repaying the advances that armadores had extended. This was a punishable crime—as was the concealment of any pearls they harvested—yet divers had few other alternatives to escape from low salaries and brutal working conditions.

The armadas traveled from Cape Pulmo to the Bay of Mulegé and Island of San Marcos. The dexterity of the divers—mestizos descended from Californio and Yaqui Indians—was widely admired. Most of them could venture as deep as five or six fathoms and stay underwater up to two minutes, up to forty times a day. The armador provided board for the divers and their families in the form of corn and dry meat equivalent to one real per day per person, all of which was considered a form of advance payment.[27] At the end of the season, divers repaid their debts by giving armadores a portion of their capture. Divers theoretically had the right to sell their catch freely, but their patrons had the right of first refusal. In practice, it was nearly impossible for divers to fulfill their debts, meaning that the armadores kept the entire harvest.[28]

During the nineteenth century, employment in pearl fisheries played a major role in regional economic development, albeit at a devastating environmental cost. Lack of knowledge about the local species and an increasing level of exploitation led to the gradual decline of natural stocks. In 1857, a local official named José María Esteva recognized that this trend would be disastrous for La Paz and proposed to regulate fisheries in the Gulf of California—one of the first such management efforts anywhere in the world. Esteva was most concerned about the welfare of divers, so his regulations dealt primarily with working conditions rather than the conservation of natural beds, which therefore continued to collapse. Only once exhaustion appeared imminent were several more laws enacted. An 1874 act divided the coasts into four sections in which fishing was allowed every other year. Another law in 1878 increased the rotation to once every four years. However, pearl merchants rarely obeyed these restrictions and continued to overfish. The government had virtually no capacity to apply the law in the first place, since doing so entailed the surveillance of vast and uninhabited areas. Only the cultivation work that Gaston Vives in the CCCP undertook in 1903 held any hope of replenishing the pearl oyster populations. This may explain why the stocks were not completely exhausted until the middle of the 1930s.[29]

These measures were only incompletely applied but allowed the pearling industry to last a few more decades. Around 1870, the beds showed signs of exhaustion again, and the armadas became less profitable. The introduction of diving suits in 1874 was a boon for the pearl companies. The gear increased immersion time, range, and depth. Officials responded by granting pearling companies concessions of vast marine areas between 1884 and 1912.[30] Only wealthy merchants could afford the costs of this new technology; the wages they paid were a pittance compared to the profits they reaped and the damage they wrought. Moreover, this innovation took place during the regime of Porfirio Díaz (1876–1911), which encouraged investment—particularly of foreign capital—and colonization. Both foreign and Mexican entrepreneurs, sometimes in association, received territorial concessions to exploit natural resources in Baja California. The whole peninsula was soon divided into zones that included both terrestrial and marine spaces. Yet the marine concessions for the exploitation of pearl oysters violated the first two articles of the 1874 regulation, which had declared these areas open to public fishing. Officials received constant complaints from local armadores who lost the right to work in areas that the government had granted to big pearling companies.

The inclusion of pearl oysters in the Porfirian policy of concessions took place a decade after the introduction of diving suits, which forced pearl merchants to reorganize labor in the fisheries. Divers in suits had access to deep sea bottoms and eventually displaced the naked divers (who continued to work in shallow areas for a time). This transition occurred because, while pearling companies still paid low wages and offered terrible labor conditions, suited divers had more economic security than those who worked as naked, independent divers. Moreover, natural beds in shallow areas had declined alarmingly, while those in deeper areas beyond the reach of naked divers remained untouched. The yield in deep water was therefore far higher, though much less sustainable.

Porfirian concession contracts all followed the same basic format. They had an initial term of six years, with the possibility of extension for another ten years or more. Taxes on the catch were to be paid to Marine Customs at a rate of eight pesos per metric ton of shells during the first two years and ten pesos per ton thereafter. Concessionaires were also required to cultivate pearl oysters and guarantee their reproduction; they could not harvest juvenile oysters or damage marine habitats. The government had the right to cancel any concession that did not fulfill these stipulations, although the pearling companies (aside from the CCCP) simply ignored them without consequence. The contracts also stipulated that foreign concessionaires were legally considered Mexican entities required, among other things, to give employment and training to Mexican workers. Another clause exempted the pearl merchants from import taxes on any articles that the armadas used for the fishery, in exchange for which pearl fleets were expected to help patrol against smuggling. Pearling companies were forbidden to sell, give, or mortgage their concession to another party without federal approval.

The federal government hoped to achieve several goals with these contracts. Above all, it sought to conserve the resource while generating revenue through taxes. Although the pearling companies were excused from the marine customs tax, they did pay two to three pesos per ton of fished oysters to the Junta de Fomento as a "contribution to the general development of the country." The policy also sought to modernize the regional economy and the society of Baja California overall.

Of the twenty-six pearling concession contracts signed over twenty-two years, only ten actually went into operation. Aside from the CCCP, the other four big pearling companies included the González and Ruffo Association, the Compañía Perlífera del Golfo de California, the Compañía Perlífera de San José, and the Compañía Perlífera de Baja California, the last of which was the largest and had several corporate incarnations. It was founded in 1885 in San Francisco, California, with an initial capital of 100,000 pesos. The associates were the North American Herman Levison (55 percent) and the Mexicans Juan Hidalgo (30 percent) and Maximiliano Valdovinos (15 percent). Its concession covered all the western coasts of the Gulf of California, from Cabo San Lucas to the outlet of the Colorado River, and from Acapulco to the frontier with Guatemala. The Cerralvo, Espíritu Santo, and San José Islands were not included in the agreement, because they had been granted a year before to the González and Ruffo company. The Compañía Perlífera de Baja California therefore had tremendous regional importance due to the scale of its capital investment in ships and equipment and the number of workers it employed (four hundred to five hundred). In 1893, several local armadores affiliated with this company and created the Compañía Perlífera del Golfo de California, but British capitalists soon acquired its concession rights and founded the Mangara Exploration Limited Company.[31]

Several factors contributed to the birth of the Mangara. The Porfirian policy of exclusive territorial concessions to pearl fishing companies favored the participation of big firms capable of imposing their conditions on the other armadas. This concentration of capital deprived the local armadores and divers from direct access to the marine resources on which they subsisted. In 1893 several small-scale armadores responded by forming the Compañía Perlífera de Baja California Sucesores, which ended its short life by transferring its rights to the Mangara later that year. This turn of events gave the English nearly absolute control of Mexican pearl resources. Mangara never fulfilled its obligation of cultivating at least ten thousand pearl oysters every year or of performing repopulation, as its contract required. On the contrary, it used ruinous techniques such as dynamite to harvest pearl oysters. Surprisingly, the newborn revolutionary government extended its concession for a period of sixteen years beginning in 1916. The Mangara jealously protected its rights while imposing deplorable working conditions on its laborers. Divers struggled unremittingly to fulfill mandatory quotas of capture under deficient food and unsanitary housing, all in a high-risk work environment that forced them to confront sharks, stingrays, moray eels, barracudas, jellyfishes, and diving accidents such as the bends and perforated eardrums. They lacked employment security, incurred debts they could never hope to repay, and worked at the whim of their supervisors. The Mangara was accused of many irregularities, but instead of receiving any kind of punishment, the British owners turned the tables and accused the marine customs officers in La Paz of negligence for failing to investigate "crimes" against the exclusivity of the company's concession. They had considerable influence with the federal and local governments and at one point even brought their complaints to Gastón Vives, who served as mayor of La Paz from 1893 to 1911. Ironically, however, the same Porfirian economic policies that made the Mangara's abuses possible also allowed Vives to form the first company that tried to conserve pearl oysters.

The World's First Successful Pearl Oyster Farm

The constantly growing demand for pearls and nacre, as well as the advent of diving suits, increased the exploitation of pearl oysters on a global scale. Merchants and entrepreneurs urged the scientists to develop cultivation of pearl oysters and avoid the extinction of the most valuable marine resource known to humanity. The expectation was expressed by Sir Alexander Lyster Jameson: "The man that solves the problem of pearl oyster cultivation, will have not only the privilege of making science and industry to progress, but rather his name will deserve the honor of being included among the founders of empires."[32] In other words, the search for sustainability was not a conservationist movement but based instead on commercial incentives.

Gastón Vives was the first person to achieve the large-scale cultivation of a pearl oyster (*P. mazatlanica*) and fulfill Jameson's prophecy. Born in La Paz to French parents, Vives studied the culture of edible oysters and mussels in France and adapted his experiences to the Gulf of California.

In 1903, after several years of research, he founded the CCCP, which became the world's most successful purveyor of nacre and pearls thanks to his expertise, management skills, political relationships, and deep concern for the conservation of pearl oyster populations.[33] This company eventually managed more than ten million animals in huge installations located at San Gabriel Bay at Espiritu Santo Island, in the Bay of La Paz. The central office occupied a large space in what today is the La Paz pier. The CCCP's cultivation process demanded complex operations at three specific phases of the pearl oyster's life cycle (spat collection, nursery culture, and late culture). Vives developed impressive artifacts for the farming procedures and built large facilities in the operation center at San Gabriel, where hundreds of workers lived almost permanently. The CCCP's technology has been thoroughly studied elsewhere,[34] but it bears repeating that the three basic stages of extensive culture developed by Vives are broadly the same used today in modern pearl farms, as seen in figure 11.1. Thanks to the reproduction activity of the millions of adult pearl oysters in the CCCP's farm, the natural replenishment of wild stocks was ensured.

Within nine years of its foundation, the CCCP had multiplied its capital and became the largest exporter of high-quality nacre and natural pearls in the world. Vives employed about 16–18 percent of the economically active population at La Paz in labor directly related to farming. Its economic activity indirectly generated still more employment in service and commercial sectors. It achieved the highest production levels of cultivated pearl oysters in history before being plundered and destroyed in 1914, during the Mexican Revolution. Vives never recovered his company despite protests to the postrevolutionary government and reminding officials how much the region would benefit from reestablishing production. The pearl oyster fishery resumed in 1916, but without the reproductive contribution of the CCCP it collapsed within two decades and marked the end of Mexico's storied pearl wealth.

From the Liberation to the Exhaustion of Fisheries

Although the destruction of the CCCP coincided with the Mexican Revolution, it did not result from the revolutionary movement itself. Instead, a staunch enemy of Vives took advantage of the violence and confusion to annihilate the man and his work. The CCCP was not the first casualty of the revolution, however. Two years earlier, revolutionary officials had rescinded the Mangara's Porfirian-era concession. Fishermen and armadores of La

Figure 11.1. Pearl oyster culture, nineteenth century and today. *Source*: Photographs by Mario Monteforte or in his possession, with the exception of photos 1, 4, 5, and 7, which are courtesy of the Gaston Vives Papers in the possession of Mario Monteforte and Micheline Cariño. The first three columns of photographs correspond to the three main stages that are used in pearl oyster farming (spat collection, nursery culture, and late culture); the last depicts the pearl culture stage. Photo 1 shows workers of the Compañía Criadora taking ashore a spat collecting device. Photo 2 shows modern spat collectors made with plastic materials. Photo 3 depicts how the spat of pearl oysters is detached from collectors and gathered in buckets or some container filled with clean seawater. Photo 4 depicts "incubators" used by the Compañía Criadora; these were placed at the bottom of the enclosure channels at the entrance of the San Gabriel coastal lagoon (photo 5). Photo 6 depicts the modern nursery culture stage in a galvanized iron pipe structure. Photo 7 depicts the tin armor used by the Compañía Criadora during the late culture stage to protect each individual against predators. Photos 8 and 9 show two different techniques of late culture used today. Photo 10 depicts Mario Monteforte performing surgery on an adult *P. mazatlanica*. Photo 11 shows an experimental harvest of round pearls, thirty months after surgery. Photo 12 shows modern Mabé pearls attached to a valve of *P. mazatlanica* that will be cut off with lapidary tools for use in jewelry.

Paz (including Gaston Vives) had complained to government officials and staged public protests against the company beginning in 1910. Further fanning these flames, some of the Mangara's workers accused the company of treating them as virtual slaves. In response, the Mangara actually increased its abuses. In June 1911 an enormous protest led local officials to petition president-elect Francisco I. Madero to annul the concession; one of his first acts in office was to order its cancellation, on May 28, 1912. Because the original concession extended through 1932, the federal government made a reparation payment of 300,000 pesos, in exchange for which it received the Mangara's infrastructure and equipment. These were later auctioned for a mere 70,000 pesos.[35]

It is important to emphasize that the complaint was directed against the Mangara rather than the CCCP, which most people understood as a benefit to the natural beds. However, in the "revolutionary" mind, Vives and the CCCP were another legacy of the Porfirian regime and deserved a similar fate. The political offices that Gaston Vives had held in La Paz during the Porfiriato did not help his case either. In the end, the government ordered the closure of his company, too.

With the end of pearling concessions, management policies reverted to what they had been before the establishment of the concessionary regime in 1875: fishermen only needed to request a license from port authorities and agree to abide by basic regulations. Local fishermen, divers, and armadores raced out to the newly accessible waters until pearl oyster populations began to plummet about two decades later. The wages paid to divers of the armadas were the same as or even less than the big pearling companies had paid, although the divers did gain the right to work for any armador they chose. Working conditions were also better than before. Moreover, income from the exploitation of pearl oysters after 1913 was invested in the regional economy instead of flowing to foreign-owned companies. Positive effects on the regional economy began to appear as domestic capital accumulation resumed. In terms of employment and wealth generation, then, the revitalization of pearl oyster fisheries once again achieved an important multiplier effect on the regional economy.[36]

The 1884–1914 boom initially masked the exhaustion of wild stocks, in part because of changing productive strategies. Most operations sought to collect nacre rather than pearls, which allowed the armadas to reduce the intensity of their pearl operations and still make a profit. Nevertheless, in the southern half of the Gulf of California, the end of the CCCP's activities had devastating consequences for the resource, the full extent of which was not evident until 1930. Once it became known, Andrés Almazán, a high functionary in the central government, proposed a ban on pearling for a period of three or four years; however, the fishermen refused, and the overexploitation of natural beds increased rapidly from that moment on. By the end of the decade, divers captured a mere two hundred to three hundred pearl oysters a day, whereas twenty years earlier, catches had numbered in the thousands. By the middle of the 1930s, divers and *armadores* could no longer disregard that depleted stocks had placed the entire pearl industry on the verge of collapse. To make a bad situation worse, manufacturers began to replace natural nacre with plastic.

The degradation of natural beds also compromised pearl oysters' resistance to an epizootic plague that broke out between 1938 and 1939. Fishermen found thousands of dead oysters in the bottom of fishing grounds with their valves open. Popular tradition attributed this phenomenon to Japanese sabotage, because they were thought to have poisoned the water to avoid competition from the Mexican product. A more likely explanation is that construction of Hoover Dam (in the United States), which opened in 1935, diminished the outflow of fresh water from the Colorado River and changed the ecological balance of the Gulf of California. Other possible causes include climatic events such as El Niño or La Niña, red tides, or some combination of all of these factors. Whatever the case, a boom of opportunistic microorganisms arrived as deadly guests of the already weakened populations of pearl oysters. Environmental change undoubtedly harmed several less commercially valuable marine species as well, although we cannot know for sure because their decline or possible exhaustion has gone unrecorded. The volume of fresh water provided by the Colorado River continued to diminish over time, even as it has become an increasingly dangerous source of pesticides, fertilizers, and sewage in the Gulf.

In 1940, the federal government finally declared pearl oysters an endangered species and established a complete and open-ended ban on their exploitation in Mexican territorial waters. However, these measures did not lead to the recovery of natural populations. Clandestine fisheries and occasional rushes have continued to the present. Irresponsible elements in the tourist industry have clandestinely promoted "find your pearl" trips, and piles of pearl oyster shells are a common sight on the beaches during holidays. Calculating the volume of these extractions is difficult.

Based on this evidence, however, we can affirm that in the face of intensive and sustained harvesting—legal or otherwise—massive cultivation represents an efficient way to avoid the exhaustion of pearl oyster beds. Several decades of successive failures have demonstrated that there is no substitute for the techniques and strategies that Vives pioneered, nor for ongoing basic research into the particular environment and ecology of local species, combined with site-specific scientific and technologic innovation.

The Resurgence of the Myth

World War II had a major influence on the expansion of pearl farms. At the request of the American military command in Japan, A. C. Cahn published a long report in 1949 about the techniques of pearl oyster culture and pearl production of P. martensii.[37] Soon thereafter, the National Pearl Research Laboratory was founded in Japan. Besides perfecting the surgical technique for inducing round pearls, the Japanese scientists studied and published widely in international journals such as the *Bulletin of the National Pearl Research Laboratory*. Their work touched on all scientific and technical aspects of the integral aquaculture and pearl production of their native species.[38] Ever since the middle of the twentieth century, world production and marketing of round pearls have been subject to Japanese patents, which until recently constituted a de facto barrier to technology transfer. Japanese businesses that received the right to use Mikimoto's patent soon moved into Australia, French Polynesia, Indonesia, Thailand, and elsewhere, giving Japanese technicians access to producing round pearls in a variety of species. Japanese cartels grew rapidly and forged agreements with pearl farms throughout the world that gave them privileged access to pearl harvests.[39]

In Mexico, the progression from irrational fisheries to applied scientific research has been slow, discontinuous, and complex. The example of Gastón Vives and the CCCP, as well as the increasing social and economic importance of pearl farms elsewhere in the world, stimulated several attempts in the Gulf of California between 1939 and 1989 (table 11.1). Most of these projects sought purely commercial benefits, and all of them failed for a variety of reasons, among which we can underline the following common factors:

1. Lack of knowledge about the bioecology of the species and pearl oysters' response to handling and treatment with pearl culture stimulations.
2. Lack of interest in applying scientific research to technological and commercial development, especially in regard to the repopulation of natural beds.
3. Application of incorrect methods and techniques, either of Japanese origin or adapted from other commercial bivalve species (edible oysters

Table 11.1. Pearl oyster culture projects in and around La Paz, 1939–88

Actors	Date and location	Actions	Outcome
Y. Matsuii (Mexico/Japan agreement)	Bays of La Paz and Loreto, 1939	Prospecting natural beds; pearl culture trials on wild individuals	Few natural populations found; high postsurgery mortality
Secretaría de Pesca	Bay of La Paz, 1961–62	Spat collection trials and extensive culture	Unsatisfactory results
Agapito Martínez (CRIP); Secretaría de Pesca	Bay of La Paz, Loreto, and nearby islands, 1962	Prospecting and transplants	Exhausted populations and poor transplant
Denis George; Secretaría de Pesca	Bay of La Paz, 1969	Spat collection and culture; Mabé and round pearl in wild individuals	High mortality and rates of rejection postsurgery; promising extensive culture
José Juan Díaz Garcés and Manuel Antonio Gallo	Bay of La Paz, 1970–71	Extensive culture and Mabé implants	Good results
Delegación de Acuacultura, BCS	Bay of La Paz, 1976–78	Extensive culture testing	Acceptable results
Shohei Shirai and Yoshiyasu Sano (Japan); Secretaría de Pesca	Bay of La Paz, 1979	Prospecting natural beds and sites; attempt to install pearl farm	Mischaracterization of natural beds as abundant; high mortality and rejection postsurgery
Takeshi Yamamoto and Yoshiyasu Sano (Japan); private interests in La Paz	Bay of La Paz, 1979–80	Culture in wild individuals	Major failure; natural beds plundered
Jaime Singh (CRIP-BCS)	Bay of La Paz, 1981–82	Extensive culture trials; culture in wild individuals	Good results, though high postsurgery mortality and no pearls
Manuel Mazón (CRIP-BCS)	Laboratory in CRIP, La Paz, 1987	Hatchery studies in *Pinctada mazatlanica*	Acceptable gonad conditioning, larval growth, and survival, but no fixation
Fernando Bückle, CICESE, Ensenada	Los Angeles Bay (Gulf of California), 1988	Extensive culture studies of *Pteria sterna*; round pearl induction trials	Excellent results in extensive culture only

and clams), that are inappropriate for native pearl oyster species or ignore local environmental conditions.
4. Pursuit of cultured pearls and economic rents at the expense of wild stocks. Amateur technicians often used wild pearl oysters in trials of pearl culture techniques even though an individual learning curve needs more than seven thousand fatalities to achieve, eventually, a success rate of at least 35 percent. This resulted in very high mortality without obtaining any pearls and put further stress on the already exhausted pearl oyster beds.

Other causes for the failures in pearl production included ongoing changes in local and national policies, financial limitations, excessive logistical complexity, competition and rivalries between groups and individual actors, vandalism, and technological failures. The production of seed in the laboratory was tried on one occasion, but funding for the project dried up despite promising results.[40] More common were disastrous missteps by entrepreneurs such as Yoshidi Matsuii in 1939, Shohei Shirai and Yoshiyasu Sano in 1979, and Takeshi Yamamoto and Sano in 1980. The participation of Japanese specialists in these three major fiascos is noteworthy. In contrast, José Juan Díaz Garcés and Manuel Antonio Gallo, assisted by Denis George, started a commercial farm in 1969 that successfully adapted the Australian techniques on the local species. In 1971 the farm was preparing to harvest its first Mabé pearls when the federal government arbitrarily closed the enterprise and seized its installations and pearl oysters. The fate of the pearls and shells remains unknown.

The long record of failures that beleaguered pearl aquaculture in the Gulf of California left policy makers skeptical about financing any further initiatives or planning for regional development based on pearl oyster farms. Ever since the 1970s, most such projects have focused on edible species that can be raised with technologies more easily adapted to local conditions. Farms of white shrimp (*Litopenaeus vannamei*) and Japanese edible oyster (*Crassostrea gigas*) were created. Both cases involved proven technologies that worked on a commercial scale and could be adapted to Mexican ecological conditions. Mexico soon became one of the world's leading exporters of white shrimp and Japanese oyster (shipping annually approximately forty thousand metric tons and twenty thousand metric tons, respectively), most of which was and continues to be produced on the eastern coast of the Gulf of California.[41]

In October 1985, the authors of this chapter approached what was then known as the Centro de Investigaciones Biológicas (Center for Biological

Research, or CIB) to propose another project for the integral pearl oyster culture and production of pearls at the Bay of La Paz.[42] The need for alternative models of regional development, the CCCP experience, and the wealth of information available from Japan, India, Sudan, Australia, and French Polynesia all undergirded our new version of the Pearl Myth. However, the CIB board of directors rejected the project—an understandable attitude in light of so many negative antecedents. Therefore, we used the help of family and friends to make independent studies on pearl oyster beds in the bay of La Paz, as well as to carry out some spat collection and cultivation tests on the local species.

Wild populations showed an alarming scarcity.[43] Moreover, existing knowledge about the biology and the ecology of both species was extremely limited. Apart from the information generated by Gastón Vives, our only guide was the partial and incomplete results of the projects that had preceded ours, which had left few theoretical or practical traces. At the end of 1988, however, we obtained the first positive results, and with a handful of juveniles of *P. mazatlanica*, we once again tried to convince the CIB of the project's feasibility. Finally, the institution endorsed our request for financing, and the National Council on Science and Technology (CONACYT) approved a grant in 1988. This marked the launch of a shoestring project that led to the constitution of the Pearl Oyster Research Group (Grupo Ostras Perleras, or GOP).[44] For the next fifteen years, the GOP systematically studied every stage of extensive aquaculture and the repopulation of natural beds (table 11.2). Along the way, the research program needed to overcome biological, ecological, and technical hurdles. Research projects included thirteen experiments, eight financed by domestic funders and the remaining five by international ones.[45] The fundamental vision of the project was to design several models for installation and operation of pearl farms able to offer a viable development alternative for the coastal communities of the region, while improving the environmental conditions of pearl oysters in the Gulf of California. The message was intended to reach the communities of the Mexican Pacific coast and eventually extend to other pearling regions.

The GOP made significant contributions to scientific knowledge and technological practices regarding pearl oyster farming. Three years after CONACYT's first grant, a variety of experiments had succeeded in establishing a regime of extensive aquaculture (table 11.2). The first pilot harvest of Mabé pearls cultivated from both species was collected in the winter of 1992–93. Concurrent studies on the procedures for producing round pearls also yielded positive results, and students working on these projects received

Table 11.2. Research by the Grupo Ostras Perleras (GOP), 1987–97

Research focus	1987	1988	1989	1990	1991	1992	1993	1994	1995
Resource surveys	M, C	M, C	M, C	M, C	M, C	M, C	M, C	M, C	M, C
Oceanography				M, C			M, C		
Evaluation of sites		M	M, C			C		M, C	
Ecology of spat collection			M, C					M, C	M, C
Temporal analysis			M, C	M, C	M, C	M, C	M, C	M, C	
Vertical repartition			M, C	M, C	M, C	M, C	M, C	M, C	
Evaluation of substrates			M, C		M, C	M, C		M, C	
Collection testing		M	M		M	M		C	C

Model design	1988	1989	1990	1991	1992	1993	1994	1995	1996	1997
Nursery:										
General	M	C	C							
Sites		M	M	M, C	M, C					
Depth			M	M, C				M, C		
Density			M	M	M,	M, C	M, C			
Duration				M	M	C	C			
Artifacts		M, C	M, C	M, C						
Ecology				M, C	M, C	M, C	M, C		M, C	M, C
Late culture:										
General			M							
Sites				C	M, C	M, C				
Depth			M	M	M, C					
Artifacts			M, C	M, C	M, C	M, C				
Ecology			M, C	M, C	M, C	M, C	M, C	M, C	M, C	M, C

Key: M = *Pinctada mazatlanica*; C = *Pteria sterna*.

degrees at the undergraduate and graduate levels. By the year 2000, nearly 70 percent of the total information on historical, biological, and technical aspects of *Pinctada mazatlanica* and *Pteria sterna* had been published in scientific journals, book chapters, books, specialized magazines, conference proceedings, research reports, and theses; further diffusion of this information took place at courses and seminars given at local, national, and international venues.[46] Perhaps a landmark in the development of modern pearl farms in Mexico and the onset of interests in other coastal countries of Latin America (including Ecuador, Panama, Costa Rica, Guatemala, Colombia, and Venezuela) was the international congress "Pearls '94," held in Hawaii in 1994, where the GOP made eleven presentations on the results of nine years of research. Some scientists who attended (including Mexicans) had developed incipient research projects by then, but only Venezuela and, more recently, Panama, began programs of their own.

The presentations at Pearls '94 and cumulative information from the GOP and other sources were also used by the Guaymas (Sonora) campus of the Instituto Tecnológico de Estudios Superiores de Monterrey (ITESM) to found a private pearl farm. Perlas del Mar de Cortés now produces excellent round pearls from extensive culture of *Pteria sterna*—which is more abundant there than *Pinctada mazatlanica*—and stands out as one successful model in the pearl world. Another academic institution, the Universidad Autónoma de Baja California Sur (UABCS) in La Paz, founded two enterprises. The first was created in 1998 and obtained modest results before closing its doors the following year. The second, named Perlas del Cortés, began operations in 2002 and, like the Sonora institution, relies on the extensive culture of *Pteria sterna*. It has achieved success in the production and marketing of Mabé jewelry, with staff also acquiring skills in the surgery to produce round pearls.

Science, Technology, and Utopia in the Gulf of California

Many Mexicans worry that the increasing commodification of knowledge has influenced the agendas of research centers and the ethos of scientists themselves. Romualdo López Zárate, professor of sociology at the Universidad Autónoma Metropolitana (Azcapotzalco, Mexico), has argued that neoliberalism constitutes "an obstacle for the development of science in Mexico," in which "scientific capitalism" has led to "the reconceptualization of knowledge as one more commodity that is for sale to the highest bidder, the loss of academic freedom by submission to the desires and interests of the market, the convenient obedience to the [goal of] generating revenue from knowledge, the restriction of the flow of information and the hiding of results of research; in brief: the privatization of knowledge for the benefit of a few."[47]

In these conditions, many academic researchers who developed marketable production technologies became independent brokers with the private sector. No administrative structures regulated the agreements between research centers and investors, leading to ambiguous relationships between academic institutions and corporate interests. This situation created incentives for researchers to pursue private business opportunities without resigning from their home institutions; in some cases, they even used institutional resources for personal profit. Administrators at most institutions simply accepted, tolerated, or ignored the situation. Wishing to solve the problem, or at least to regulate it, different institutional structures were created.

The CIBNOR was no exception. The Office of Technological Management was created in early 1994 to promote and oversee agreements between potential investors and the academic staff who had developed a marketable technology. This new entity's first action was to establish procedures for obtaining patents on behalf of CIBNOR for all technologies either completed or in the process of development by its personnel. Because of its high potential profit and overall efficiency, the GOP project was first on the list. However, the GOP declined to seek a patent for its processes, which are based on technologies that have been known and practiced throughout the world since the early twentieth century. Many other pearl farms had achieved success using virtually the same technique that this CIBNOR office wanted to protect. Moreover, we considered that the knowledge inherently belonged in the public domain regardless of its economic value. The GOP had already published a great body of information and continued to do so. Unfortunately, this philosophy did not fit within CIBNOR's regulations.

The authors of this chapter along with other members of the GOP therefore decided to found a corporation in 1995 called Perlamar de La Paz to promote pearl farms as an alternative model of regional development. Perlamar planned to open an experimental farm to provide training, technical assistance, and consulting services—including the marketing of pearl products—in an effort to attract interest from the state government of

Baja California Sur and local communities. It also attempted to recruit support from local businessmen interested in the development of economically sustainable pearl farms. Perlamar's project was also presented to funding agencies dedicated to regional development, and even to Mexican and foreign investors who shared its business philosophy. The idea was not new: something similar had been successfully applied thirty years earlier by the Groupe d'Intérêt Économique in French Polynesia.

The GOP and Perlamar established clear parameters on technology transfer and other elements of the relationship between researchers and private enterprises. It also insisted that wild stocks should not be touched and that cultured young adults be planted in natural beds to help ensure repopulation. Business partners were expected to share a commitment to regional socioeconomic development and community welfare. Nevertheless, all of the proposals we received entailed the extraction of wild adults to use them in pearl culture, and none of them offered options for regional community development. The authors of this chapter, along with other members of the GOP-Perlamar, opposed these proposals, which arrived from both Mexican and foreign investors.

Neither the institutional autonomy of GOP-Perlamar nor its emphasis on public service and concept of sustainability fit within existing institutional structures. After the dismemberment of the GOP in 1998, some studies continued through external projects, including one that led to the first successful experimental hatchery of *P. mazatlanica*,[48] and further research on pearl culture techniques. Perlamar was finally compelled to close in 2000 without ever really having begun operations. It had been a daring adventure that sought to leverage the Pearl Myth as a model of sustainable community development that could have given fishing people along the coast a new source of competitiveness and diminished the pressure they exerted on marine resources.

In 2004, after nineteen years of insistence, CIBNOR finally declared that research on hatchery techniques for pearl oysters had become an institutional priority. The authorities appointed a new project manager (an alumnus of the GOP), who continued the preliminary experiments. This new stage built upon the scientific and technological advances of the GOP's original project but downplayed the principles of social service and the production of practical knowledge that could be useful for the sustainable development of coastal communities. Instead, the research aimed at specialized aspects of biotechnology intended for publication in peer-reviewed scientific journals, along with a new emphasis on graduate education with an "efficient" time-to-degree program.

Conclusions and Perspectives

The environmental history of fisheries and husbandry of nacre and pearls is a vivid example of the transformation from overexploitation of a valuable natural resource to its sustainable management. This change was forced by the exhaustion of pearl oyster natural stocks in the face of growing demand — or the commodity fetishism — that is characteristic of capitalist society. More than a conservationist movement, this transformation came about through the efforts of scientists pressed into service by economic interests of governments and merchants who feared the extinction of their business.

The present situation is the result of important historical, scientific, and technological transformations, which we have only briefly described here.[49] However, it is important to bear in mind that nearly every pearl sold on the global market since the middle of the twentieth century has been obtained from cultured pearl oysters and pearl induction procedures. No pearling company can succeed without managing the scientific and technological aspects of aquaculture, bringing to bear the requisite expertise in pearl culture, and paying attention to environmental conditions. In addition, meeting the demands of the market is essential: cultured pearls must have the size and quality that buyers expect and be mounted in finely crafted jewelry (which represents the most lucrative segment of the market). The development of a successful pearl farm is not simple; moreover, the economic and human costs of pearl farms must take into account the social and environmental benefits they render.

The Gulf of California has a long heritage as a global pearling center and is uniquely suited to the needs of contemporary pearl farms. It possesses all the required scientific, technological, environmental, and economic resources. No less important is the fact that the collection and marketing of nacre and pearls has a long history as the region's most important economic activity.

Yet these economic considerations are only part of the equation. Pearl aquaculture presents not only a priceless opportunity for sustainable development but also a logical way to address the social and environmental challenges the region now faces. The Gulf of California has not escaped the global crisis in fisheries; indeed, the decline of fishing as a viable subsistence strategy has had and will continue to have serious consequences for coastal communities. Fishing people constitute one of the region's most vulnerable social groups, and development planners must somehow improve on current strategies of subsidizing fishing supplies, granting cooperatives special permission to fish for specific species, and the fruitless struggle against clandestine fishing. Worse still, the recent boom in real estate development for the tourist market and second homes for North Americans means that fishing communities now face a threat as intractable as the collapse of fisheries. Many have lost access to the beaches that are their workplaces. This panorama threatens communities with the permanent loss of their employment, of their culture, and of their contribution to Mexico's food supply.

It is both necessary and urgent to recognize the social, economic, and cultural values these communities possess and to propose sustainable and productive means of guaranteeing their survival. In this context, the husbandry of pearl oysters to produce nacre and cultured pearls has special relevance. This long-term historical review has demonstrated the importance of valuable marine resources in the Gulf of California and argues for the importance of aquaculture in maintaining the ecological and productive balance of commercial species throughout the region. The GOP's efforts to develop local aquaculture technology suggest that it is possible to independently manage pearling resources while creating means for regional, sustainable development of coastal communities.

Planning for alternative futures necessarily involves uncertainties, even in the short term, as we adjust to the realities of the twenty-first century. Nacre and pearls are luxury goods for an upscale market that few can afford. The current challenges imposed by economic crisis, the shortage of water and energy, pollution, declining tourist revenues, changes in sea level and global warming, violence and poverty, and the loss of ethical values (just to name a few) complicate the search for a sustainable and socially just version of the Pearl Myth. Nevertheless, as social and natural scientists, we need to seek this new version and provide the society with better ideas of competitiveness. We hope that the rich environmental history of pearl oyster fisheries and aquaculture will contribute in its own way to the quest for these alternatives.

Notes

1. Pearl oysters are marine bivalve mollusks of the family Pteriidae, which comprises more than three hundred species distributed into four genera. Natural populations generally inhabit rocky-coral biotopes along tropical and subtropical coasts of the world, from depths of a few meters to seventy meters or more depending on the species. Only six species of the genus *Pinctada* (including *P. mazatlanica*) and two species of the genus *Pteria* (including *Pt. sterna*) are actually managed in farming and pearl culture, either commercially (*P. maxima, P. margaritifera, P. martensii, P. chemnitzi, Pt. sterna*) or in pilot-experimental scale (*Pt. penguin, P mazatlanica, P. imbricata, Pt. colymbus*).

2. Micheline Cariño and Mario Monteforte, *El primer emporio perlero sustentable del mundo: La Compañía Criadora de Concha y Perla de la Baja California S.A. y perspectivas para Baja California Sur* (Mexico City: UABCS, CONACULTA-FONCA, 1999), 118–237.

3. Extensive culture starts with deployment of artificial collectors in the sea to capture wild "spat," or "seed," of target species. After a couple of months of immersion, the collectors are washed to recover tiny individuals measuring two to four millimeters in diameter. In the intensive culture, hatchery laboratories generate the spat. The CCCP developed its own extensive culture technology but did not produce cultured pearls.

4. Gastón Vives (1903–14) with *P. mazatlanica* in La Paz; William Saville-Kent (1904–24) with *P. maxima* in Australia; and Cyril Crossland (1905–22) with *P. margaritifera* var. *erythraensis* in Dongonab Bay, Sudan.

5. Micheline Cariño, "The cultured pearl polemic," *World Aquaculture* 27, no. 1 (1996): 42–44.

6. Y. Hisada and T. Fukuhara, *Pearl Marketing Trends with Emphasis on Black-Pearl Market*, FAO Corporate Document Repository, Field Document no. 13 (n.p.: FAO Fisheries and Aquaculture Dept., 1999); A. Müller, "Cultured Pearls: Update on Global Supply, Demand and Distribution (Focusing on the Pearls from the White-Lipped Oyster *Pinctada maxima*)," presentation at Gemmo Basel 2005 International Colloquium on Gemmology, Basel, Switzerland, 29 April–2 May 2005; C. Tisdell and B. Poirine, "Economics of Pearl Farming," in *The Pearl Oyster*, ed. Paul Southgate and John Lucas, 473–98 (Amsterdam: Elsevier, 2008); http://web-japan.org/atlas/nature/nat28.html.

7. M. M. Cariño Olvera, "Le mythe perlier dans l'histoire coloniale de Sudcalifornie, 1530–1830," Mémoire de Maîtrise d'Historie, Université de Paris VII Jussieu, Paris, 1987.

8. The geographic distribution of *P. mazatlanica* and *Pt. sterna* actually reaches Peru and Ecuador. The first pearls and nacre the Spanish knew about were found along the Antilles coasts, where *P. imbricata* and *Pt. colymbus* (small-sized species of commercial Pteriidae) produce small natural pearls.

9. Miguel Mathes, "Asesinato y descubrimiento: El motín de Fortún Ximénez y la incorporación de California al Imperio Español," *Meyibó* (Instituto de Investigaciones Históricas UNAM-UABC) 1 (July–December 1990): 31–49.

10. Hernán Cortés, "Cuarta Carta de Relación, 1534," in *Cartas de Relación*, 22nd ed. (Mexico City: Porrúa, 2007).

11. An *armada* was a fishing fleet formed by a large ship such as a brig, frigate, or sloop, complemented by a certain number of canoes carrying one or two naked divers. Generally, the *armador* was not the proprietor of the fleet but an employee of richer entrepreneurs.

12. "Establecimiento de veda permanente sobre la pesquería de ostras perleras en la costa mexicana y declaración de las mismas como especies en peligro de extinción," *Diario Oficial de La Federación* (Mexico City), 24 October 1939; "Cambio de estatus de las ostras perleras de especies en peligro de extinción a especies bajo protección especial," *Diario Oficial de la Federación* (Mexico City), 16 May 1944.

13. Denis George, "Debunking a Widely Held Japanese Myth: Historical Aspects on the Early Discovery of the Pearl Cultivating Technique," *Pearl World* 10 (2008): 1–7; Joseph Taylor and Elizabeth Strack, "Pearl Production," in *The Pearl Oyster*, ed. Paul C. Southgate and John S. Lucas, 273–302 (Oxford, U.K.: Elsevier, 2008).

14. *P. chemnitzi*, the Japanese *P. martensii*, the Indian *P. fucata*, and the Caribe-Antillian *P. imbricata* and *Pt. colymbus* represent the small-size sector of commercial pearl oysters (Akoya-type pearls).

15. Miguel del Barco, *Historia natural y crónica de la Antigua California (1780)*, ed. Miguel León-Portilla (Mexico City: Instituto de Investigaciones Históricas, Universidad Nacional Autónoma de México, 1988), 171. On population, see Homer Aschmann, *The Central Desert of Baja California: Demography and Ecology* (Berkeley: University of California Press, 1959), 133–44; and Julia Bendimez-Patterson, "Antecedentes históricos de los indígenas de Baja California," *Estudios Fronterizos* (Revista del Instituto de Investigaciones Sociales, Universidad Autónoma de Baja California) 5, no. 14 (September–December 1987): 12–27.

16. José Francisco Castellanos and Arturo Cruz, "Aprovechamiento de los moluscos en la dieta aborigen," in *Ecohistoria de los Californios*, ed. Martha Micheline Cariño-Olvera, 61–80 (La Paz: Universidad Autónoma de Baja California Sur, 1995).

17. The first cartographic charts labeled this bay "Bahía de Las Perlas." The island now known as Espíritu Santo was likewise labeled "Isla de Las Perlas." See Mario Monteforte and Micheline Cariño, "El Mar de Cortés no existe," *Biodiversitas* (CONABIO) 86 (September–October 2009): 12–15.

18. The term *placer* is used mostly in the mining industry to denote sites having important deposits of precious metals or minerals. A pearling placer refers to the richest natural beds of pearl oysters where adults are abundant, sometimes coinciding with excellent nacre quality and high incidence of natural pearls.

19. Micheline Cariño, "Exploraciones y descubrimientos, 1533–1678," in *Sudcalifornia: De sus orígenes a nuestros días*, ed. Micheline Cariño and Lorella Castorena, 55–84 (Mexico City: Gobierno del Estado de Baja California Sur, SIMAC-CONACYT, UABCS-SEP, 2007).

20 Micheline Cariño, "Mito y perlas en California (1530–1830)," *Sociales-Humanidades: Revista del Área Interdisciplinaria de Ciencias Sociales y Humanidades de la UABCS* 2 (Second Semester 1990): 53–59.

21. Jorge Luis Amao-Manríquez, *El establecimiento de la comunidad minera en la California jesuítica* (La Paz: Colección Cabildo, Serie Premios, 1981), 25–55.

22. Archivo Histórico de Baja California Sur "Pablo L. Martínez," La Paz BCS (hereafter AHPLM), "Cádiz 1811"; Archivo del Juzgado de Distrito de Baja California Sur, Procesos Penales, exp. 3/918.

23. Biblioteca Nacional de México, *Junta de Fomento de las Californias 1828*, Colección Lafragua, 2 vols.: LAF 437 and LAF 31.

24. Archivo General de la Nación (Mexico City) (hereafter AGN), Archivo Histórico de Hacienda, leg. 117, México 1830.

25. M. M. Cariño Olvera, "Les mines marines du golfe de Californie: Histoire de la région de La Paz à la lumière des perles," PhD diss., École des Hautes Études en Sciences Sociales, Paris, 1998, p. 549.

26. The information from the years 1838 to 1853 was obtained from several documents of the AHPLM: legajos 36, 36', 42, 43, 44, 47, and 48. For the next years, the information comes from Adrián Valadés, *Temas históricos de la Baja California* (Mexico City: Jus, 1963).

27. Probably equivalent to a *real de vellón*. In the seventeenth and eighteenth centuries, 2.5 *reales de vellón* was equivalent to one *real de plata* (whole silver coin).

28. José María Esteva, "Memoria sobre la pesca de la perla en la Baja California (1857)," in *Las perlas de Baja California*, 30–45 (Mexico City: Departamento de Pesca, 1977).

29. Mario Monteforte and Micheline Cariño, "Exploration and Evaluation of Natural Stocks of Pearl Oysters *Pinctada mazatlanica* and *Pteria sterna* (Bivalvia: Pteriidae): La Paz, South Baja California, México," *AMBIO: A Journal of the Human Environment* 21, no. 4 (1992): 314–20; Micheline Cariño and Mario Monteforte, "History of Pearling in the Bay of La Paz, South Baja California, México (1533–1914)," *Gems and Gemology* 31, no. 2 (1995): 108–26.

30. Cariño, "Les mines marines," 636.

31. *Diario Oficial* (Mexico City), 28 December 1901. The Companía Perlífera del Golfo de California remained active for ten years. Owners Adolfo Schirabe and Edmundo Vives (Gastón Vives's brother) received a concession for a portion of the peninsular coasts between the parallels 24°N and 29°N.

32. Alexander Lyster Jameson, "On the Identity and Distribution of the Mother-of-Pearl Oysters; With a Revision of the Subgenus *Margaritifera*," *Proceedings of the General Meetings for Scientific Business of the Zoological Society of London* 1 (January–April 1901): 372–94.

33. Cariño and Monteforte, *El primer emporio perlero*.

34. See Cariño, "Les mines marines"; Cariño and Monteforte, *El primer emporio perlero*; and Cariño and Monteforte, *History of Pearling*.

35. AHPLM, La Paz 1912, Fomento, vol. 578, exp. 33.

36. AHPLM, La Paz 1913, Fomento, vol. 595, exp. 246.

37. A. C. Cahn, *Pearl Culture in Japan*, Natural Resources Section, Report 122 (Tokyo: General Headquarters, Supreme Commandant for the Allied Powers, 1949).

38. S. Mizumoto, "Pearl Farming in Japan," in *Report of the FAO Technical Conference on Aquaculture, Kyoto, Japan, 26 May–2 June 1976*, 381-85 (Rome: Food and Agriculture Organization of the United Nations, 1976).

39. One exception to this pattern was India, where scientists associated with CMFRI (Central Marine Fisheries Research Institute) developed their own technology for the hatchery and round pearl production in *P. fucata*, an industry that has played a key role in developing the Tamil-Nadu region. See K. Alagarswami, "Pearl Culture," in special edition of *Bulletin of the Central Marine Fisheries Research Institute* (India) 39 (1987): 115.

40. José Manuel Mazón-Suastegui, "Evaluación de cinco dietas microalgales en el crecimiento larval de *Modiolus capax* (Conrad, 1837) y *Pinctada mazatlanica* (Hanley, 1856), Mollusca: Bivalvia," master's thesis in marine resource management, CICIMAR-IPN, La Paz, Baja California Sur, 1987.

41. Food and Agriculture Organization, *State of World Fisheries and Aquaculture* (Rome: FAO Fisheries Department, 2004), 155.

42. The Centro de Investigaciones Biológicas de Baja California Sur (CIB) was created in 1975. In 1992 the institution extended its area of influence and increased its functions, at which point it changed its name to Centro de Investigaciones Biológicas del Noroeste (CIBNOR).

43. Monteforte and Cariño, "Exploration and Evaluation."

44. Grupo Ostras Perleras, a research group in CIB-CIBNOR, was led by Mario Monteforte and included students from Mexican universities such as Universidad Autónoma de Baja California Sur, the Centro Interdisciplinario de Ciencias Marinas of the

Instituto Politécnico Nacional, Universidad Autónoma de Baja California, and Universidad Nacional Autónoma de México.

45. Domestic funding includes grants from CONACYT (3), SIMAC (2), SIBEJ (1), FMCN (1), CONABIO (1), and ANUIES (1). International funding includes grants from IFS in Sweden (2), ECOS in France (1), and two US organizations, Hubbs-McBean Foundation (1) and Island Pearls (1).

46. For instance, the international congresses "Pearls '94" in Hawaii, "World Aquaculture Society (WAS) '95" in San Diego, and "WAS '96" in Bangkok, Thailand.

47. Romualdo López Zárate, "El modelo neoliberal, obstáculo para la ciencia en México," *La Jornada* (Mexico City), 21 September 2009; Romualdo López Zárate, "Los valores de la ciencia en el subsistema de educación superior," in *Cultura mexicana: Revisión y prospectiva*, ed. Francisco Toledo, Enrique Forescano, and José Woldenberg, 31–60 (Mexico City: Taurus, Santillana Ediciones Generales, 2008).

48. Presentations at the XII Congreso Nacional de Oceanografía, Huatulco, Oaxaca, Mexico, 2000, including M. Robles, P. Monsalvo, T. Reynoso, and M. Monteforte, "Desarrollo larvario de la Madreperla de Calafia, *Pinctada mazatlanica* (Hanley 1856) en condiciones controladas"; H. Bervera, J. J. Ramírez, and M. Monteforte, "Evaluación de substratos artificiales e influencia del color de los materiales en la fijación de semilla de la Madreperla de Calafia, *Pinctada mazatlanica* (Hanley 1856)"; and P. Saucedo, C. Aldana, C. Rodríguez, P. Monsalvo, T. Reynoso, and M. Monteforte, "Acondicionamiento gonádico y maduración de la Madreperla de Calafia, *Pinctada mazatlanica* (Hanley 1856), bajo dos tratamientos de temperatura."

49. For further discussion, see Micheline Cariño and Mario Monteforte, *Histoire mondiale des perles et des nacres* (Paris: L'Harmatan, 2006), 251; Micheline Cariño and Mario Monteforte, "De la sobreexplotación a la sustentabilidad: Nácar y perlas en la historia mundial," *El Periplo Sustentable*, no. 12 (2007): 81–131; and Mario Monteforte, "Ecología, biología y cultivo extensivo de la madreperla de calafia, *Pinctada mazatlanica* (Hanley, 1856) y la concha nácar arcoiris, *Pteria sterna* (Gould, 1852), en Bahía de La Paz, Baja California Sur," postdoctoral thesis in biological sciences, Faculty of Biological Sciences, University of La Habana, Cuba, 2005.

CHAPTER TWELVE

Conclusion
Of the "Lands in Between" and the Environments of Modernity

Cynthia Radding

The contributions to the present volume on the environmental history of modern Mexico bring innovative research on a variety of topics spanning over two centuries. Taken together, they honor the deeply rooted traditions of environmental history for Mexico, based on cultural geography and the ecological heritage of Mesoamerican agricultural technologies and landscapes. At the same time, they transcend the boundaries of these disciplines to bring to the discussion of history and the environment such pressing topics as urban water supplies and sewage, irrigation, health and environmental contaminants, the conservation of fisheries and forests, and the establishment of national parks and reserves. The integrative themes that weave these chapters together include water, technology, labor, the governance of common resources, and the production of knowledge. This concluding chapter summarizes the salient points of each chapter and closes with some final reflections on modernity and the intersections between environmental and social history as we seek to understand more fully the relationships between nature and culture in the formation of modern Mexico.

Chapter 2, by Angus Wright, demonstrates that Mexico constituted one of the global staging grounds for the Green Revolution, a mid-twentieth-century cluster of technological innovations that expanded grain production in wheat, rice, and corn in South Asia, Latin America, and other world regions with ancient traditions of cultivation. Wright takes the themes of technology and modernity into an overview of twentieth-century Mexican agriculture, focusing on the Green Revolution and efforts to eradicate centuries-old tropical diseases such as malaria. The Green Revolution in

Mexico centered the attention of international agencies such as the Rockefeller Foundation, private funds, and the Mexican national government on the humid tropics and the northern deserts, considered to be heretofore sparsely populated with low productivity in foodstuffs. Paralleling the findings that Luis Aboites presents later in the book on urban water and sewage, the targets set by the proponents of the Green Revolution for expanding agricultural production in what were considered to be new frontiers depended heavily on chemical fertilizers and deep well drilling, both closely linked to the petroleum industry and the expanding grid of rural electrification.

The principal critiques of the Green Revolution supported by Wright's analysis center on the assumption of "exhausted soils" in the traditional areas of Mexican agriculture; the dependency on synthetic fertilizers and consumption of electricity to advance cultivation techniques in the desert and tropical regions; and widening regional and social inequalities in postrevolutionary Mexico resulting from selective investment in tropical and arid-lands agronomic technologies. What emerges from Wright's nuanced discussion of soil fertility and its susceptibility to erosion or degradation in relation to population densities and levels of civilization and agricultural sophistication is the importance of understanding the ecological complexity of anthropogenic landscapes, as evidenced in terracing and irrigation systems.

Wright's long-term historical view of the Mixtec region of Oaxaca links the ingenuity of pre-Hispanic systems for capturing soil runoff and controlling irrigated water flow in a stratified and increasingly oppressive society dominated by the Mixtec nobility to modern scenarios of poor soils and low agricultural output under the weight of mid-twentieth-century population growth and minimal technological innovation. While comparing evidence of soil erosion in other regions of central and western Mexico—pointing plausibly to substantial degradation under the techniques of milpa cultivation—Wright argues that fertility losses in Mexico stemmed largely from the colonial and postcolonial periods of agricultural development in both the heartland of Mesoamerica and its borderlands. He observes that Mexican national policies recognized the problems of soil degradation and deforestation, but (as in the United States) efforts at conservation focused primarily on conserving forest cover.

This summary of the deep roots of Mesoamerican agricultural traditions leads to his principal topic of the Green Revolution and its consequences for sustainable systems of food production and rural peasant livelihood. The goals of the Green Revolution resonated with postrevolutionary policies of land reform, which stressed expansion of cultivated areas and bringing new lands into production over the redistribution of existing farmlands in response to the agrarian movements that had fueled the popular forces of the Revolution of 1910. Although the Cardenista reforms brought the agrarian movement into national policies of land redistribution, soil conservation, and forestry, their radical edge was blunted by the conservative turn in Mexican national politics after 1940. Technological solutions for perceived problems of poor soil management and low productivity in traditional farming areas provided the scientific rationale for large-scale, private and *ejidal* agricultural enterprises in northern Mexico, based on irrigation and industrial fertilizers. Chemical pesticides followed closely the use of fertilizers to counteract the spread of plant diseases and parasites, strengthening clusters of industrial and commercial agribusiness interests and changing the structure of late twentieth-century agriculture in Mexico.

The initial stages of the Green Revolution focused on producing grains for human consumption in Mexico, but profitability turned the later stages of high-tech production systems toward tropical fruits and vegetables for export, cotton, and tobacco. The price to pay for the chemically expanded production fostered by the Green Revolution included leached and saline soils, devastating consequences for farmworkers' health, and contamination of waterways as irrigation runoff flowed into the rivers and reservoirs of northwestern Mexico. Health and labor intersect with agricultural environmental issues in ways that compare Wright's critical assertions with Myrna Santiago's analysis of the protracted struggle of labor movements in the petroleum industry and Aboites's discussion of the flaws inherent in the federal government's reliance on technology to resolve the problem of water distribution in Mexico's major cities. Wright closes with a question on sustainability that points to the restoration of traditional farming methods in Oaxaca, under local initiative and modest levels of funding, as an alternative to the modernist reliance on synthetic fertilizers and pesticides that became the visible legacy of the Green Revolution.

Martín Sánchez Rodríguez centers his study on water and technology for the late colonial Bajío on the following questions: What were the environmental consequences of the expansion of irrigation, increased cereal production, and the transformation of agrarian landscapes in the basin of the Río Lerma? Did the adaptation of Old World technologies for water storage in the Bajío contribute to sustainable systems of economic production in the face of population growth and rising demand for food? Taking the *Relaciones Geográficas* of 1570 as a baseline for his description of the Bajío, Sánchez provides a clear portrait of the semiarid climate of this portion of western Mexico, which corresponded to the core of the fifteenth-century

Purépecha sphere of influence. The region was characterized by irregular rainfall, concentrated in the summer months, thus providing sufficient moisture for agriculture but subject to the risks of drought and frost. The surfeit of precipitation in only one season created both the conditions and the necessity for water storage, and the network of streams and tributaries that fed into the Lerma basin served as a grid for building systems of irrigation as the colonial economy unfolded.

Sánchez re-creates the landscapes of the Bajío from the literature in geography, archaeology, and history. Vegetation biomes combined the varied species of low forest, including thorn forest species, and semideciduous tropical forests, along with marshes and grasslands surrounding the lakes and wetlands produced by the river systems of the region, notably the Laguna de Chapala and Lake Cuitzeo. Pre-Hispanic settlement, with agricultural villages and a fairly dense population, flourished for over a millennium, from approximately 800 BC to 900 CE. For reasons that may be attributed to drought, the sedentary landscapes of this period gave way to seminomadic indigenous landscapes, which Spaniards would first describe in the sixteenth century. The precontact history of the Bajío illustrates, significantly, that the oft-cited frontier between the Mesoamerican civilizations of the subtropical highlands and the Gran Chichimeca of the northern steppes did not describe a clear demarcation—even when following natural features such as the Río Lerma—but rather constituted a series of moving borders and changing clusters of sedentary and nomadic modes of subsistence over time and space.

Colonial landscapes evolved with Spanish colonial endeavors in both livestock breeding, adapting Andalusian methods, and agriculture. Irrigation methods required construction of dams and networks of canals; of these, Sánchez describes in particular the creation of the Yuriria Lake, by widening existing swamps and lagoons and diverting river flow into them to sustain large reservoirs. Cereal production and grain milling developed in the Bajío to supply the mining *reales* of Zacatecas and Guanajuato, which began to attract large populations by the mid-sixteenth century. Nevertheless, the breeding and migratory grazing of cattle and livestock proved to be the main factor in the transformation of Indo-Hispanic landscapes in the Bajío during the first two centuries of Hispanic presence there.

Eighteenth-century expansion of agricultural production opened a new phase of landscape transformations in the Bajío. Sánchez departs from the well-established literature for this region and period to argue that the changes in political ecology, so widely observed in the increased production of cereals and modified terms of land tenure and rents, proceeded not only from the expansion of cultivated fields but also from significant technological innovations. He further argues that the relative aridity of western Mexico and the extreme variability of rainfall regimes in different regions made "the administration of water" the most significant factor for the growth of agriculture in New Spain, implying the physical distribution of fluvial streams to planted fields as well as the social distribution of water among different groups and classes of landholders and farmers. In the Bajío, as in other regions of northern and western Mexico, the most formidable technical challenge that cereal growers faced concerned harnessing the torrential rains that swelled rivers during one or two summer months, and then left their streambeds dry. While previous authors have noted the investments by *hacendados*, *rancheros*, tenants, and even sharecroppers in catchment dams and diversion canals, Sánchez turns our attention to the adaptation of Egyptian and Indian techniques of water storage to the Mexican countryside, known as *entarquinamiento* or *cajas de agua*. These complex systems of water storage and redistribution through dikes, holding tanks, and canals were identified during colonial times in the Canary Islands and in Mexico. Their adaptation in the Bajío arose from the social and economic processes that transformed the internal markets of New Spain and its factors of production. Cereal production expanded; significantly, milpas traditionally dedicated to maize were transformed into fields reserved for wheat, responding to market demand and rising prices in late Bourbon New Spain. Thus, the hydraulic infrastructure imprinted upon the landscape undergirded the increased agricultural production even as it symbolized the uneven distribution of benefits socially across the Bajío. Sánchez's work implicitly raises the intriguing question—which would make an ideal topic for further study—of whether the cajas de agua could attenuate the devastating consequences of crop failures, such as the "great hunger" of the mid-1780s, which have had such dramatic effects on the environment and the history of Mexico.

Rick López, writing on the same period of transition from the late colonial to the early republic, turns our attention from the direct application of technology for production to the production of knowledge, as shaped by viceregal policies and Enlightenment concepts of science in the Iberian imperial sphere and its legacy for Mexican nationalist renderings of nature. Specifically, the Royal Botanical Garden of Mexico, founded as part of the Royal Botanical Expedition to New Spain (1787–1803), represents a significant attempt to collect and systematize knowledge through scientific and bureaucratic networks that linked the metropolis and its colony in the late Bourbon regalist vision of empire. Yet this vision and the scientific methods it espoused came into conflict with indigenous and Creole taxonomies, challenging the values ascribed to natural specimens by native herbalists and—more broadly—the cultural meanings associated with nature in its different formations throughout New Spain.

The Bourbon regalist projects focused on the botanical expedition and the garden set out to hierarchize science according to the European concepts of the period. Scientific research and the governance of nature became gendered as "masculine," cosmopolitan, and subject to royal standards of accountability, in contrast to the "feminine," local, and traditional norms and practices regarding the knowledge of nature and the uses and conservation of its resources. López illustrates these epistemological and conceptual contradictions, with political and cultural repercussions, through the historiography surrounding the work of the *protomédico* Francisco Hernández. Sent by King Phillip II during the 1570s, his efforts to catalog, describe, and collect a wide variety of botanical specimens produced *The Natural History of New Spain*, written in Latin and consisting of six folio volumes filled with descriptions of over three thousand plants previously unknown in Europe (along with descriptions of a much smaller number of animals and minerals) and ten folio volumes of paintings by Mexican indigenous artists illustrating the plants and animals he described. Steeped in the Nahuatl language of his informants and collaborators, Hernández promoted Mesoamerican systems of classification as the organizing vehicle for his empirical data.

The reception of Hernández's prodigious efforts in Spain, upon his return, were blunted by the imperial crises of the late sixteenth century and the Habsburg obsession with maintaining a guarded control over the knowledge produced in and about its overseas colonies. Although portions of the *Natural History* were transcribed and copied, as in the *Rerum medicarum Novae Hispaniae Thesaurus* of 1628, and a selection of its contents was published in reference to its medical uses by the Dominican Francisco Jiménez, in New Spain (1615), the work in its entirety was never published. Tragically, the original folio volumes perished in the great fire that swept the Escorial in 1671, destroying much of the text and the irreplaceable paintings by the Mexican artists.

Two centuries later, the botanical expedition of 1787 to New Spain acknowledged the traditional schema of learning and classification derived from Nahuatl but rendered them as inferior to the Linnaean nomenclature that would shape the science of botany. The imposition of Latinate terminology over the vocabularies derived from Nahuatl, which had given continuity to medical treatises and manuals in use in New Spain, both symbolized and implemented the valuation of European scientific principles over local epistemologies. Furthermore, this hierarchization of knowledge that presumed the superiority of European conceptual frameworks clashed with the eighteenth-century *novohispano* or Creole sense of identity and pride in the natural wealth and intellectual traditions of Mexico as a cultural space and an imagined community. Two leading exponents of this Mexican tradition of science and learning, José Antonio Alzate y Ramírez and José Ignacio Bartolache, played a major role in this and in other endeavors of late Bourbon cultural flowering in New Spain.

The Royal Botanical Garden laid out in Mexico City was linked closely to the Botanical Garden of Madrid in both its inception and the administrative efforts to complete its construction. Thus, Alzate y Ramírez and Bartolache joined Casimiro Gómez Ortega, director of the Madrid garden, and Martín Sessé y Lacasta, a medical doctor with ample experience in the Caribbean and Mexico City, in what was ultimately an unsuccessful search for copies of the Hernández *Natural History* in Mexico. While Gómez Ortega and Sessé y Lacasta conceived of the botanical project in broad terms for coordinating a colony-wide effort at research, collection, and classification—directed from the imperial center of Madrid—their Mexican counterparts sought to meld a hybrid of local knowledge and European scientific principles in this early stage of national consciousness grounded in the material reality of New Spain.

José Juan Juárez Flores's dramatic analysis of deforestation on the legendary volcanic peak of Malintzin (Matlalcuéyetl: the blue-skirted goddess) constitutes a compelling and complex story that provides a representative case study for the national picture of accelerated deforestation in the final decades of the nineteenth century. The interlinking volcanic peaks of Pico de Orizaba (Citlaltepec: the starred hill), Popocatépetl (the steaming mountain), Iztaccíhuatl (the woman in white), and Malintzin had undergone repeated episodes or phases of intervention to cull resources from their forests for fuel and construction, for gathering and hunting, and for ritual renovation during the pre-Hispanic and colonial periods. Importantly, these mountainous forests constituted reservoirs of water and their root systems formed barriers of defense against erosion, retaining the texture and nutrients of the soils that eventually flowed into the alluvial valleys of Tlaxcala and Puebla. During the nineteenth century, especially between 1860 and 1910, in the period chosen for this study, the exploitation and destruction of the forests on Malintzin reached alarming proportions, turning her blue skirts into scarred swaths of denuded slopes from which the once-verdant tropical woods have retreated.

Juárez integrates his history around the themes of deforestation, water, and technology in the second half of the nineteenth century. These strands of environmental history take on clearer meanings, as Juárez demonstrates, when they are analyzed in the light of political economy and of fundamental changes in the legal and social structures of Porfirian Mexico, which altered the governance and use of both water and woodlands. The quantitative evidence for the accelerated destruction of forests—in this case, in the ecosystem of Malintzin and the volcanic chain of which it is a part—makes sense when we view these figures in terms of secular processes of population growth, increased consumption, and the interlinking of national and international markets. Not only did the extraction and use of woods and grasses increase (because of the growth of urban settlements surrounding Malintzin) but also new uses for the products of the forests developed in the late nineteenth century, centered on kerosene for household and urban lighting, paper, and other chemical products derived from resins; these were consumed in Mexico and exported to the rapidly expanding global markets of the second industrial revolution. Industrialization consumed these wood products as well as *zacatón*, a grasslike palm that was used traditionally for thatch, fodder, and local manufacture of brooms. Furthermore, the accelerated deforestation documented for this period owed its rhythm to radical changes in the judicial regimen governing the possession and use of woodlands and waterways in Porfirian Mexico, emanating from the political economy of liberalism.

One of the most pervasive themes of nineteenth-century Mexican history highlights the shift from communal land tenure to private property. As in Tortolero's subsequent chapter on Morelos, Juárez shows that the privatization of basic agricultural resources involved not only the land but also the water necessary for peasant livelihood. Forests and grasslands, traditionally managed on a communal basis, entered as well into the secular process of privatization that so altered the landscapes of modern Mexico. Juárez emphasizes that this was not a simple transference of valuable resources from communities to private landowners; instead, it was a process that entangled local villages, municipal corporations, and the federal government. During the colonial period and the early nineteenth century, the forests of the volcanic slopes did not belong to the indigenous communities settled in the piedmont; rather, they were part of the corporate patrimony of the city of Tlaxcala. The municipal government rented the forests to the communities in return for labor and regular supplies of timber for public works. Beginning with the famous *desamortización* of all corporate lands decreed in 1856, and continuing with the constitutional reforms and the radical privatization of land, water, and subsoil under Porfirian legislation, the federalization of waterways and forests disrupted regional processes of negotiation between local communities and municipal and state-level authorities.

In terms that seem to echo the research presented by Sánchez for the Bajío, Alejandro Tortolero Villaseñor returns to the classic sources from the late Porfiriato and early postrevolution to emphasize the importance of water for the agricultural valleys south of Mexico City. Tortolero's focus on the hydraulic works for irrigation explains the strategies developed by sugar planters to expand their holdings and increase their levels of production during the late nineteenth century as well as the momentum of the peasant movement that swept Morelos in the first decade of the twentieth century.

Meticulously researched in both archival and printed sources, Tortolero's contribution to this volume deconstructs the simplification of the *agrarista* movement in conventional histories of the revolution, to remind us that the peasant movement to which Emiliano Zapata gave his name was not limited to land but also encompassed water and the means to irrigate the crops. Tortolero's analyses, based on ample quantitative data and qualitative contemporary texts, present us with an innovative critique of the historiography and new insights into the divergent policies that were proposed for agricultural development in postrevolutionary Mexico. Tortolero confirms the dramatic expansion of acreage in the hands of private hacendados at the turn of the twentieth century but reminds us that the majority of these lands, ostensibly dedicated to sugar and other kinds of commercial agriculture, were turned over to sharecroppers because of the limited supply of water for irrigation. Directing our gaze to the villagers and landless day laborers of Morelos, Tortolero shows us that the apparent simplicity of the political tenets of Zapatismo masked the thorny issues surrounding the redirection of technological advances and the redistribution of vital resources such as irrigable land, which would define the environmental component of social revolution.

Deforestation, the underlying theme of Juárez's chapter on the volcanic slopes of Malintzin, links the destruction of tropical woods and zacatón and their repercussions in the valley of Tlaxcala-Puebla to Sterling Evans's reflections on "King Henequen," but for a very different geographical region and agricultural environment. In the peculiar ecological conditions of Yucatán, with its semiarid limestone karst, uncertain rainfall, and poor soils for traditional crops, the *bosque seco*, or subtropical dry forest, characterized most of the central and northern portions of the peninsula. Paradoxically, these very harsh conditions provided an ideal environment for the development of plantations based on two species of agave, henequen and sisal, whose fibers bound Yucatán (and Mexico) in concentric webs of dependency during the nineteenth-century industrialization of grain harvesting in Canada and the United States. Thus, the ecological consequences of large-scale henequen and sisal planting, harvesting, and processing in Yucatán—from roughly the 1850s to the 1930s—cannot be explained without reference to the global markets that reshaped national economies, created new international financial systems, and altered the meanings of both nationalism and imperialism.

Both of these fibrous succulents are native to Yucatán and thus could be considered "natural" species on which to build a regional economy. For centuries, the Maya of Yucatán had lived with the agave plants and converted their fibers into clothing, bedding, roofing materials, and tools; the thorns became needles for both utilitarian and ritual purposes. Spanish colonial authorities and (the relatively few) settlers in Yucatán maintained henequen production and harvesting at a low scale, used primarily for naval rigging and cordage.

The conversion of the Yucatecan agaves from wild vegetation or small-scale garden plantings to large-scale plantations, invading communal lands under municipal administration, began as early as 1830. This process, although interrupted by the crisis of the Caste War and its aftermath, was reinforced and accelerated after 1860, due in part to the collapse of the sugar industry in Yucatán. During the boom years of 1880–1915, henequen production increased exponentially, as did the extension of acreage dedicated to henequen and sisal plantations, eventually absorbing over two-thirds of all cultivable land in the peninsula.

As Evans explains, Yucatecan geology and soil composition protected henequen plantations from the fungal infections and parasites that so often bedeviled large-scale tropical agriculture in other parts of the Americas. Despite these apparently ideal environmental conditions for the unlimited expansion of henequen plantings and production, the accelerated clearing of *monte* (the semiarid scrub forest) and the conversion of communal lands (prerevolution *ejidos*) into henequen plantations under the Porfirian political economy had far-reaching social and ecological repercussions for the region. Deforestation followed the spectacular boom in henequen, even if in a different dynamic than in the volcanic slopes and valleys of central Mexico. Expanded milpa clearings followed the march of henequen, in response to the increased demand for food (maize) from a rapidly growing population, including the semiservile labor force recruited for the grueling labor of the henequen plantations, but whether the milpas contributed as heavily as did the plantations to the rhythm of deforestation in Yucatán is questionable.

The henequen boom of Yucatán preceded, yet paralleled, the early phase of the petroleum industry in the transitional rainforests of the Huasteca. Myrna Santiago presents contrasting views of nature and the environment from different key actors in the drama of the exploitation of Mexico's "black gold," whose riches depended, as in the case of Yucatán, on the often-volatile currents of the world market. In her discussion of nature and labor in the petroleum industry, we find the theme of resource extraction extended from the forested slopes of Malintzin to the rainforests and petroleum fields of the Huasteca. Several of the chapters discussed thus far raise the issue of opposing values ascribed to basic natural resources by different groups of peasant producers and private landowners, whether as sustenance, as communal reserves, or as the raw materials for capitalist profit. Tortolero, for example, makes clear that the court litigations and direct confrontations that ensued between sugar planters and peasant villagers in Morelos reveal distinct and contradictory claims to both use rights and ownership of land and water in the agricultural landscapes dominated by the sugar industry. Santiago explicitly turns to the analysis of class to argue that the economic and social structures erected by the oil magnates, sharpened and distorted by race, and challenged by Mexican labor organizations, shaped the concepts of nature that governed the relationships among these social groups through the environmental transformations of the Huasteca in turn-of-the-century Mexico.

The oil barons, personified by Edward Doheny and Weetman Pearson, hailing from the United States and Great Britain, respectively, reconstructed in Mexico the privileged standards of food, shelter, recreation, and sociability to which they considered themselves entitled in their countries of origin. Foreign administrative employees and skilled workers, while enjoying privileged status within the labor force, experienced the tropical biota of the Huasteca less as a recreational space and reservoir for the collection of exotic species than as a formidable and unfamiliar set of surroundings. In greatest contrast, however, the Mexican workers—consigned to the lowest ranks of laborers, forced to live in substandard housing, and exposed to flooding, toxic gases, and malarial infection—confronted the natural environment as a series of dangerous challenges to their physical survival through injury, disease, and exposure. Most of the labor force that was recruited for the rapidly expanding petroleum industry came from other regions of Mexico, and they found the tropical climate of this region forbidding.

This class-stratified orientation to nature and the environment takes its meaning primarily from the fact that all of these groups of laborers and managers in the petroleum industry—as well as the remnant population native to the Huasteca—lived in an environment that underwent dramatic and violent transformation. Mexicans and foreigners alike did not merely live in a tropical forest, they were actively engaged in its destruction through fire, construction, population growth, and the exposure of soils, plants, and animals to the eruptions of petroleum wells ("blowouts"), ruptured pipelines, and the runoff of *chapopote*. The expansion of the petroleum industry and its environmental consequences were not limited to the land base of Veracruz and Tamaulipas; rather, it spread to the sand dunes, beaches, estuaries, and mangroves along the coast of the Gulf of Mexico.

Within this context of dangerous work and rapid, visible, environmental change, Santiago revisits the well-known history of the labor conflicts that fueled social unrest in the region and led to the 1938 nationalization of the petroleum industry, interpreting these events in terms of class struggle over nature. Mexican workers who led the efforts to organize unions and demand collective bargaining with their foreign employers did not espouse an "environmentalist" discourse that lamented the destruction of the forest; rather, their environmental awareness focused on health issues and the glaring disparities created by the oil industry. Striking workers who achieved the first collective contracts, in 1926, did compose a broad agenda for occupational health with demands that included compensation for loss of limbs and injuries, prostheses, and a well-equipped hospital. Although these contracts were rendered largely ineffective by the companies' retaliatory efforts, the unified labor movement's demands reiterated, in 1936, a comprehensive set of conditions linking compensation (wages) to the dangers that workers experienced daily in the extraction, storage, and processing of petroleum as well as the basic health necessities of laborers and their families. Labor history and environmental history together clarify the troubled sequence of events surrounding the petroleum industry in the postrevolutionary political context, even as they open new dimensions to the complex story of modernity in contemporary Mexico.

Emily Wakild brings together the themes of water, conservation, and the intersection of tradition and modernity in her overview of the development of the Mexican system of national parks and natural reserves. Her cultural and political history of the creation of parks shows them to be planned and controlled spaces within the contradictions of rapid population growth, urbanization, and economic expansion as well as the seemingly chaotic conditions of Mexico's urban environments. If, as Wakild argues, the creation of reserves and parks is not at odds with the objectives of national development and modernization emanating from the Revolution of 1910 and postrevolutionary policies, nevertheless the history of setting aside reserves did not follow a consistent trajectory; rather, it depended on the vagaries of overall political and economic policies. Moreover, the overlap of institutional jurisdictions and contradictory property and usufruct claims on the lands reserved for parks has historical roots in the long-term traditions of communal land use and access to resources. Wakild's discussion of the chronological (and ideological) phases of Mexico's "hybrid" system of parks offers new insights into questions concerning the tension between nature and culture, between conservation and modern development.

In her overview of the literature regarding the concepts and practices of conservation in different national and colonial settings, Wakild reminds us that the designation of rural or "wilderness" areas as parks or reserves set aside from production but compatible with different kinds of land use and cultural values enacted by indigenous inhabitants does not constitute a fundamental or unbridgeable contradiction; the more telling conflicts arise from the conversion of rural areas with multiple uses (and landscapes) into urban and industrial areas, undergoing transformations that seem to be irreversible. Ironically, the International Union for the Conservation of Nature, an organization that has its roots in an international meeting in Mexico City in 1947, has categorized Mexico's national parks at a relatively low level of preservation because of the heterogeneity of their contours and uses.

The initiatives for designating certain areas as reserves date from the late nineteenth-century Porfiriato, stemming from concerns about deforestation—a theme that resonates in the chapters by Juárez Flores on the slopes of Malintzin, Sterling Evans on Yucatán, and Angus Wright on the Green Revolution—focused on the Desierto de los Leones (on the southern margins of Mexico City) and the mining concession of El Chico, Hidalgo. In addition, early conservation efforts involved reforestation of wetlands and denuded hillsides. The institutional structures for a national system of parks did not take shape, however, until the constitutional changes put into place at the close of the Revolution of 1910, undergirded by the nationalist and restorative principles of Article 27 of the Constitution of 1917, the enactment of the Forestry Code of 1926, and the integration of conservation as a component of land reform. The presidential term of Lázaro Cárdenas (1934–40), outstanding among all postrevolutionary administrations for its policies in land reform, labor rights, and the nationalization of the petroleum industry (as cited in the chapters by Myrna Santiago and Angus Wright) provided the leadership to put in place the foundations for a national system of parks in Mexico. Indeed, the guidelines developed by the Department of Forestry, Fish, and Game envisioned the national parks and reserves as complementary to and supportive of farming, ranching, and extractive subsistence strategies through the conservation of woodlands and water sources. Their conservation through the creation of national parks, as Wakild explains, served at this time as a way of restoring the "commons," whereby the ejidos controlled impressive extensions of forested lands and, in turn, were entrusted with the responsibility for their maintenance. Somewhat parallel to the marked change in course for agrarian reform after 1940, as remarked by Angus Wright, the post-Cardenista period seemed abruptly to change the course of conservation policies and nearly close the program for creating national parks.

The period of the "Mexican Miracle" (broadly, 1940–80) favored industrial development and import substitution over conservation in the national goals for modernization, spurred by the international market for both raw materials and finished goods produced in Mexico. This was also a period of accelerating population growth and rural-to-urban migration, enclosing and absorbing traditional rural spaces. The miracle of nearly four decades of sustained economic growth and tangible benchmarks of social welfare began to crumble in the 1980s, leading to a shrinkage of governmental regulation and the severe diminishment of both the public and social sectors—epitomized by the dismantlement of the ejido and the neoliberal reforms leading to NAFTA. At the same time, however, conservation efforts were renewed, but with a different rationale than the policies put in place by the Cárdenas regime. The biosphere reserves created during the late twentieth and early twenty-first centuries encompassed a larger land mass than the parks of the earlier conservation efforts; however, their importance lay not in communal usufruct but in their recognition by the international scientific community and the nascent ecotourism industry for their particular biological and geological features. The quantitative evidence summarized in the graphs included in Wakild's chapter support her argument that the rhythm of park creation and the extension of national territory protected as either parks or reserves varied considerably over nearly a century, as did the policy objectives that governed conservation measures, responding to different historical circumstances and overall governmental strategies. National parks in Mexico, in contrast to those in the United States, did not represent "nature in the wild" but nature as linked to its social context for access to and governance of communal resources. Furthermore, conservation developed as a set of articulated principles and policies shaped by the content of Mexican historical traditions.

Luis Aboites Aguilar's discussion of mid-twentieth-century policies to federalize the task of supplying urban populations with piped water and wastewater removal elaborates further on the idea of health as integral to environmental issues and brings our consideration of histories of the environment and of nature firmly into the urban milieu. As Aboites explains, the "national model" of providing credit to municipalities to undertake the construction of urban water systems arose in part from the fiscal constraints of the municipal corporations to carry out major public works and in part from the agendas set by the national party and its governing bodies to assert its presence in the provincial cities. Furthermore, federal programs to finance local systems for potable water served to promote population growth and enhance the health of the Mexican population by preventing disease and providing this vital resource.

If, indeed, the national model fell short of its goals, and the uneven distribution of clean water and sewage systems exacerbated further the inequalities that marked Mexican society, the effort represented an impressive commitment to investment in basic infrastructure paralleling similar federal programs to extend electrification in both rural and urban areas, expand irrigation, and build public housing. To be sure, the federal bureaucracy expanded through the "national model," as evidenced by the creation of the Banco Nacional de Obras y Servicios Públicos, the Banco Hipotecario, and the Secretaría de Recursos Hidráulicos. Federal investment in public housing supported, as well, private associations such as the Ingenieros Civiles Asociados. Thus, public housing, electricity, and potable water formed three major strands of the national model for postrevolutionary Mexico, making technology a necessary part of national development.

This formula, based on substantial federal investment and the confidence in technological innovation to resolve basic infrastructural and social problems, began to erode in the 1970s. The reasons that Aboites adduces for the retreat of the national model concern financing, as the federal government began to diminish its rate of investment in potable water systems; technology, as the consequences of deep-water pumping became evident in problems of sinking and flooding in Mexico City; salinization of agricultural wells in northwestern Mexico; and perceived inequalities in the distribution of water and the targets of investment in different regions and localities. In ways that bear comparison with other Latin American nations, the federal government returned the functions and responsibilities for the provision of potable water to the municipalities in 1983. As Mexico entered the twenty-first century, controversies arose over the contracts extended to private companies (including foreign firms) for building water pipelines, managing the distribution of potable water, and charging the citizenry for its consumption.

Mario Monteforte and Micheline Cariño turn the discussion of technology and communal resources to the fisheries and development of aquaculture in contemporary Baja California. The arid coastal plains west of the San Pedro Mártir mountain range that frame the Gulf of California seemed forbidding to European settlement. Yet historically and ecologically, the Gulf of California constituted a rich site of biodiversity; its estuaries provided important spawning grounds for a broad variety of maritime species, including the famed pearl oysters that provided the initial impetus for commercial exploitation along the Gulf coast of the arid peninsula. Global demand for the beauty of nacre shells and pearls spurred the exploitation of pearl fisheries in major oceans and bays from Australia and the Philippines to Panama and the Caribbean. The Gulf of California entered the pearl trade from the early eighteenth century onward, and as occurred with pearl fisheries around the world, its wild stocks of oysters were susceptible to periodic overexploitation that resulted in depletion. Cariño and Monteforte build their chapter on the transition from pearl fishing to pearl culture, based on their research and documented experimentation, to underscore the environmental consequences of both extraction and "farming the sea."

The commercial success of the Compañía Criadora de Concha y Perla de Baja California constituted an important enterprise in its time and served as a memorable example of how technology could be brought to bear on the commercial production of a natural resource benefited by a stable market. By the mid-twentieth century, saltwater farms had supplanted the extensive cultivation of wild pearl oysters; with Japanese technology and commercial expansion, cultured pearls came to dominate the international market for nacre and pearls. Monteforte and Cariño show how the case study of pearl fishing and cultivation in the Gulf of California illustrates the basic ecological issues of sustainable production versus overexploitation and exhaustion of a theoretically renewable resource. Furthermore, they show that environmental problems, and their scientific solutions, cannot be separated from the social and political contexts in which they arise. In Baja California, as elsewhere, the debates over the use of different technologies for the production of pearls and the conservation of their natural base became struggles for power with multiple polarities among local community efforts, research centers, and governmental bureaucracies.

Employing a multidisciplinary approach to their subject, Monteforte and Cariño allude to the legends and the lure of wealth and beauty centered on pearls arising from both Iberian and Mesoamerican traditions. They underscore the cultural values ascribed to pearls and precious shells, which antedated commercial exploitation in the webs of imperialism that encircled the globe beginning in the sixteenth century. Ethnohistory of the Californio Indians and their adaptive strategies in the arid conditions of Baja California, informed by ethnoarchaeological research, illuminates current debates over sustainable resources. Labor constitutes an important theme for the history of pearl fishing, referring to the recruitment of indigenous divers in colonial times and to the contentious dependency between *armadores* and divers in the harsh labor conditions prevailing under modern practices, as in the chapters by Myrna Santiago on the petroleum industry and by Angus Wright on large-scale commercial farming, but in different environmental and economic frameworks. The story of pearls and nacre contributes to the history of science, relating to theories about the formation of pearls in oysters and to the technologies for controlling the process through pearl induction; in addition, it raises issues of scientific and public policies concerning endangered and strategic species within the bounds of national jurisdictions and territorial seas.

The collapse of the Compañía Criadora during the Revolution of 1910, followed by multiple small-scale endeavors at pearl fishing, brought the natural population of oysters to a critical point in the 1930s. Depleted oyster stocks were further endangered by a series of epizootic blights and the contamination of the Gulf waters due to runoff from the chemically polluted Colorado River and the construction of dams and agricultural development in the United States. Monteforte and Cariño build their history of pearl cultivation for the contemporary period from their extensive research in the literature and from their own experience as scientists and direct agents in their attempts to establish sustainable commercial operations for cultivating pearls and to conserve this valuable natural resource. Their work through the Grupo Ostras Perleras from 1989 to 2004 amassed important information and established the techniques for successful pearl cultivation, even in the altered ecological conditions of the Gulf of California. Monteforte and Cariño conclude their chapter by raising significant questions concerning the ethical bounds of protecting scientific knowledge and the principle of open access to the results of publicly funded research.

All of these chapters raise issues concerning the governance of communal resources at the same time as they explore the cultural and scientific themes that are an integral part of environmental history. The studies included in this volume offer complementary interpretations of the cultural values ascribed to nature in the production of knowledge and in the materiality of living in and transforming the landscapes that provide physical sustenance and meaning. Rick López's presentation of the contradictions inherent in the viceregal project to catalog the biota of New Spain and to establish the Royal Botanical Garden in Mexico City as well as Emily Wakild's research on the creation of parks and natural reserves in modern Mexico both contribute important insights into the ideological frameworks for developing science about nature and for the conservation of communal resources through the mediation of the state.

The majority of the chapters—in particular, the contributions by José Juan Juárez, Alejandro Tortolero, Sterling Evans, Luis Aboites, Angus Wright, and Myrna Santiago—highlight eloquently the parallel tropes of hierarchical systems of knowledge and the codification of social inequalities. Across diverse themes of deforestation, water for agricultural and urban consumption, health, labor, and the contamination of water and soils, they demonstrate the necessary reinforcement of social and environmental histories to comprehend the formation of Mexico as an "imagined community" and, moreover, as a constellation of anthropogenic landscapes.

Equally important for understanding the relationships between people and the environment in "the land between the waters," the questions raised concerning the conservation and governance of the commons serve to integrate the different regional and temporal focal points of all the chapters in this book. Community jurisdiction over property, productive assets, and the "intangible" resources of both "folk" and scientific knowledges constitute key themes for Martín Sánchez Rodríguez's study of water and its reservoirs in western Mexico and for the chapter by Mario Monteforte and Micheline Cariño on the conservation of pearl oyster beds, their cultivation, and the access to knowledge created through scientific research. Emily Wakild's summary of the "tragedy of the commons" debate—usually framed in terms of material resources of water, fisheries, forests, and grazing lands— highlights the emphasis on Mexican traditions of communal property and resource use, values that suffused the early stages of park development in Mexico. Monteforte and Cariño offer a nuanced response to the alleged "tragedy" of overconsumption of common goods by insisting on the public nature of knowledge and its open access to diverse practitioners of science and to the communities that depend on both gathered and cultivated resources to survive. Together the contributing authors show that the commons, when governed through locally nurtured institutions or governmental legal frameworks and policies, constitute sustainable resources with requirements for their conservation and possibilities for their development.

Notwithstanding the denunciations evident in many of the chapters of wanton waste and social injustice through the mismanagement of communal resources and the poor implementation of conservation policies, this volume on the liminal landscapes and highly diverse peoples of Mexico is not merely or principally a declensionist history of deterioration and destruction of Mexico's natural patrimony. Rather, it provides a series of windows through which to view the entwined histories of environmental change, society, political economy, and science in the multistranded production of Mexico as a modern nation. These authors of diverse nationalities and working within different epistemological frameworks have all contributed to a dynamic vision of environmental history that informs the topic of modernity through a careful appreciation of Mexico's pasts in both time and space.

The field of environmental history opens new windows in the corridors and chambers of the edifice of Mexican historiography. On the one hand, it brings to the forefront of historical scholarship topics that might have been consigned to the fields of agro-ecology or archaeology, such as water storage, deforestation, and pearl fisheries; on the other, it takes the staple themes of Mexican history in new directions. The land tenure debates that gave rise to extensive and well-documented histories of haciendas and rural communities provided the foundation for several of the studies included here at the same time that the analyses by Sánchez for colonial Guanajuato, Juárez on Puebla-Tlaxcala, and Tortolero on Morelos enrich the concepts of land tenure and resource usufruct. Not parenthetically in light of the 2010 commemorations of independence (1810) and the revolution (1910), the environmental content of these studies sharpens the focus on peasant movements, but it also suggests the resilience of power structures that governed access to water and basic resources. Luis Aboites's contribution on urban water distribution, Angus Wright's study of migrant labor and the Green Revolution, Sterling Evans's work on henequen, and Myrna Santiago's chapter linking health and environmental issues with the labor movements of the Huasteca—on very different topics—all open new research paths for the postrevolutionary period of twentieth-century Mexico.

In parallel streams emerging from the social histories that contribute in important ways to our understanding of Mexico and Mexicans, recent historiographical innovations for modern Mexico have developed the fields of gender, sexuality, and culture—including art, film, and literature—as well as history of science and environmental history. Rick López and Emily Wakild contribute chapters on the late colonial and postrevolutionary periods, respectively, that conjoin the history of science and the cultural values attached to the environment. In different ways they point to the contradictions of specimen collection, competing knowledge systems, and state-sponsored efforts to monumentalize nature within the constraints of imperial and national political economies. All of the authors included in this volume illustrate that environmental history, as practiced for Mexico and other regions of the Americas, raises new and compelling questions concerning the asymmetrical and contested relations between civil society and the governmental institutions of the state. They bring the environment into the historical narrative and, at the same time, historicize nature.

About the Contributors

Luis Aboites Aguilar is a research professor at the Centro de Estudios Históricos of El Colegio de México. His research centers on the use of water in Mexico in the nineteenth and twentieth centuries, northern Mexico, and public finance in the twentieth century. His most recent publication is *La decadencia del agua de la nación: Estudio sobre desigualdad social y cambio político en México (segunda mitad del siglo XX)*, published in 2009.

Christopher R. Boyer is an associate professor of history and Latin American and Latino studies at the University of Illinois at Chicago and coeditor of the Latin American Landscapes series. His first book, *Becoming Campesinos: Politics, Identity, and Agrarian Struggle in Postrevolutionary Michoacan, 1920–1935*, was published in 2003. He is currently finishing a book on community and scientific management in Mexican forests between 1880 and 2000.

Micheline Cariño received her PhD from the École des Hautes Études en Sciences Sociales and is currently a research professor at the Universidad Autónoma de Baja California Sur, where she has taught since 1989. She specializes in the world history of fisheries, commerce, and the technology of pearl oysters, nacre, and pearls, especially as it relates to the Gulf of California and Baja California. She is a founding member of the Sociedad Latinoamericana y Caribeña de Historia Ambiental (SOLCHA) and the author of ten books and more than ninety articles.

Sterling Evans holds the Louise Welsh Chair in History at the University of Oklahoma, where he teaches environmental and borderlands history. His research interests include North American transnational history and ecosystems, or landscape histories that transcend national boundaries. He is the author of *Bound in Twine: The History and Ecology of the Henequen-Wheat Complex for Mexico and the American and Canadian Plains, 1880–1950* (2007) and *The Green Republic: A Conservation History of Costa Rica* (1999).

José Juan Juárez Flores is a doctoral student at the Universidad Autónoma Metropolitana–Iztapalapa in Mexico City and is currently a professor of history at the Universidad Autónoma de Tlaxcala. He is the winner of the Francisco Javier Clavijero Prize from the Instituto Nacional de Antropología e Historia (INAH) and the Premio Banamex Atanasio G. Saravia for regional Mexican history. His research centers on the economic, fiscal, urban, and ecological history of the Puebla-Tlaxcala region in the eighteenth and nineteenth centuries, and his writing has appeared in several journals and edited volumes.

Rick A. López is an associate professor of history at Amherst College. He is the author of *Crafting Mexico: Intellectuals, Artisans, and the State after the Revolution* (2010). Currently, he is working on a manuscript titled "Science, Nationalism, and Aesthetics in the Shaping of Mexico's Environmental Imagination," which analyzes the development of Mexico's nationalist ecological imagination from the 1780s through the 1910s.

Mario Monteforte is a specialist in biological oceanography. He is currently a titular scientist at the Centro de Investigaciones Biológicas del Noroeste (CIBNOR) and a faculty member of Escuela Superior de Ciencias Marinas at the Universidad Autónoma de Baja California. He holds PhDs in biological oceanography from the Université Pierre et Marie Curie (Paris), in marine ecology from the École Pratique des Hautes Études (Paris), and in biological sciences from the Universidad de La Habana, Cuba. His primary research interests center on the scientific and technological development of pearl oysters and pearl culture, as well as the planning of coastal development based on mariculture. He is the author of seventy articles, ten book chapters, and three books.

Cynthia Radding is the Gussenhoven Distinguished Professor of Latin American Studies and History at the University of North Carolina, Chapel

Hill. Her most recent books are *Wandering Peoples: Colonialism, Ethnic Spaces, and Ecological Frontiers: Northwestern Mexico, 1700–1850* (1997) and *Landscapes of Power and Identity: Comparative Histories in the Sonoran Desert and the Forests of Amazonia from Colony to Republic* (2005). Her current research project focuses on ethnic identity, "wild" and cultivated landscapes, and different pathways to knowledge in northern New Spain.

Martín Sánchez Rodríguez is the president of El Colegio de Michoacán, where he is also a professor of history. His books include *"El mejor de los títulos": Riego, organización social y administración de recursos hidráulicos en el Bajío mexicano* (2005) and two coauthored books on the hydraulic cartography of Michoacán and Guanajuato (2005, with Brigitte Boehm, and 2007, with Herbert H. Eling, respectively). His current research focuses on the relationship between the social uses of water and landscape transformation in Jacona, Michoacán.

Myrna I. Santiago received her PhD in history from the University of California, Berkeley, and is now professor of history at Saint Mary's College of California, where she teaches Latin American and world history. Her scholarly interests include environmental and labor history in Latin America, as well as workers' occupational safety and environmental health. Her book, *The Ecology of Oil: Environment, Labor, and the Mexican Revolution, 1900–1938* (2006), won the Bryce Wood Book Award from the Latin American Studies Association and the Elinore Melville Memorial Prize from the Conference on Latin American History.

Alejandro Tortolero Villaseñor received his PhD from the École des Hautes Études en Sciences Sociales and currently holds the position of professor titular at the Universidad Autónoma Metropolitana, Iztalapa. He has held visiting professorships in Berlin, Paris, Seville, Costa Rica, and Buenos Aires, as well as at Harvard University, and has been awarded both a Guggenheim fellowship and the National Research Award (Premio Nacional de Investigación) in Social Sciences from the Mexican Academy of Sciences. His specialties include the agricultural, environmental, and economic history of Mexico. He is the author of three monographs, including, most recently, *Notarios y agricultores: Crecimiento y atraso en el campo mexicano, 1780–1920* (2009).

Emily Wakild is an assistant professor of history at Wake Forest University in Winston-Salem, North Carolina. Her book-length examination of the convergence of social reforms and nature protection in Mexico's national parks, *Revolutionary Parks: Conservation, Social Justice, and Mexico's National Parks, 1910–1940*, was published by the University of Arizona Press in 2011. Her research interests include the social and environmental history of revolution in Latin America, the comparative history of conservation, and cultural understandings of climate history.

Angus Wright earned his doctorate in Latin American history at the University of Michigan and is a professor emeritus of environmental studies at California State University, Sacramento. His work concerns the social and environmental consequences of agriculture and land tenure, primarily in Latin America. He is the author of *The Death of Ramon González: The Modern Agricultural Dilemma* (2nd ed., 2005) and coauthor with Ivette Perfecto and John Vandermeer of *Nature's Matrix: Linking Agriculture, Conservation, and Food Sovereignty* (2009).

Index

Acámbaro, Gto., 51, 60, 225
agrarian reform, 10–11, 30, 32, 166–67
Aguascalientes, Ags., 225–26, 232, 239, 240(t)
Ajuría, Miguel, 142–44
Alamán, Lucas, 92
alcohol production, 106–7, 115
Alemán, Miguel, 230
Alzate, José Antonio de, 7, 77, 79, 283
Amilpas River, 129
Anáhuac, 1, 16n1, 16n2
Anisz, Enrique, 200
aquaculture. *See* pearl oysters, aquaculture of
aquifers, 1, 9, 224–25, 233–34; depletion, 234–35, 238–39
Atlampa Pasture (Mexico City), 82, 88; flooding in, 84–85
Ávila Camacho, Manuel, 33–38, 44, 205
Aztecs, 1, 16n2, 25–26, 28, 51, 204, 241; botanical knowledge of, 74–76, 79–80, 97n22, 282; population, 7. *See also* Tenochtitlán

Baja California, 6, 9, 36, 292–93; aquaculture research in, 263–66, 264(t); colonialism in, 251–53; geography of, 251. *See also* Northwestern Center for Biological Research
Baja California Pearl and Shell Breeding Company (CCCP), 246, 259, 261, 263; and aquaculture, 246–47, 255, 259, 260(f), 263, 266
Bajío, 36, 39, 50–60, 63, 68–69
Balsas River, 51, 125
Banco Nacional Hipotecario Urbano y de Obras Públicas (later, BANOBRAS), 222–24, 230–32, 235, 238, 266
Beltrán, Enrique, 203
Benítez, Fernando, 162
Bennett, Vivienne, 236
Bigot, Raoul, 124–25, 144
binder (agricultural implement), 150, 167
biosphere reserves, 201–4, 202(t)
Boehem, Brigitte, 14
Borah, Woodrow, 25
botanical garden. *See* Mexican Botanical Garden
botany, 75–76, 81–82, 90–91, 93–95
Bourbon Reforms, 4, 5(t), 6–7, 90, 253, 281–82
Bourlag, Norman, 37
Brading, David, 59, 63, 65

301

Branfield, Richard, 35. *See also* Rockefeller Foundation
Briones Rodríguez, Cruz, 173
broom root, 107–9, 110, 111(t)
Bustamante, Miguel, 93

cajas de agua, 50–51, 57–60, 67–68; in Cañada de Negros, 63–65; in Irapuato, 62; in Jalpa, 65–67
Californio Indians, 249, 251–52, 255
Calles, Plutarco Elías, 30–32
Cañada de Negros Hacienda, Gto., 63–65, 64(f), 67
canals, 54, 56, 59, 62, 163(t)
Cárdenas, Juan de, 75, 80
Cárdenas, Lázaro, 5, 32–33, 37, 44, 205; conservationist policies, 198–99; land reform, 10–11, 32, 166–67; nationalization of petroleum, 186–88; national parks under, 199–200; rural development policies of, 11, 32; water infrastructure, 223, 227
Carranza, Venustiano, 198
cattle. *See* livestock
Cedillo, Saturnino, 227
Celaya, Gto., 53–54, 57, 59–60
Center for Biological Resarch (CIB). *See* Northwestern Center for Biological Research
Cervantes, Vicente, 74, 78–82; after independence, 89–93; and Mexican Botanical Garden, 82–87
Chapultepec, 15, 83, 85, 192–93, 204–8, 205(f), 206(f), 210–12, 277; ahuehuete trees in, 192, 206–7, 210; amusement park, 208–10; and the Mexican Botanical Garden, 88, 94; zoo, 208–10
charcoal, 102–3, 104, 106, 112, 115
Charles II, 252
Charles III, 6, 73, 76–77, 79, 253
Chávez, Diego de, 55
Chevalier, François, 25
Chiautempan, Tlax., 106, 108, 117
Chichimecs, 51, 53
Chihuahua, 9–10, 220, 221, 226, 236, 240(t)

chinampas, 26, 28
cholera, 220–21
científicos (during Porfiriato), 8, 94–95, 157–58, 198
class: as analytical category, 173, 186–87; and environment, 174–75, 181, 183–84; and ethnicity, 177–80, 183, 187
Coahuistla hacienda, Morelos, 137(t6.2), 139–42
Colima National Park, 200
colonization projects (twentieth century), 11, 26–27, 31
Columbian exchange, 3, 14, 20n40
Conchos River, 226
conservation, 9, 32, 45, 115, 118–20, 164–65; and national parks, 193–94, 198–202, 211; and pearl oysters, 249–50, 253, 255, 262; of water, 136, 138. *See also* forest management; soil, management
Cook, Sherburne, 14–15, 25
corn, 2, 24; production in the Bajío, 57, 60–62; production in Yucatán, 166
Cortés, Hernán, 249, 252
Cosío Villegas, Daniel, 31–32, 44
Covarrubias, José, 127
credit: rural, 38; for water infrastructure, 223–25, 230, 232, 238
Crosby, Alfred, 3, 14, 28
Cuernavaca, Mor., 129
Culiacán Valley, 34, 39–41

dams, 5, 11, 33, 35, 45; and cajas de agua, 54, 56, 62–63, 66; pre-Columbian, 27; and water infrastructure, 221, 224, 236–37
DDT, 38, 41–42
Deep Drainage Project, 234
deforestation, 54, 60; in the Huasteca, 181; in Tlaxcala, 100–102, 114; in Yucatán, 150, 161–64. *See also* Malinzin
de la Madrid, Miguel, 210
de la Peña, Guillermo, 129
Desierto de los Leones, 198
Díaz, Porfirio: and henequen boom, 160–62; and pearl boom, 255–58,

261–62; presidency of, 4, 5(t), 6–7, 94, 124–26, 157–59, 205. See also científicos
Díaz Garcés, José Juan, 264(t), 265
Diez, Domingo, 126, 145
dispossession: of communal land, 9–10, 29, 102, 126–29, 145–46, 159; of water, 141–42, 145–46
Doheny, Edward, 173, 176
Dolores hacienda, Morelos, 142–44
drought, 2, 7, 26, 52, 165, 280

Egypt, 58–59
ejidos, 34
electricity generation, 219–20
el niño (ENSO), 7, 262
El Puente hacienda, Mor., 132, 134–37, 134(f), 137(t6.1), 137(t6.2)
Enlightenment: European, 73–74; Mexican, 74
erosion, 24–29, 43–44, 107–8

farm workers, 39–41
fertilizers, 4, 23, 35, 37, 168; and cajas de agua, 59–60, 69
firewood. See wood
flood farming. See cajas de agua
food security: in Yucatán, 166
forest management, 105–7, 115, 118–20, 164–65, 198
forests, 3, 5, 10, 12–13, 27; dry tropical, 151–52, 161, 164–65; ecological services of, 101, 199; in the Huasteca, 176–77; industry, 102–3, 105–7; and national parks, 200; social conflicts over, 114–19. See also deforestation; Malinzin; reforestation; tree tapping
Fox, Vicente, 39
Frey, Eugenio, 164
Fuerte Valley, 34

Gallagher, Kevin, 43
Gallo, Manuel Antonio, 264(t), 265
García Zamora, Guillermo, 59
George, Denis, 250, 264(t), 265
Gibson, Charles, 25
glass production, 106

Gómez Ortega, Casimiro, 77
González Navarro, Moisés, 166
González Roa, Fernando, 127
Goubert, Jean Pierre, 220
Graff, Santiago, 221
Green Revolution, 4, 11, 22–24, 34–39, 44–45, 48n48, 277–79
Guadalajara, Jal., 50, 54, 225, 240(t)
Guanajuato, 3, 36, 55, 93, 280, 295. See also Bajío
Guanajuato River, 51–54, 62–63
Gulf of California, 9, 15, 41, 43, 245–48, 292–93; fisheries in, 271–72

Haber–Bosch process, 35
Havana, Cuba, 90–91
henequen, 153, 154(f), 160–61; cordage industry, 154–55, 163–64, 166–68; in Maya mythology, 154–55; plantations, 156–57, 161–66; –wheat complex, 150, 161
Hernández, Francisco: botanical expedition, 75–76, 78–79, 94; manuscript destroyed, 77; and Nahua science, 75–76, 78–79
Higuerón River, 138, 139(t), 146
historiography: cajas de agua, 58–59; "hydraulic" vs. "agrarian" 125–29, 144–46; Mexican environment, 13–15, 293–96
Hobsbawm, Eric, 218
Hoover, Herbert, 31
Huamantla, Tlax., 105–6, 107–13; social conflict in, 114–19. See also Zitlaltepec resin refinery
Huasteca, 175–76; rainforest in, 176–77
Huastecos, 175
Huerta, Victoriano, 220
Humboldt, Alexander von, 7, 56, 59, 90
hunting, 158, 179, 196, 254, 293
Hunt Institute (Pittsburgh), 94

Independence, war of, 7–8, 68, 89–90, 253
influenza pandemic of 1918, 5, 10
International Harvester, 161
International Union for the Conservation of Nature (IUCN), 196, 197

Irapuato, Gto., 54, 61–62, 67, 68(f), 225
irrigation, 36–37, 44, 50–51, 230, 238; in Morelos, 131–34, 132(f), 133(f); as national problem, 124, 144–45. *See also* cajas de agua; technology, hydrological; water

Jalisco, 51, 63, 65, 153
Jalpa Hacienda, Gto., 65–67, 66(f)
Japan, 247, 263
Jesuits, 6, 253

Kaerger, Karl, 59
Kew Botanical Garden, 73, 86
Koch, Robert, 220

Laguna district, 9, 58, 67–68, 224, 237
Lake Chapala, 55
Lake Cuitzeo, 53
Lake Texcoco, 1, 26, 28
Lake Yuriría, 53, 54, 280
La Laja River, 51–53, 53(f), 68(f)
Landa, Diego de, 154
Lejeune, Luis, 124, 144
Lerdo Law, 104, 141–42, 284–85
Lerma Project (Mexico City), 230–31, 234
Lerma River, 50, 52–54, 53(f), 55, 58, 65, 68–69, 68(f), 237, 279–80
Lerma River Basin, 14, 50–54, 199, 230–31
liberalism, 8–9, 29–30, 102–5, 106–7, 115–16, 140–42, 157–59, 284; and henequen export boom, 157. *See also* Lerdo Law
Linnaeus, Carolus, 74, 164, 250; taxonomic system of, 77, 80, 81–82
Lipsett-Rivera, Sonia, 7
livestock, 54, 107, 116; estates, 6, 9–10, 54–56, 61–63, 65–67, 161; in national parks, 199
Lockhart, James, 25
López, George, 62
López, Jacinto, 86, 89–90
López Zárate, Romualdo, 268–69

Madero, Francisco, 261
Magaña, Gildardo, 128

maize. *See* corn
malaria, 23, 42, 82, 180, 182, 186, 277–78
Malgelsdorf, Paul, 35. *See also* Rockefeller Foundation
Malinzin (Tlaxcala): etymology, 100, 104–5; forest history, 100–104; forest species, 104. *See also* broom root
Mangara (pearling company), 249, 257–58, 261
marshes, 53–54
Matsuii, Yoshidi, 264(t), 265
Maximilian, 94
Maya, 6, 9, 24, 286; collapse, 26–27; common lands, 158–59; henequen mythology, 154–55
Melville, Elinor, 14, 25, 27–28, 32
Mendoza Cortina, Manuel, 139–42
mesquite, 54
Mexica. *See* Aztecs
Mexican Botanical Garden, 73–74, 81–82, 87–89; Atlampa site, 84–85; funding of, 82–84; as national institution, 93; during war of independence, 90–91; Zócalo site, 85–87
Mexican Federation of Workers (CTM), 227
Mexican miracle: 4, 5(t), 11–12, 200–201, 290
Mexican revolution, 5, 94, 260–61
Mexico City, 74, 81–82, 201, 211–12, 230; earthquake, 2; population growth, 207, 212, 218–19; public housing in, 228, 236; water infrastructure in, 221–24, 229–31, 234. *See also* Chapultepec; Tenochtitlán
Michoacán, 39, 51; bishopric of, 54, 60–61
migration, 5, 13; to United States, 13; urban, 23
Mikimoto, Kokichi, 247, 250–51, 263
mining, 3; silver, 6, 8–9, 29, 55, 253; technological innovations, 8–9
Mixteca, 25–26, 43, 45, 278
Mixtecs, 27–28, 40–41, 43, 45
Mociño, José Mariano, 80, 82, 90, 92, 94
Molina, Olegario, 160–61

Molina Enríquez, Andrés, 125, 144
monoculture, 162, 165–66
Monterrey, NL, 221, 229, 236, 240(t); water shortages, 237–38
Morelia, Mich., 220, 225, 232, 240(t)
Morelos, 10, 125, 203, 285; conflicts over water in, 138–43; geography of, 129–30; irrigation in, 131–34, 132(f), 133(f); sugar haciendas, 129–32
mother of pearl. *See* nacre
Murphy, Michael, 59

nacre, 245–46, 254–55. *See also* pearls
Nahuas, 7; botanical knowledge of, 75–76, 79–80, 97n22, 282. *See also* Aztecs
National Action Party (PAN), 44–45
nationalism, 74, 93–95, 174–75, 184, 187–88
national parks, 288–90, 198–201, 199(f); and conservation, 195–97, 199–200; multiple use in, 197–99, 202–3
neoliberalism, 12–13, 43–44, 201, 209–10, 239, 268–69
Nile River, 58–59
North American Free Trade Agreement (NAFTA), 43–44, 201
Northwestern Center for Biological Research (CIBNOR), 265–66, 269–70

Oaxaca, 25, 26–28, 29, 45
Obregón, Álvaro, 30–31
Ocio, Manuel de, 253
Otomís, 28
oysters. *See* pearl oysters

Pachuca, Hgo., 3, 225, 240(t)
Pani, Alberto, 223
parks, 11, 79, 86, 193–95, 211–12. *See also* national parks
Party of the Institutionalized Revolution (PRI), 11–12, 44–45
Pasteur, Louis, 220
pearl divers, 255–56, 258, 261, 265
Pearl Oyster Research Group (GOP), 266, 269, 272; aquaculture research by, 267(t), 266–69

pearl oysters, 9, 245, 253; aquaculture of, 246–47, 250, 263–68, 264(t); concessions, 256–58, 261; management of, 255, 266, 280, 282; overfishing of, 253, 255, 257–58, 261–62, 271
pearls: jewelry, 245–46, 268; markets for, 247–48, 251, 281; merchants (*armadores*), 252–53, 255–58; myth of, 245, 248–50, 254, 266, 280, 282
Pearson, Weetman, 176
PEMEX, 38
pesticides, 4, 23, 37, 39–41, 43, 168; poisoning, 40–42
petroleum industry, 9, 176–77; nationalization of, 33; occupational safety, 180, 182–86; strikes, 184, 186; workers, 178, 179–81
Philip II, 75
Picó, Fernando, 62
Plan de Pitic, 7
pollution, 12–13, 26, 177, 262, 272; of air, 196, 210; of drinking water, 247–48; and social class, 183, 187
Popocatépetl, 101, 129
population, 200; in the Bajío, 55, 60; and Chapultepec, 207; demographic collapse, 2, 25, 28, 156; growth, 60, 102, 200–201; on Malinzin, 102; in Mexico City, 207, 218–19, 230; pre–Columbian, 24–25; in Yucatán, 159
Porfiriato. *See* Díaz, Porfirio
Powell, Philip, 54
presidios, 56
public housing, 228
public works committees (*juntas de mejoras materiales*), 225–26, 231
Puebla: 7, 125, 229, 235–36, 239, 240(t); market for forest products, 101, 105–6, 109–10, 113–14, 122n50; valley of, 101, 183
Purépechas, 51

Quéretaro, 36, 220, 225, 226, 227–28, 240(t)
Quevedo, Miguel Ángel de, 119–20, 198–200

Radding, Cynthia, 14
railroads, 8, 102–3, 105, 163–64
rain: and disease, 182; and erosion, 24; seasonality of, 2, 51–52, 56–57, 59, 144; and sugar cane, 135–65; torrential, 57, 59, 131, 151, 175
rancheros, 7–8, 9, 59, 175, 281
ranches. *See* livestock, estates
reaper/binder. *See* binder
recreation, 179; in Chapultepec, 206
reforestation, 167–68
resin. *See* tree tapping
revolution. *See* Mexican revolution
Rivera, Diego, 230–31
river basin commissions, 11, 36
Robles, Gonzalo, 223
Rockefeller Foundation, 34–35, 37
Rodríguez, María Guadalupe, 63
Roosevelt, Franklin, 32
Roosevelt, Theodore, 195
Róvalo, Luis, 140–41
Royal Botanical Expedition to New Spain (1787–93), 73–74, 77, 80–81, 94
Royal Botanical Garden (Madrid), 73–74, 77, 84
Royal School of Botany (Mexico), 81–82
Ruiz de Velasco, Felipe, 127–28, 131–36, 138, 145

Sahagún, Bernardino de, 75, 80
Salamanca, Gto., 57
San Luis Potosí, 3, 221, 222–24, 227, 236, 239, 240(t)
Sano, Yoshiyatsu, 264(t), 265
San Pedro Apatlaco, Mor., 141–42
Santa Inés hacienda, Mor., 137(t6.2), 139–40
Santos, León, 45
Sauer, Carl, 25
science: enlightenment, 74–78; historiography of, 294; nationalist, 93; royal patronage of, 77, 79, 81–84. *See also* pearl oysters, aquaculture of
Sea of Cortez. *See* Gulf of California
Secretariat of Agriculture and Water Resources (SARH), 37

Secretariat of Human Settlements and Public Works (SAHOP), 235
Secretariat of Hydrological Resources (SRH), 129–31, 134–35, 238
Sessé y Lacasta, Martín, 7, 74, 77, 89–92, 94, 283; botanical expedition of, 73–74, 77, 80–81; on Hernández, 78–79; and the Mexican Botanical Garden, 82–87
sewerage, 220, 225, 229–30
Shirai, Shohei, 264(t), 265
shrimp farming, 41, 43, 265
Silao River, 51–53, 53(f), 62–63, 68(f)
silver. *See* mining
Simpson, Lesley Bird, 25
Sinaloa, 34, 36, 39–41, 44
sisal. *See* henequen
soil, 24–26, 108–9; management, 26–27, 29, 32, 34–35, 37, 45; as "naturally" poor, 22–24, 31–32, 44; in Yucatán, 152, 157; volcanic, 24–25, 103. *See also* erosion
Sonora, 7, 30, 34, 36, 37, 166, 268
Sonoran Dynasty, 30–32, 34, 44–45
Spanish Resin Union (LURE), 109, 110, 114
Stackman, Elvin, 35. *See also* Rockefeller Foundation
sugar: cane, 3; industry in Morelos, 129–30; plantations, 6, 10, 115, 125, 144–45, 155–56; water requirements of, 131–37, 137(t6.2)

Tamaulipas, 6, 175, 222, 288
Tamayo, Jorge, 57
Tampico, Tamps., 173, 185
Tarascans. *See* Purépechas
taxes: on forest use, 102, 105–7, 115; protests against, 229; for water infrastructure, 221–22, 229, 231
Taylor, James, 25
technology: hydrological, 55–56, 68–69, 224; and urbanization, 219–20
Téenec. *See* Huastecos
Tenochtitlán, 26, 28, 85
terraces, 24, 26
Tetlanohcan, Tlax., 114–19

Texas, 6, 176
Tlaxcala (City), 104, 109; city council of, 104–5, 116, 118
Tlaxcala (State), 115, 118–19
Tlaxcalans, 101, 104
Toluca, Mex., 221, 232, 240(t)
Toluca river basin, 50, 55, 234, 240(t)
Tortolero, Alejandro, 14
tragedy of the commons, 206–7
tree tapping, 105, 109–13, 116–17
Turibio River, 51–54, 53(f), 64–65, 67, 68(f)
turpentine, 102, 114; and Zitlaltepec refinery, 109–13, 111(t)
Tutino, John, 54, 61
typhus, 2, 42, 220–21, 227

United Civil Engineers (ICA), 228, 234
United Nations: FAO, 43; REDD+, 13
United States: agricultural policies of, 22, 29–30, 34–35; markets in, 39; migration to, 13; national parks, 195–96; war on drugs, 12, 41; war against Mexico, 7

Valle del Mezquital, 27–28
Valley of Mexico, 1
Van Young, Eric, 50, 55
Veracruz, 173–75, 198
Versailles, 73, 86
Villa, Francisco, 30
Vives, Gaston, 246, 255, 258–59, 261–63, 266. *See also* Baja California Pearl and Shell Breeding Company
volcanoes, 2, 200, 283. *See also* Malinzin; Popocatépetl

Wallace, Henry, 34
War of the Reform, 7, 68
water, 1–2, 10; aquifers, 9, 19, 224–25, 233–34; competition over, 138–42; management, 56; "national model" of, 223–24, 230, 234–35; politics of, 7, 140–42; potable, 12, 220, 227–29; and public health, 222, 227–28; rights to, 138, 139(t), 141–42; shortages of, 207, 221; and social inequality, 233, 235–36; urban infrastructure, 218–19, 228–29, 232–33, 233(t), 235–37, 240(t). *See also* irrigation; technology, hydrological
wheat, 55, 57, 60–62, 67
wilderness, 176, 195–97, 211
Womack, John Jr., 128–29
wood: firewood, 102–3, 104–7, 110–12, 111(t), 115, 119; for railroad ties, 105. *See also* forests
World War II, 11, 22–23, 33, 41–42, 167, 200, 230, 263

Yaquis, 255
yellow fever, 23, 180, 182
Yellowstone, 195
Yucatán, 6, 9, 286–87; Caste War of, 8, 155–56; geography, 151–54, 153(f), 157; henequen boom in, 156–61
Yuriría, Gto., 51, 57. *See also* Lake Yuriría

Zacatecas, 3, 55, 280
Zacatepec hacienda, Morelos, 132, 135–37, 137(t6.1), 137(t6.2)
Zapata, Emiliano, 30–31, 44, 127–28, 285
Zapatismo (Chiapas), 12
Zapatismo (Morelos), 10, 125–29, 138, 145–46, 285
Zapotecs, 43
Zitlaltepec resin refinery (Tlaxcala), 109–12; destroyed by fire, 113–14, 119; workers in, 112–13, 113(t)